GLOBAL DYNAMICS

Wiley Series in Computational and Quantitative Social Science

Computational Social Science is an interdisciplinary field undergoing rapid growth due to the availability of ever increasing computational power leading to new areas of research.

Embracing a spectrum from theoretical foundations to real world applications, the Wiley Series in Computational and Quantitative Social Science is a series of titles ranging from high level student texts, explanation and dissemination of technology and good practice, through to interesting and important research that is immediately relevant to social / scientific development or practice. Books within the series will be of interest to senior undergraduate and graduate students, researchers and practitioners within statistics and social science.

Primary subject areas within the scope of the series include mathematical sociology, economic sociology, social simulation / agent-based social simulation, social network analysis, social complexity, social & behavioural dynamics, social contagion, demography, causality, data mining & analysis, data privacy and security, analytical sociology, econophysics/ sociophysics, (evolutionary/ algorithmic) game theory, and computational/experimental social sciences.

Behavioral Computational Social Science

Riccardo Boero

Tipping Points: Modelling Social Problems and Health

John Bissell (Editor), Camila Caiado (Editor), Sarah Curtis (Editor), Michael Goldstein (Editor), Brian Straughan (Editor)

Understanding Large Temporal Networks and Spatial Networks: Exploration, Pattern Searching, Visualization and Network Evolution

Vladimir Batagelj, Patrick Doreian, Anuska Ferligoj, Natasa Kejzar

Analytical Sociology: Actions and Networks

Gianluca Manzo (Editor)

Computational Approaches to Studying the Co-evolution of Networks and Behavior in Social Dilemmas

Rense Corten

The Visualisation of Spatial Social Structure

Danny Dorling

GLOBAL DYNAMICS

APPROACHES FROM COMPLEXITY SCIENCE

Edited by

Alan Wilson

WILEY

Library of Congress Cataloging-in-Publication Data applied for

ISBN: 9781118922286

A catalogue record for this book is available from the British Library.

Set in 10/12pt, TimesLTStd by SPi Global, Chennai, India.

Printed and bound in Singapore by Markono Print Media Pte Ltd

1 2016

Contents

Notes on Contributors

Peter Baudains is a Research Associate at the UCL Department of Security and Crime Science. He obtained his PhD in Mathematics from UCL in 2015 and worked for five years on the EPSRC-funded ENFOLDing project, contributing to a wide range of research projects. His research interests are in the development and application of novel analytical techniques for studying complex social systems, with a particular attention on crime, rioting and terrorism. He has authored research articles appearing in journals such as *Criminology*, *Applied Geography*, *Policing* and the *European Journal of Applied Mathematics*.

Janina Beiser obtained her PhD in the department of Political Science at University College London. During her PhD, she was part of the security workstream of the ENFOLDing project at the UCL's Centre for Advanced Spatial Analysis for three years. Her research is concerned with the contagion of armed civil conflict as well as with government repression. She is now a Research Fellow in the Department of Government at the University of Essex.

Jyoti Belur is a Senior Research Associate and Senior Teaching Fellow. She served as a senior officer of the Indian Police Service for several years. Her experience and understanding of policing has contributed to her research interests in various aspects of policing, counter-terrorism, crime prevention in the United Kingdom and overseas. She has conducted research on a wide variety of topics including police use of deadly force, police investigations, police misconduct, policing left-wing extremism and crimes against women and has published a book titled *Permission to Shoot? Police Use of Deadly Force in Democracies*, as well as a number of journal articles and book chapters.

Steven Bishop is a Professor of Mathematics at UCL where he has been since arriving in 1984 as a post-doctoral researcher. He published over 150 academic papers, edited books and has had appearances on television and radio. Historically, his research investigated topics such as chaos theory, reducing vibrations of engineering structures and how sand dunes are formed, but has more recently worked on 'big data' and the modelling of social systems. Steven held a prestigious, 'Dream' Fellowship funded by the UK Research Council (EPSRC) until December 2013 allowing him to consider creative ways to arrive at scientific narratives. He was influential in the formation of a European network of physical and social scientists in order to investigate how decision-support systems can be developed to assist policy-makers and, to drive this, has organised conferences in the United Kingdom and European Parliaments. He has been involved in several European Commission funded projects and has helped to forge a research agenda which looks at behaviour of systems that cross policy domains and country borders.

Alex Braithwaite is an Associate Professor in the School of Government and Public Policy at the University of Arizona, as well as a Senior Research Associate in the School of Public Policy at University College London. He obtained a PhD in Political Science from the Pennsylvania State University in 2006 and has since held academic positions at Colorado State University, UCL, and the University of Arizona. He was a Co-Investigator on the EPSRC-funded ENFOLDing project between 2010 and 2013, contributing to a wide range of projects under the 'security' umbrella. His research interests lie in the causes and geography of violent and nonviolent forms of political conflict and have been published in journals such as *Journal of Politics*, *International Studies Quarterly*, *British Journal of Political Science*, *Journal of Peace Research*, *Criminology* and *Journal of Quantitative Criminology*.

Simone Caschili has a PhD in Land Engineering and Urban Planning, and after being a Research Associate at Centre for Advanced Spatial Analysis (UCL) and Senior Fellow of the UCL QASER Lab, he is currently an Associate at LaSalle Investment Management, London. His research interest covers the modelling of urban and regional systems, property market, spatio-temporal and economic networks and policy evaluation for planning in both transport and environmental governance.

Adam Dennett is a Lecturer in Urban Analytics in the Centre for Advanced Spatial Analysis at University College London. He is a geographer and fellow of the Royal Geographical Society and has worked for a number of years in the broad area of population geography, applying quantitative techniques to the understanding of human populations; much of this involving the use of spatial interaction models to understand the migration flows of people around the United Kingdom, Europe and the world. A former secondary school teacher, Adam arrived at UCL in 2010 after completing a PhD at the University of Leeds.

Robert J. Downes is a MacArthur Fellow in Nuclear Security working at the Centre for Science and Security Studies at the Department of War Studies, King's College London. Trained as a mathematician, Rob received his PhD in mathematics from UCL in 2014; he studied the interplay between geometry and spectral theory with applications to physical systems and gravitation. He also holds an MSc in Mathematics with Theoretical Physics awarded by UCL. As a Postdoctoral Research Associate on the ENFOLDing project at The Bartlett Centre for Advanced Spatial Analysis, Rob studied the structure and dynamics of global socio-economic systems using ideas from complexity science, with particular emphasis on national economic structure and development aid.

Shane Johnson is a Professor in the Department of Security and Crime Science at University College London. He has worked within the fields of criminology and forensic psychology for over 15 years and has particular interests in complex systems, patterns of crime and insurgent activity, event forecasting and design against crime. He has published over 100 articles and book chapters.

Rob Levy is a researcher at the Centre for Advanced Spatial Analysis at University College London. He has a background in quantitative economics, database administration, coding and visualisation. His first love was Visual Basic but now writes Python and Javascript, with some R when there is no way to avoid it.

Elio Marchione is a Consultant for Ab Initio Software Corporation. Elio was Research Associate at the Centre for Advanced Spatial Analysis at University College London (UK). He obtained his PhD at the University of Surrey (UK) at the Centre for Research in Social Simulation; MSc in Applied Mathematics at the University of Essex (UK); MEng at the University of Naples (ITA). His current role consists, among others, in designing and building scalable architectures addressing parallelism, data integration, data repositories and analytics while developing heavily parallel CPU-bound applications in a dynamic, high-volume environment. Elio's academic interests are in designing and/or modelling artificial societies or distributed intelligent systems enabled to produce novelty or emergent behaviour.

Francesca Romana Medda is a Professor in Applied Economics and Finance at University College London (UCL). She is the Director of the UCL QASER (Quantitative and Applied Spatial Economics Research) Laboratory. Her research focuses on project finance, financial engineering and risk evaluation in different infrastructure sectors such as the maritime industry, energy innovation and new technologies, urban investments (smart cities), supply chain provision and optimisation and airport efficiency.

Pablo Mateos is Associate Professor at the Centre for Research and Advanced Studies in Social Anthropology (CIESAS) in Guadalajara, México. He is honorary lecturer in the Department of Geography, University College London (UCL), in the United Kingdom where he was Lecturer in Human Geography from 2008 to 2012. At UCL, he was a member of the Migration Research Unit (MRU) and Research Fellow of the Centre for Research and Analysis of Migration (CReAM). His research focuses on ethnicity, migration and citizenship in the United Kingdom, Spain, the United States and Mexico. He has published over 40 articles and book chapters, and a book monograph titled *Names, Ethnicity and Populations: Tracing Identity in Space* published by Springer in 2014.

Thomas Oléron Evans is a Research Associate in the Centre for Advanced Spatial Analysis at University College London, where he has been working on the ENFOLDing project since 2011. In 2015, he completed a PhD in Mathematics, on the subject of individual-based modelling and game theory. He attained a Masters degree in Mathematics from Imperial College London in 2007, including one year studying at the École Normale Supérieure in Lyon, France. He is also an ambassador for the educational charity, Teach First, having spent two years teaching mathematics at Bow School in East London, gaining a Postgraduate Certificate in Education from Canterbury Christ Church University in 2010.

Alan Wilson FBA, FAcSS, FRS is Professor of Urban and Regional Systems in the Centre for Advanced Spatial Analysis at University College London. He is Chair of the Home Office Science Advisory Council and of the Lead Expert Group for the GO-Science Foresight Project on the Future of Cities. He was responsible for the introduction of a number of model building techniques which are now in common use – including 'entropy' in building spatial interaction models. His current research, supported by ESRC and EPSRC grants, is on the evolution of cities and global dynamics. He was one of two founding directors of GMAP Ltd. in the 1990s – a successful university spin-out company. He was Vice Chancellor of the University of Leeds from 1991 to 2004 when he became Director-General for Higher Education in the then DfES. From 2007 to 2013, he was Chair of the Arts and Humanities Research Council. He is a Fellow of the British Academy, the Academy of Social Sciences and the Royal Society. He was knighted in 2001 for services to higher education. His recent books include *Knowledge Power* (2010), *The Science of Cities and Regions* and his five volume (edited) *Urban Modelling* (both 2013) and (with Joel Dearden) *Explorations in Urban and Regional Dynamics* (2015).

Belinda Wu holds a PhD in Geography and has a broad interest in modelling and simulating complex systems in socio-spatial dimensions, with a focus on quantitative strategic decision-support systems. Currently, she works as a Research Associate on the development aid workstream of the ENFOLDing project in CASA, UCL. She was appointed as the main researcher on a series of transport planning and policy research projects in Northern Ireland, before working as a main researcher at two nodes of UK National e-Social Science Centre: Genesis (Generative Simulation for the Social and Spatial Sciences) and MoSeS (Modelling and Simulation of e-Social Science). In 2012, she became the named researcher of the ESRC project SYLLS (Synthetic Data Estimation for the UK Longitudinal Studies) to produce the synthetic microdata to broaden the usage of the valuable UK Longitudinal Studies data for ONS (Office of National Statistics). She is also a Fellow of Royal Geographical Society (FRGS) and a Member of Institute of Logistics and Transportation (MILT).

Acknowledgements

I am grateful to the following publishers for permission to use material.

INDECS, Interdisciplinary Description of Complex Systems, Scientific Journal: A Review of the Maritime Container Shipping Industry as a Complex Adaptive System, INDECS, 10(1), 1–15, used in Chapter 2.

CASA, UCL: Shipping as a complex adaptive system: A new approach in understanding international trade, CASA Working Paper 172, used in Chapter 2; Global migration modelling: A review of key policy needs and research centres, CASA Working Paper 184, used in Chapter 5.

Pion Ltd: A multi-level spatial interaction modelling framework for estimating inter-regional migration in Europe, Environment and Planning A 45: 1491–1507, used in Chapter 6.

Springer: Space-time modelling of insurgency and counterinsurgency in Iraq, Journal of Quantitative Criminology, 28(1), 31–48, used in Chapter 12.

I am very grateful to Helen Griffiths and Clare Latham for the enormous amount of work they have put into this project. Helen began the process of assembling material which Clare took over. She has been not only an effective administrator but an excellent proof reader and sub-editor!

I also acknowledge funding from the EPSRC grant: EP/H02185X/1.

Part One

Global Dynamics and the Tools of Complexity Science

1

Global Dynamics and the Tools of Complexity Science

Alan Wilson

The populations and economies of the 220 countries of the world make up a complex global system. The elements of this system are continually interacting through, for example, trade, migration, the deployment of military forces (mostly in the name of defence and security) and development aid. It is a major challenge of social science to seek to understand this global system and to show how this understanding can be used in policy development. In this book, we deploy the tools of complexity science – and in particular, mathematical and computer modelling – to explore various aspects of change and the associated policy and planning uses: in short, global dynamics.

What is needed and what is the available toolkit? Population and economic models are usually based on accounts. Methods of demographic modelling are relatively well known and can be assumed to exist for most countries. In this case, we will largely take existing figures and record them in an information system. An exception is the task of migration modelling. National economic models are, or should be, input–output based. We face a challenge here, in part, to ensure full international coverage and also to link import and export flows with trade flows. In the case of security, there are some rich sources of data to report; in the case for development aid, the data are less good. In each case, we require models of the flows – technically, models of spatial interaction.

Flow models represent equilibria or steady states. Our ultimate focus is dynamics. There will be imbalances in the demographic and economic accounts, and these become the drivers of change in dynamic models. Typical combinations of systems and models that we explore are as follows:

- multi-layered spatial interaction models of trade flows – in the context of rapidly changing ship, port and route 'technologies';

Global Dynamics: Approaches from Complexity Science, First Edition. Edited by Alan Wilson.
© 2016 John Wiley & Sons, Ltd. Published 2016 by John Wiley & Sons, Ltd.

- dynamic models of trade and economic impact using a variant of spatial Lotka–Volterra;
- input–output models linked by spatial interaction models of imports and exports;
- spatial interaction models of migration combined with biproportional fitting;
- models of riots (i) using epidemiological and spatial interaction modelling, (ii) using discrete choice models, (iii) using spatial statistics and (iv) using diffusion models;
- models of piracy (i) using agent-based models and (ii) using spatial interaction models with 'threat' as the interaction;
- models of ethnic contagion using spatial statistics;
- modelling the impact of development aid through input–output models;
- spatial Richardson (arms race) models;
- Colonel Blotto game-theoretic security models.

We introduce each of these in a little more detail, noting the actual or potential planning and policy applications of each.

In the case of shipping (Chapters 2 and 3), we can use the models we develop to explore the consequences of changing patterns of trade and changing transport technologies. There are rich, albeit disparate, sources of data. The global trade system is complex – through the variety of goods, commodities and services that are carried and through the set of transport modes deployed – sea, air, rail, road and telecommunications. This means that we have to choose levels of resolution at which to work and particular systems of interest on which to focus. In making these decisions, we are, to an obvious extent, constrained by the availability of data. We also wish to connect – and make consistent – any predictions from a model of trade with the import and export data which form part of the input–output tables to be outlined in the next chapter. We focus on a coarse level of aggregation based on seven economic sectors, and we present these sectors and volumes of trade in money terms. We focus mainly on 'container shipping', though container routes usually include road and/or rail elements as well as sea. This covers 80% or more by volume of trade flows. We proceed in two stages beginning with a review of the evolution of the container shipping system (Chapter 2) and then by building a multi-layered model of international trade (Chapter 3).

A key component of an integrated global model will be a submodel that gives us the state of economic development of each country. The ideal model for each country is an input–output model and these, of course, would be linked through trade flows. It seems appropriate, therefore, to report our response to this challenge in this section along with trade (Chapter 4). The basis of this development has to be the existence of national input–output tables. WIOD (2012) provides an excellent source for 40 countries. However, there are enormous gaps of course – 40 out of 220 – and these gaps embrace the whole of Africa. We have sought to handle this situation by developing new tools, based on high-dimensional principal components' analysis, which enable us to estimate the missing data. The detail of this method is presented in Chapter 5 of our companion book *Geo-Mathematical Modelling* (Wilson, ed., 2016).

The policy challenges facing governments associated with migration are essentially of three kinds: the effective integration of in-migrants; limiting the inflows of some types of migrant; encouraging inflows of others. There are forces driving migration which, from governmental perspectives, are more or less controllable in different circumstances. It is important, as ever, to seek to provide a good analytical base to underpin the development of policy. There has been extensive research on migration, and we first provide a background to our own work by surveying this research in the context of the policy questions that arise (Chapter 5).

A typical problem facing the global systems' modeller is the situation in which the data available are not sufficiently detailed. In this case, bearing in mind the nature of the policy challenges, in Chapter 6 we take on the task of estimating flows at a regional (sub-national) level. We do not have the data to achieve this on a global scale, but we have good European data and so we develop the methodology on this basis. This is a classical biproportional fitting problem. Migration data have to be assembled from a variety of sources and different ones are more or less reliable. In order to build as complete a picture as possible of global migration, we explore a variety of sources and seek to integrate them (Chapter 7).

Security challenges vary in scale from the urban – even street level – to the international, for example, through the global deployment of a country's military forces. These different scales, in general, demand different modelling methods and we seek to illustrate a range of these. Security has rich but disparate data. We have developed a two-pronged approach: first to develop some new theoretical models by taking some traditional ones and adding spatial structures, and second, we have assembled a wide range of data that has allowed us to carry out some preliminary tests. We recognise that in this case, there will be government agencies around the world who are modelling these systems with far richer resources than we can bring to bear. What we hope to have achieved is to demonstrate some new approaches to security modelling that may be taken up by these agencies. In this case as well, we have been able to develop models at finer scales in relation to riots, rebellions and piracy. A key concept in this work is the representation of 'threat' and in particular, threat across space. We introduce this in broad terms in Chapter 8.

We then present five distinct applications which between them offer a wide range of methods. In some cases, we can apply different methods to the same problem and so discover the strengths and weaknesses in a comparative framework. Chapter 9 offers a variety of approaches to the London riots of August 2011. We built a three-stage model – propensity to riot (from epidemiology), where to riot (a version of the retail model) and the probability of arrest. We use Monte Carlo simulations to determine whether the counts of observed patterns are more or less frequent than might be expected under conditions in which the extent of spatio-temporal dependency of offences is varied.

In Chapter 10, we shift scale and location again and examine the Naxalite rebellions in India. The data on Naxalite terrorism include the date on which events took place and the district (of which there are 25) within which they occurred. Events include Naxalite attacks and police responses. A key idea in the insurgency literature concerns the contagion of events. This can occur for a number of reasons. For example, conflict may literally spillover from one locality to a nearby other, leading to an increase in the area over which the conflict occurs or moving from one location to those adjacent. In this case, we explore a number of hypotheses. We can test whether there are non-spatial effects of police action on insurgent activity. Moreover, we may test the hypothesis that police action is triggered by insurgent activity. If only the latter is observed, this would suggest that police action is reactive but has little effect on insurgent actions (at least on a short-time scale). For such models, the count of attacks per unit time is described by two components: (i) the first is a baseline risk – which may be time invariant or not, but where it changes it will tend to do so over a relatively long-time scale; and (ii) a self-excitation process, whereby recent events have the potential to increase the likelihood of attacks today considerably.

In Chapter 11, we explore a very different system of interest: piracy in the Gulf. An important question is the security of shipping in relation to pirate attacks. There are two possible approaches to this problem: first, to develop an agent-based model with a given (and realistic) pattern of shipping, and pirates as agents; and secondly, as adopted in this Chapter, to develop a spatial interaction model of 'threat' and to use this to explore naval strategies.

In Chapter 12, we explore a different kind of security issue with a different method: the impact of IEDs (improvised explosive devices) in Iraq. The null hypothesis is that they are independent in time and space. We use Knox's method of contagion analysis to seek evidence of clustering – an important issue in the assessment of response to this kind of threat – and find that there is evidence for clustering in space, time and space-time.

Another kind of security issue is posed in nearby countries where there is a threat of cross-border contagion fuelled, for example, by social networks and this is the subject of Chapter 13. We consider whether ethnic conflict is contagious between groups in different countries and if so, how? And then, whether governments react pre-emptively to potential conflict contagion by increasing repression of specific groups? The argument to be tested is that ethnic groups that are discriminated against in a society identify with groups fighting against the same grievance in other countries and become inspired by their struggle to take up arms against their own domestic government as well. For this process, information about foreign struggles is important, not geographic proximity as such. The empirical test involves using a statistical model on country-years from 1951/1981 to 2004 and this gives some support for the argument. The test will be repeated using data on the analytic level of ethnic groups in different years and improving on the measures of information flows. In the case of government reaction, the argument to be tested is whether governments pre-emptively increase repression against ethnic groups they expect to become inspired by foreign conflicts in order to deter them from mobilising. The empirical test in this case is through a strategic model using data on the behaviour of governments towards domestic ethnic groups.

Development aid (Chapters 14 and 15) offers different challenges: first, defining categories; and then assembling data from very diverse sources. In this case, the ultimate challenge is to seek to measure the effectiveness of aid, and a starting point is to connect aid to economic development. This creates a demand to 'measure' development, and we have done this by constructing input–output models for each country which can then be integrated with our trade model. It then becomes possible to compare the magnitudes of different kinds of aid flows with other trade flows and with flows within national economies. Not surprisingly, aid is much more significant in developing countries than in those with advanced economies. The value of the global input–output model now becomes apparent: in a selected country, we can compute the multiplier effects of increased demand or investment in particular sectors and then begin to address the question of whether investment aid is most effectively targeted. We model aid allocation in Chapter 16.

We finally seek to move beyond our investigation of the impact of aid on development in particular countries and explore the extent to which it has any impact on trade, on migration flows or on helping to maintain security. It has been necessary to drive our work in developing particular submodels by assembling relevant data in each case. Our global input–output and trade system then provides the basis for integrating the main submodels so that we explore the interdependencies which make the global system so complex. It is foolish to think that we (or anyone else) can offer a detailed and convincing 'model of the world' in all its aspects. But what we can do is to offer a demonstration model that reveals some of the complex system consequences of interdependence and points the way to further research, possibly to be carried out by government and inter-governmental agencies that can bring far greater resources to bear. In Chapter 17, therefore, we draw together the different submodels into a comprehensive model which enables us to incorporate the key interactions. Some of the most obvious interdependencies to be picked up are as follows:

- the impacts of net migration on economic development through the labour elements of the national input–output models;
- security-led pushes in outward migration;
- security-led changes in economic development – whether from damage from attack or because of more intensive development of the arms industry;
- many aspects of changing trade patterns – for example, from investment in new ports as well as changes in economic development levels;
- the re-targeting of development aid.

We proceed by establishing a base model and year – taking 2009 as the latest year for which input–output data are available at the time of writing. As noted in Chapter 4, the model is rooted in WIOD (2012) data but then enhanced through a principal components' technique to cover all countries. The import and export flows are integrated with those from a trade model by a biproportional fitting procedure. We assemble base year data and models (as appropriate) for the flows of migrants, military dispositions and development aid. These become drivers of change for subsequent time periods (which we take to be years). At each year end, a number of indicators are calculated and particular attention is paid to imbalance as these will provide the basis for driving the system dynamics. Each year end 'model run', for this reason, is likely to involve iterations driving the system to a new equilibrium.

Reference

WIOD (2012) The World Input–Output Data Base: Content, Sources and Methods, Technical report Number 10.
Wilson, A. (ed) (2016) *Geo-Mathematical Modelling*, Wiley, Chichester.

Part Two

Trade and Economic Development

2

The Global Trade System and Its Evolution

Simone Caschili and Francesca Medda

2.1 The Evolution of the Shipping and Ports' System

Shipping volumes have grown dramatically in recent decades, and this growth has been coupled with changing technologies – particularly through larger ships and improved port logistics – and with changing geographies. At present, many authors estimate that maritime shipping ranges between 77% and 90% of the intercontinental transport demand for freight by volume compared to shipping in the 1980s when it was around 23% (Rodrigue et al., 2006, 2009; Glen and Marlow, 2009; Barthelemi, 2011). The total number of twenty-foot equivalent units (TEU) carried worldwide ranged from 1,856,927 in 1991 to 7,847,593 in 2006 (Notteboom, 2004), and the average vessel capacity has grown from 1900 TEU in 1996 to 2400 TEU in 2006 (Ducruet and Notteboom, 2012). Kaluza et al. (2010) attribute this substantial increase to the growth in trans-Pacific trade. The lower cost per TEU-mile in long distance transport for large quantities of goods has also driven this growth (Rodrigue et al., 2006) coupled with significant technical improvements in size, speed and ship design as well as automation in port operations (Notteboom, 2004; Rodrigue et al., 2009). For instance, in 1991, the use of vessels larger than 5000 TEU was unheard of, but by 1996 large vessels constituted about 1% of the world's fleet, increasing to 12.7% in 2001 and 30% in 2006.

A variety of independent agents (shipping companies, commodity producers, ports and port authorities, terminal operators and freight brokers) play roles in this process, and through their mutual interactions, they determine the patterns of development and growth. It is helpful to view shipping as a complex system of relatively independent parts that constantly search, learn and adapt to their environment while their mutual interactions shape hidden patterns with

Global Dynamics: Approaches from Complexity Science, First Edition. Edited by Alan Wilson.
© 2016 John Wiley & Sons, Ltd. Published 2016 by John Wiley & Sons, Ltd.

recognisable regularities that continuously evolve. In this context, the science of Complex Adaptive System (CAS) provides a useful framework to analyse a shipping system (Arthur, 1997; Dooley, 1996; Gell-Mann, 1994; Gell-Mann, 1995; Holland, 1992; Holland, 1995; Holland, 1998; Levin, 1998; Levin, 2003), a field of study in which strategic analysis is based on what is essentially a bottom-up perspective. CASs are generally assumed to be composed of a set of rational, self-learning, independent and interacting agents whose mutual interrelations generate non-linear dynamics and emergent phenomena.

Since the 1980s, maritime agents have continuously evolved in their organisation in response to both external stimulus and market competition. In the logistics and management perspective, a new form of inter-firm organisation has emerged in the shipping industry. Rodrigue et al. (2009) succinctly explain how this change has occurred:

[…] *many of the largest shipping lines have come together by forming strategic alliances with erstwhile competitors. They offer joint services by pooling vessels on the main commercial routes. In this way they are each able to commit fewer ships to a particular service route, and deploy the extra ships on other routes that are maintained outside the alliance. […] The 20 largest carriers controlled 26% of the world slot capacity in 1980, 42% in 1992 and about 58% in 2003. Those carriers have the responsibility to establish and maintain profitable routes in a competitive environment* Rodrigue et al. (2009).

The evolution of the shipping industry has gone hand in hand with the changes in port organisation. According to a study for the European Parliament (2009), from the growth of containerisation to what is known as the terminalisation era, where ports have become multifunctional operations through the development of highly specialised terminals, ports have undergone a major transformation in their management structures. The role of port authorities has also changed as the shipping system has evolved. Their main duties now involve the optimisation of process and infrastructures, logistics performance, promotion of intermodal transport systems and increased relations with their hinterlands.

In light of these observations, our objective in this chapter is to examine how maritime shipping can be modelled through the use of CAS theory. If we assume that emergent global behaviour such as international trade can be explained through bottom-up phenomena arising from the local interaction among individual agents, our main goal is to understand how patterns emerge in the global shipping system. The argument of our analysis is organised as follows. In Section 2.2, we summarise the main features regarding the worldwide movements of goods. Section 2.3 provides a detailed discussion of the CAS methodology for maritime trade, and in Section 2.4 we discuss the opportunity to apply CAS modelling to the maritime system and conclude with a research agenda for future studies.

2.2   Analyses of the Cargo Ship Network

Two recent articles (Ducruet and Notteboom, 2012; Kaluza et al., 2010) examine the main characteristics of the Global Cargo Ship Network (GCSN). Other studies focus on sub-networks of the GCSN. Ducruet et al. (2010) have analysed the Asian trade shipping network, McCalla et al. (2005) the Caribbean container basin, Cisic et al. (2007) the Mediterranean liner transport system and Helmick (1994) analysed the North Atlantic liner port network. But the two studies of Kaluza et al. (2010) and Ducruet and Notteboom (2012) are particularly interesting because they examine the complete global network and give us a view of its macroscopic properties. However, one drawback is their inability to forecast future trends or changes in the

Table 2.1 Overview of the main features of the GCSN as proposed in the studies of Kaluza et al. (2010) and Ducruet and Notteboom (2012)

	Kaluza network (Kaluza et al., (2010))	GAL (Ducruet and Notteboom, (2012))	GAL (Ducruet and Notteboom, (2012))
Main features	Asymmetric (59% connections in one direction); structural robustness (densely connected)	Weighted indirect network; small network	Weighted indirect network; small network
# Vessels	Total = 11,226; container ship = 3,100; bulk dry carrier = 5,498; oil tankers = 2,628	Container ship = 1,759	Container ship = 3,973
Weights	Sum of cargo capacity between port i and port j	Not specified	Not specified
# Nodes	951	910	1,205
# Links	36,351	28,510	51,057
Min shortest path	2.5	2.23	2.21
Clustering coef.	0.49	0.74	0.73
Average degree; Max degree	76.5; -	64.1; 437	87.5; 610
$P(k)$	Right skewed but not power law	−0.62	−0.65
$P(w)$	Power law (1.71 ± 0.14)	–	–
Betweenness centrality	Strong correlation between degree and centrality with some exceptions	Suez and Panama canals have high centrality (vunerability of the GCSN)	

networks. Nevertheless, for our purpose, the aim of both studies is to characterise the global movements of cargo in order to define quantitative analyses on existing structural relations in the rapidly expanding global shipping trade network.

Table 2.1 highlights similarities and differences between the two studies on the GCSN. Kaluza et al. use the Lloyd's Register Fairplay for the year 2007, while Ducruet and Notteboom utilise the data set from Lloyd's Marine Intelligence Unit for the years 1996 (post-Panamax vessels period) and 2006 (introduction of 10,000+ TEU vessels).

Both studies apply different approaches to the network analysis and sometimes reach different conclusions. Ducruet and Notteboom built two different network structures: the first (Graph of Direct Links – GDL) takes only into account the direct links generated by ships mooring at subsequent ports, and the second (Graph of All Linkages – GAL) includes the direct links between ports which are called by at least one ship. Kaluza et al. (2010) differentiate between the movements according to the type of ship. Four networks are subsequently constructed: all available links, sub-network of container ship, bulk dry carriers and oil tankers. Despite the clear differences between the approaches adopted in the two studies, in order to compare them we have considered the complete network of ships' movements by Kaluza et al. and the GAL network of Ducruet and Notteboom.

All the networks are quite dense (on average, the ratio between number of edges and nodes is 37.2). Some network measures indicate a tendency of the GCSN to belong to the class of small world networks[1] given the high values of the Clustering Coefficient.[2] Small world networks are a special class of networks characterised by high connectivity between nodes (or, in other words, low remoteness among the nodes). In an economic setting, this property has an underestimated value, for example, in the retail market, and connections among firms can create clusters of small specialised firms that gravitate around a big firm (hub). This large firm uses small sub-peripheral firms to sub-contract production. Thus, both firms (the hub and peripheral ones) reach their goals and increase the economic entropy of the system (Foster, 2005).

The degree distribution $P(k)$[3] shows that "*most ports have few connections, but there are some ports linked to hundreds of other ports*" (Kaluza et al., 2010). However, when the authors examine the degree distribution in detail, they find that the GCSN does not belong to the class of scale-free networks.[4] Both studies show low power law exponents or right-skewed degree distributions. In this sense, the degree distribution analysis would have had a higher significance if the authors had showed a ranking of the ports over time. This would have allowed them to understand if there was ever a turnover of dominant hubs that in turn had led to the detection of competitive markets in maritime shipping. Opposite results would have depicted a constrained market.

Kaluza et al. (2010) also studied the GCSN as a weighted network where the distribution of weights and Strength[5] displays a power law regime with exponents higher than 1. These findings agree with the idea that there are a few routes with high-intense traffic and a few ports that handle large cargo traffic. The detection of power law regimes is often associated with inequality (i.e. distribution of income and wealth) or vulnerability in economic systems (Foster, 2005; Pareto, 1897). The correlation between Strength and Degree of each node also fits a power law; this implies that the amount of goods handled by each port grows faster than the number of connections with other ports. Hub ports do not have a high number of connections with other ports, but the connected routes are used proportionally by a higher number of vessels.

Unfortunately, Ducruet and Notteboom's work (Ducruet and Notteboom, 2012) does not provide results of the weighted network analysis over the years 1996 and 2006. Such as analysis would have allowed us to discuss relevant facts about the dynamics of flows in the main transatlantic routes and also address constructive criticism of the influence of the introduction of large loading vessels (post-Panamax era) on specific routes.

The centrality of ports in a network (i.e. the importance of a node) can be inspected with other topological measures rather than the crude number of connections per node (degree k).

---

[1] For an extensive review of complex networks, see Watts and Strogatz (1998), Albert and Barabási (2002) and Newman (2003).
[2] A measure of the tendency of nodes to cluster together.
[3] In network theory, the degree k is meant to be the number of connections of every node.
[4] For an extensive review of complex networks, see Albert and Barabási (2002), Newman (2003) and Hubner (2003).
[5] In the case of ship networks, Strength represents the sum of goods that pass through a port in a year, in other words, the sum of the links' weights that converge on a node.

In the case of GCSN, both studies use measures of the betweenness centrality.[6] Kaluza et al. (2010) highlight a high correlation between the degree k and betweenness centrality and then validate the theory that hub ports are also central points of the network. Ducruet and Notteboom detect interesting anomalies in the centrality of certain ports. Large North American and Japanese ports are not on the top rank position in terms of network centrality despite their traffic volume. The most central ports in the network are the Suez and Panama canals (as they are gateway passages), Shanghai (due to the large number of ships that "visit" the port) and ports such as Antwerp (due to its high number of connections.)

Although maritime shipping has been going through a tremendous period of expansion in the last decade, the underlying network has a robust topological structure that has not changed in recent years. Kaluza et al. (2010) observe the differences "in the movement patterns of different ship types." The container ships show regular movements between ports, which may be explained by the nature of the service they provide. Dry carriers and oil tankers tend to move in a less regular manner because they change their routes according to the demand of goods they carry.

Finally, maritime shipping appears to have gained a stronger regional dimension over the years. In 1996, there was a strong relation between European and Asian basins while in 2006 these connections appear to be weaker. Ducruet and Notteboom (2012) explain this as a twofold phenomenon. Each basin has reinforced the internal connectivity while the Asian basin is witnessing a strong increase in the volume of goods shipped. The direct consequence is that Asian countries have been splitting their links with European countries. Physical proximity also helps to explain the increase in regional basins as well as the establishment of international commercial agreements such as the NAFTA and MERCOSUR between North and South America (Ducruet and Notteboom, 2012).

2.3   A Complex Adaptive Systems (CASs) Perspective

In the previous section, we have discussed two recent studies that take into account a static analysis of the global cargo shipping network. From these studies, we conclude that GCSN is a small world network with some power law regimes when it is examined as a weighted network. This evidence indicates that the underlying structure is not dominated by random rules, and that the complex organisation emerges from the interaction of lower-level entities.

In shipping, *self-organisation* is identified as a bottom-up process arising from the simultaneous local *non-linear* interactions among agents (i.e. vessels, ports, shipping alliances or nations according to the scale of analysis). This allows us to note that in GCSN, our aim is to understand why certain ports are able to play a leading role and also to estimate the shipping trade trends. Using an example from the nature, flocking birds generate patterns based on local information. Each bird learns from other birds and adapts its speed and direction accordingly in order to reach the next spot. In the same way, shipping companies compete in the market according to their own interests. The introduction of innovation makes a company more competitive and sets new rules in the market that force other companies to co-evolve accordingly in order to be profitable. This adaptive process has been witnessed in maritime shipping at different stages with the introduction of new technologies such as improvements in the fleets

---

[6] The betweenness centrality of a node is the number of topologically shortest paths that pass through that node.

(launch of post-Panamax ships) or in the management processes in the seaports (automation of loading and unloading services).

Based on the work in (Ducruet and Notteboom, 2012; Kaluza et al., 2010), our next step is to identify a set of CAS features related to shipping systems. We select 10 characteristics extracted from a number of works that have proposed applications of CAS modelling (Wallis, 2008). In Table 2.2, we relate each characteristic to Holland's classification described in Appendix and to a possible CAS modelling application for shipping systems. In the reminder of this section, we discuss how these 10 characteristics are constructive elements for a CAS shipping system.

As shown in Section 2.2, international shipping is a large collection of entities (Table 2.2 – Feature: *Many interacting/interrelated agents*) whose interactions create non-linear trends (Table 2.2 – Feature: *Non-linear/Unpredictable*). Given these two analytical perspectives, we can examine the local interactions among ships and how they are assigned to different ports according to price and demand of the goods they carry (Table 2.2 – Feature: *Goal seeking*).

However, according to the modelling proposed in Ducruet and Notteboom (2012) and Kaluza et al. (2010), seaports may be considered as agents of a CAS. In this case, the most interesting questions revolve around understanding how a shipping system evolves in relation to external shocks (Table 2.2 – Feature: *Co-evolutionary*). For instance, in cases of sudden undesired events such as terrorist attacks or extreme natural phenomena (earthquakes and hurricanes), the cargo shipping network would *co-evolve* in order to maintain the same level of provided service if a big seaport hub were to disappear or be severely damaged.

In the case of natural systems, the fundamental questions are: how would an ecosystem evolve if a species disappeared? Would it be replaced by new species and would other species be able to survive without it? We can apply those questions to the case of maritime shipping in order to forecast future configurations as well as prevent global breakdowns in national and international markets (Table 2.2 – Feature: *Self-organisation*).

The cargo shipping industry comprises several sectors such as international maritime transport, maritime auxiliary services and port services, which play a relevant role; they have a fairly long history of co-operation since the 1990s with the formation of consortia and alliances. Each co-operation is regulated by a wide range of *"national and international regulations responding to specific issues that have arisen as the international trading system has evolved"* (Hubner, 2003). The outcomes of these collaborations influence the setting of freight rates and shipping companies' tariffs. In the light of the previous remarks, co-operation among agents (shipping companies, port authorities and so on) should be included in the modelling (Table 2.2 – Features: *Distributed control* and *Nested Systems*).

In particular, the international economic alliances in trade agreements play a significant role in the definition of trade flows and development. For instance, China's admittance into the WTO has affected the bilateral negotiations between WTO countries and China itself as well as among former members (Table 2.2 – Feature: *Co-evolutionary* and *Self-organisation*), but other examples of international trade agreements show similar impacts on international trade processes (NAFTA among North American countries, MERCOSUR in South America, ASEAN-AFTA among five Asian countries and the Trans-Pacific Strategic Economic Partnership (TPP) in the Asian-Pacific region).

Based on these observations, when modelling shipping relationships, trade agreement memberships should be included for two reasons: first, in order to understand the actual effects on

Table 2.2 Comparison of Complex Adaptive Systems (CAS)s features with shipping

Features	Description	Authors	Holland basics	Maritime shipping system
Self-organisation	Formation of regularities in the patterns of interactions of agents that pursue their own advantages through simple rules	Brown. Eisenhardt; Harder, Robertson, Woodward; Lichtenstein; McDaniel, Jordan, Fleeman; McKelvey; Pascale; Tower; Zimmerman, Lindberg, Plesk; Axelrod, Cohen	Tagging, non-linearity	The GCSN is a small world network with some power law regimes when inspected as a weighted network. This evidence shows that the underlying structure is not dominated by random rules and that complex organisation emerges from the interaction of lower-level entities.
Many interacting/interrelated agents	Large number of locally interconnected and interacting rational agents that are continually pursuing their advantageous interests	Chiva-Gomez; Brown, Eisenhardt; Harder, Robertson, Woodward; Daneke; Dent; Hunt, Ropo; Lichtenstein; McGarth; McDaniel, Jordan, Fleeman; McKelvey; Moss; Olson, Eoyang; Kurtz, Snowden; Keshavarz, Nutbeam, Rowling, Khavarpour; Seel	Flows, tagging	This concept is already embodied in the definition of the trade shipping system. If we take into account the mere fleet system and the connections established between ports, we observe around 10,000 vessels, 1,000 ports and 50,000 connections (see Table 2.1 for details).

(continued)

Table 2.2 (*continued*)

Features	Description	Authors	Holland basics	Maritime shipping system
Distributed control	The outcomes of a CAS emerge from a process of self-organisation rather than being designed and controlled externally or by a centralised body	Holland; Chiva-Gomez; Keshavarz, Nutbeam, Rowling, Khavarpour; Seel	Flows, internal model	Although there are international trade agreements that unavoidably influence maritime shipping, these pacts can be seen as external forces that increase the system's entropy and prompt more economic relationships
Non-linear/ Unpredictable	Interactions are non-linear and thus intractable from a mathematical point of view	Holland; Chiva-Gomez; Dent; Hunt; Lichtenstein; McGarth; McDaniel, Jordan, Fleeman; McKelvey; Moss; Olson, Eoyang; Tower; Stacey; Kochugovindan, Vriend; Foster	Non-linearity	The GCSN shows power law fit distributions and not random topological structures. This signals the emergence of non-linear interactions between system's agents
Co-evolutionary	The environment is influenced by the activities of each agent	Chiva-Gomez; Dent; Hunt; Lichtenstein; Pascale; Funtowitcz, Ravetz; Zimmerman, Lindberg, Plesk; Seel	Diversity, tagging	Introduction of post-Panamax and 10,000+ TEU ships change carriers routing networks and tariffs as well as the volume of transshipped cargo handled at main ports

	Description	References	Keywords	Port application
Emergence	The interplay between agents shapes a hidden but recognisable regularity (i.e. the brain has consciousness but not the single neurons)	Seel; Holland; Stacey	Aggregation, flows, internal model	Emergence of regional clusters of ports.
Goal seeking.	The agents try to adapt in order to fulfil their goals	Chiva-Gomez; Brown, Eisenhardt; Daneke; Kurtz, Snowden; Funtowitcz, Ravetz; Zimmerman, Lindberg, Plesk	Flows, internal model	Dry carriers and oil tankers tend to move in an irregular manner because they change their routes according to demand of goods they carry
Nested systems	Each agent can be considered as a system. Each system is a part of something bigger, thus each system can be a sub-system of a bigger system	Tefatsion; Seel	Diversity, internal model	Ports alliances at national or international level are nested clusters of ports. The same port may belong to a cluster of ports at the national level and to a cluster of ports at the international level, but this category may not necessary include all the ports it belongs to within the national cluster

Source: Reproduced with permission from Caschili and Medda (2012).

agents involved in the agreements, and second, to understand the effects generated on agents who are not members of a specific trade bloc. In this regard, a CAS application on maritime international trade would help us to better understand the role of alliances in trade, the effects of the establishment of new alliances or the admission of new members in existing agreements (Table 2.2 – Feature: Emergence).

The aforementioned are some of the questions a CAS application should potentially be able to answer when policy constraints are reckoned with the agents' behaviour modelling (Table 2.2 – Feature: Distributed control). Referring to Holland's classification, the modeller has to set up the internal model of each agent so that it takes into account the distinguishing factors an agent uses to direct its economic choices. For example, national and international port alliances are nested clusters of ports. A single port may belong to a cluster of ports at the national level and also belongs to a cluster of ports at the international level. Conversely, not all ports in national cluster are necessarily part of an international cluster. These structures emerge at the time of mutual interactions between the agents.

2.4   Conclusions: The Benefits of a Systems Perspective

In an article in *Nature*, Farmer and Foley (2009) call for more applications of CAS in economics, asserting that "*agent-based models potentially present a way to model the financial economy as a complex system, as Keynes[7] attempted to do, while taking human adaptation and learning into account, as Lucas advocated.*"

Unlike classic top-down approaches, whose modelling components are carefully designed and evaluated, the CAS theory proposes bottom-up methods based on the modelling of simple interactions among its components (or agents) that generate complex, robust and flexible phenomena and macro regularities. The only duty of the modeller is to set the initial rules that regulate the agent–agent interactions and report the initial conditions of the environment where the agents act.

CAS application may be useful for multiple aims. Following is the list of some of the goals that may be achieved in applying CAS theory in maritime trade:

- To test standard economic theories;
- To understand the spatial structure and organisation of economies such as the formation of regional clusters of ports, business agglomerations and industrial alliances;
- To understand why certain types of co-operation among shipping firms appear to be more adaptable than others and to know what factors regulate the stable relationship among them;
- To provide policy makers with a set of comprehensive tools able to address issues of growth, distribution and welfare connected to global trade trends;
- To put into practice an integrated multidisciplinary approach among fields belonging to social science.

Many scholars have thrown down the gauntlet to the scientific community for a multidisciplinary approach based on the CAS paradigm (Holland, 1995; Martin and Sunley, 2007). Communication across disciplines may allow discerning clues from different fields and quicken

---

[7] According to Keynes' theory, active government intervention in the marketplace and monetary policy is the best method of ensuring economic growth and stability.

theoretical and practical advances. Complexity in economics is currently lacking provision of a more general theory but is characterised by the application of ad hoc ideas and concepts. In previous paragraphs, we have argued that even in CAS theory there is not a shared theoretical framework whose scholars gather in the application of its notions. Most of these notions derive from physical, chemical and biological studies and their assumptions are "*restricted to the content of the specific scientific models used*". Economic systems are unambiguously complex systems, as they are highly interconnected, adaptive, self-organising and emergent systems. The scientific community should take steps towards the construction of a more tenable ontological framework while scholars and modellers should derive a common approach from that framework.

References

Albert, R. and Barabási, A.L. (2002) Statistical mechanics of complex networks. *Reviews of Modern Physics*, **74**, 47–97.

Anderson, P.W. (1994) The eightfold way to the theory of complexity: A prologue, in *Complexity: Metaphors, Models, and Reality* (eds G.A. Cowan, D. Pines and D. Meltzer), Addison-Wesley, USA, pp. 7–16.

Arthur, W.B. (1997) Introduction: process and emergence in the economy, in *The Economy as an Evolving Complex System II* (eds W.B. Arthur, S. Durlauf and D.A. Lane), Mass, Reading, pp. 1–4.

Barthelemi, M. (2011) Spatial networks. *Physics Reports*, **499**, 1–101.

Buchanan, M. (2002) *Ubiquity: Why Catastrophes Happen*, Weidenfeld and Nicolson, London.

Caschili S., Medda F. (2012) "A Review of the Maritime Container Shipping Industry as a Complex Adaptive System", *INDECS* **10**(1), 1–15.

Cisic, D., Komadina, P. and Hlaca, B. (2007) *Network Analysis Applied to Mediterranean Liner Transport System.* In conference proceedings of the International Association of Maritime Economists Conference, Athens, Greece, July 4–6.

Corning, P.A. (2002) The re-emergence of "Emergence": A venerable concept in search of a theory. *Complexity*, **7** (6), 18–30.

Dooley, K. (1996) Complex adaptive systems: A nominal definition. *The Chaos Network*, **8** (1), 2–3.

Ducruet, C., Lee, S.W. and Aky, N.G. (2010) Centrality and vulnerability in liner shipping networks: revisiting the Northeast Asian port hierarchy. *Maritime Policy and Management*, **37** (1), 17–36.

Ducruet, C. and Notteboom, M. (2012) The worldwide network of container shipping: spatial structure and regional dynamics. *Global Networks*, **12** (3), 395–423.

European Parliament – Directorate General for Internal Policies (2009) *The Involving Role of EU Seaports in Global Maritime Logistics – Capacities, Challenges and Strategies*, Policy Department B: Structural and Cohesion Policies.

Farmer, J.D. and Foley, D. (2009) The economy needs agent-based modelling. *Nature*, **460**, 685–686.

Foster, J. (2005) From simplistic to complex systems in economics. *Cambridge Journal of Economics*, **29** (6), 873–892.

Gell-Mann, M. (1994) Complex adaptive systems, in *Complexity: Metaphors, Models, and Reality* (eds G.A. Cowan, D. Pines and D. Meltzer), Addison-Wesley, USA, pp. 17–45.

Gell-Mann, M. (1995) *The Quark and the Jaguar: Adventures in the Simple and the Complex*, Owl Books, New York.

Glen, D. and Marlow, P. (2009) Maritime statistics: a new forum for practitioners. *Maritime Policy & Management*, **36** (2), 185–195.

Helmick, J.S. (1994) *Concentration and Connectivity in the North Atlantic Liner Port Network, 1970–1990*, University of Miami, Miami.

Holland, J.H. (1992) *Adaptation in Natural and Artificial Systems*, The MIT Press, Cambridge.

Holland, J.H. (1998) *Emergence: From Chaos to Order*, Addison-Wesley, Redwood City.

Holland, J.H. (1995) *Hidden Order: How Adaptation Builds Complexity*, Addison Wesley Publishing Company, USA.

Hubner, W. (2003) *Regulatory Issues in International Maritime Transport*. Prepared for the Organization for Economic Co-operation and Development (OECD), Directorate for Science, Technology and Industry, Division of Transport.

Kaluza, P., Kolzsch, A., Gastner, M.T. and Blasius, B. (2010) The complex network of global cargo ship movements. *Journal of the Royal Society Interface*, **7** (48), 1093–1103.

Kochugovindan, S. and Vriend, N. (1998) Is the study of complex adaptive systems going to solve the mystery of Adam Smith's invisible hand? *The Independent Review*, **1**, 53–66.

Levin, S.A. (2003) Complex adaptive systems: Exploring the known, the unknown and the unknowable. *American Mathematical Society*, **40**, 3–19.

Levin, S.A. (1998) Ecosystems and the biosphere as complex adaptive Systems ecosystems; biomedical and life sciences and earth and environmental science. *Ecosystems*, **1**, 431–436.

Markose, S.M. (2005) Computability and evolutionary complexity: Markets as complex adaptive systems (CAS). *The Economic Journal*, **115**, 159–192.

Martin, R. and Sunley, P. (2007) Complexity thinking and evolutionary economic geography. *Journal of Economic Geography*, **7**, 579–601.

McCalla, R., Slack, B. and Comtois, C. (2005) The Caribbean basin: adjusting to global trends in containerization. *Maritime Policy and Management*, **32** (3), 245–261.

Newman, M.E.J. (2003) The structure and function of complex networks. *SIAM Review*, **45** (2), 167–256.

Notteboom, T. (2004) Container shipping and ports: An overview. *Review of Network Economics*, **3** (2), 86–106.

Pareto, V. (1897) Cours d'économie politique professé a l'Université de Lausanne. Vol. I, Paris.

Rodrigue, J.P., Comtois, C. and Slack, B. (2009) *The Geography of Transport Systems*, 2nd edn, Routledge, New York.

Rodrigue, J.P., Comtois, C. and Slack, B. (2006) *The Geography of Transport Systems*, Routledge, London.

Wallis, S.E. (2008) Emerging order in CAS theory: Mapping some perspectives. *Kybernetes*, **37** (7), 1016–1029.

Watts, D.J. and Strogatz, S.H. (1998) Collective dynamics of 'small-world' networks. *Nature*, **393**, 409–410.

Appendix

A.1 Complexity Science and Complex Adaptive Systems: Key Characteristics

Some scholars (Holland, 1998; Corning, 2002; Markose, 2005) define complex systems by observing particular features within a given system such as *emergent, self-organising/adaptive, non-linear interactions* and *in evolution*. For instance, emergent phenomena are classifiable by demonstrating their unpredictable behaviours when we account for each part of the system. This concept is epitomised by the famous sentence *"the whole is greater than the sum of the parts"* (Markose, 2005; Buchanan, 2002). Recessions and financial growth are, for example, emergent phenomena of national economies.

The class of CAS is one of the conceptualisations belonging to the framework of complex systems. According to Anderson (1994), scholars have theorised approaches in order to better understand complexity: Mathematical (Turing and von Neuman), Information theory, Ergodic theory, Artificial entities (cellular automata), Large random physical systems, Self-organised critical systems, Artificial intelligence and Wetware. Anderson's classification places CASs into the class of Artificial Intelligence approach. What characterises this peculiar class of complex system are the processes of adaptation and evolution. A system is adaptive when its agents *"change their actions as a result of events occurring in the process of interaction"* (Kochugovindan and Vriend, 1998). Evolution is created by the local interactions among agents. In this sense, adaptation can be seen as a passive action in which the agents absorb information from the surrounding environment (or from previous experience).

Conversely, evolution is generated by the mutual actions among agents. Figure A.1 shows how adaptation and evolution are embedded in different classes of systems. In our view, based on the previous definitions, complex systems are adaptive and evolving systems. Unintelligent

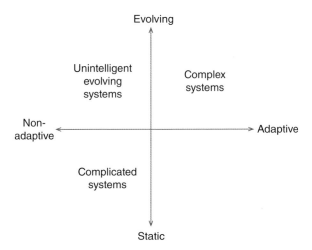

Figure A.1 Graph of systems showing their ability to evolve and adapt. *Source*: Reproduced with permission from Caschili and Medda (2012)

evolving systems evolve through interaction processes but they do not adapt. For example, a crystal is generated by the mutual interactions among atoms or molecules that have no intelligence of the process in which they are involved. Furthermore, complicated systems are made by numerous interacting elements that do not adapt or evolve in the system. Complicated artefacts such as a car engine belong to this class. The IV quadrant of the graph in Figure A.1 is empty, as no system that is adaptive shows static structures. Adaptation and evolution play off each other, and by this we mean that the adaptation process includes the concept of evolution but not the reverse.

According to Wallis (2008), there is no consensus on CAS unified theory. Holland (1992) nonetheless calls for the need for a unified theory of CAS. Although many authors developed compressive frameworks (Gell-Mann, 1994, Gell-Mann, 1995, Dooley, 1996, Arthur, 1997, Levin, 1998), in this work we focus on Holland's (1995) approach to modelling CASs. His framework is used widely in much of CAS literature, especially in economic applications. In one of the most robust works towards a theory of CAS Holland (1995) suggests four properties and three mechanisms a CAS must possess. Wallis (2008) states that Holland seven attributes for CAS are not the only ones a CAS may possess, but we agree with him "*other candidate features can be derived from appropriate combinations of these seven.*" We present below a summary of the seven basic features, and group them into properties and mechanisms.

A.1.1   Four Properties

- **Aggregation:** The concept of aggregation is twofold. The first facet involves how the modeller decides to represent a system. What features to leave in and what to ignore are of paramount importance. In this sense, elements are aggregated in reusable categories whose combinations help in describing scenes, or to be more precise, "*novel scenes can be decomposed into familiar categories.*" The second meaning we can ascribe to aggregation properties for CASs relates to the emergence of global behaviours caused by local interactions, where agents tend to perform actions similar to other agents rather than to adopt independent configuration. Furthermore, aggregation often yields co-operation in that the same action of a number of agents produces results unattainable by a single agent. We can explain this concept through the example of the ant nest. An ant survives and adapts to different conditions when its action is coordinated with a group of ants (ant nest), but it will die if it works by itself. Likewise, in a CAS, a new action will survive and induce global effects if it is adopted by a large number of agents.
- **Non-linearity**: Agents interact in a non-linear way so that the global behaviour of the system is more than the sum of its parts.
- **Flows**: Agents interact with one another to create networks that vary over time. The recursive interactions create a *multiplier effect* (interactions between nodes generate outcomes that flow from node to node creating a chain of changes) and a *recycling effect* (in networks cycles improve the local performance and create striking global outcomes).
- **Diversity**: Agent persistence is highly connected to the context provided by the other agents to define "*the niche where the agent outlives.*" The loss of an agent generates an adaptation in the system with the creation of another agent (similar to the previous) that will occupy the same niche and provide most of the missing interactions. This process creates diversity in the sense that the new specie is similar to the previous but introduces a new combination of

features into the system. The intrinsic nature of a CAS allows the system to have progressive adaptations, further interactions and be able to create new niches (the outcome of diversity).

A.1.2 Three Mechanisms

- **Tagging**: Agents use the tagging mechanism in the aggregation process in order to differentiate other agents with particular properties; this facilitates a selective interaction among the agents.
- **Internal models**: these are the basic models of a CAS. Each agent has an internal model that filters the inputs into patterns and learning from experience. The internal model changes through agent interactions, and the changes bias future actions (agents adapt). Internal models are unique to each CAS and are a basic schema for each system. The internal model takes input and filters it into known patterns. After an occurrence first appears, the agent should be able to anticipate the outcome of the same input if it occurs again. Tacit internal models only tell the system what to do at a current point. Overt internal models are used to explore alternatives, or to look to the future.
- **Building blocks**: With regard to the human ability to recognise and categorise scenes, CAS uses the building blocks mechanism to generate internal models. The building block mechanism decomposes a situation evoking basic rules learnt from all possible situations it has already encountered. The combination of the seven features has allowed analysts to define environments where adaptive agents interact and evolve. In the next section, we examine two specific studies dedicated to global cargo ship network through the lens of CAS.

3

An Interdependent Multi-layer Model for Trade

Simone Caschili, Francesca Medda and Alan Wilson

3.1 Introduction

We saw in the previous chapter how the international trade system is made up of a number of interdependent networks. We now aim to model this system by articulating the horizontal and vertical connections in these networks to generate what we call the Interdependent Multi-layer Model (IMM). The IMM allows us to estimate bilateral trade flows between countries (imports and exports). This model has a range of uses and in particular allows us to investigate the effects of exogenous shocks in the system.

Natural disasters such as Hurricane Katrina and the 2011 earthquake in Japan underline how our cities, regions and nations are composed of complex interdependent systems (Akhtar and Santos 2013; Burns and Slovic, 2012; Levin et al., 1998). We are especially interested in the consequences for trade and whether we can evaluate the ability of systems to recover after a disruption? What can we learn about their resilience capacity in order to design better and more reliable economic systems?

Recent advances in complexity science have facilitated such investigations into resilience to disruptive exogenous and endogenous events (Marincioni et al., 2013; Liu et al., 2007, Holt et al., 2011; Markose, 2005; Rosser, 1999). For example, global supply chains are characterised by interrelated organisations, resources and processes that create and deliver products and services to consumers in countries around the world. The activity of buying a product or using a service then not only impacts on the national economy where the purchase happens but also affects several other international economies (Bade and Parkin, 2007). Global supply chains are grounded in the notion that each firm contributing to the chain can provide a

Global Dynamics: Approaches from Complexity Science, First Edition. Edited by Alan Wilson.
© 2016 John Wiley & Sons, Ltd. Published 2016 by John Wiley & Sons, Ltd.

service or good at a price below the domestic price of final consumption (Teng and Jaramillo, 2005). Such an economic system is likely to be profitable when a particular country has an economic advantage due to its availability of specific natural resources, a cheap labour force or a specialised technology (Garelli, 2003). But this is not enough. An efficient distribution chain must be structured so that equal relevance be given to hard infrastructures (i.e. ports, railways, roads) as well as soft infrastructures (how infrastructures are operated) of the chain, which ultimately determine the competitive advantage of low transport costs (Gunasekaran et al., 2001). As global trade increases, firms enlist trusted partners when operating within the international trading environment. International economic agreements have in fact enabled increased volumes of capital, goods and services transacted across traditional state borders, thereby raising direct foreign investments and the level of economic engagement across countries (Büthe and Milner, 2008).

This chapter is set out as follows. In Section 3.2, the main concepts underlying the construction of the model are discussed. In Sections 3.3–3.4, the assumptions, the layers and the algorithm comprising the model are examined. In Section 3.5, our data set and calibration process are presented; in Sections 3.6 and 3.7, the results of the model at the steady state and after the introduction of system shocks are clarified. Section 3.8 concludes with comments on future extensions of the present work.

3.2   The Interdependent Multi-layer Model: Vertical Integration

Networks are ubiquitous across societies and economies in the form of physical networks, such as transportation, communication and utility services, and also as intangible economic, financial and social networks (Clauset et al., 2008; Newman, 2003; Strogatz, 2001). Whatever their nature, as technology progresses, we are becoming more dependent on networks and networks are simultaneously becoming more interdependent (Burger et al., 2014; Rinaldi et al., 2001). The organisation and growth of a network determine its success or failure and consequently influence other interdependent networks (Castet and Saleh, 2013). For instance, over the last 15 years, we have witnessed the exponential growth of the World Wide Web; this virtual network has transformed numerous other networks, including transportation, commerce and supply chains. In the case of retail business, the introduction of e-commerce as an alternative to traditional commerce has increased competitiveness, provided countless specialised services and opened access to much wider markets (Harvie, 2004).

We can conceptualise the interdependence among networks in a hierarchical structure. A high-level network such as the one carrying trade flows is underpinned by other networks (see Figure 3.1).

In Figure 3.1a, trade is depicted as a function of common currency in yellow colour, as a membership in alliances such as the EU and as a common language. Different kinds of noteworthy events are shown in Figure 3.1b that have strong impacts on trade: financial failure, migration, illicit trade and cyber attacks.

As shown in Figure 3.2, the 'layers' can be represented in functional relationships in which an upper layer is a dependent variable and the next layer is the set of associated independent variables. Any of the latter can in turn be seen as a dependent variable with its own set of independent variables – hence we have multi-layering. Our next step is to develop this argument explicitly through the use of the relationships shown in Figure 3.1. By using an aggregate picture of the system under consideration, we are able to demonstrate the idea more easily.

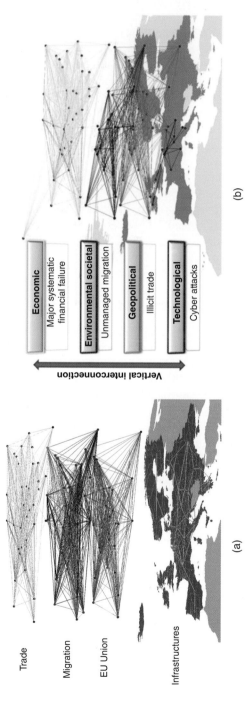

Figure 3.1 A multi-layer network's functional dependencies (a) and 'events' (b). *Source:* Reproduced from Caschili et al. (2015) with permission of Springer

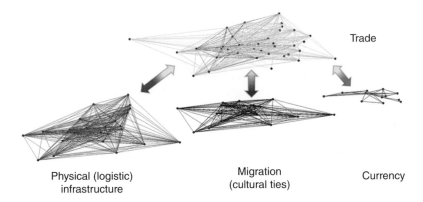

Trade

Physical (logistic)
infrastructure

Migration
(cultural ties)

Currency

Figure 3.2   Functional relationship between networks. *Source*: Reproduced from Caschili et al. (2015) with permission of Springer

These are all potentially interactive layers. Consider, for example, the generalised transport cost (i.e. physical infrastructure layer in Figure 3.2), which represents the network of multi-lateral resistance to the movement of goods between countries (Aashtiani and Magnanti, 1981). The physical layer not only represents the transport cost from country i to country j but also carries information about the underlying physical infrastructure. A disruption in the physical infrastructure (due to exogenous events such as an earthquake, hurricane or tsunami) generates impacts on transport costs. After an extreme event, companies are compelled to re-route their supply chains, resulting in added days and higher logistics costs (Tang, 2006). More specifically, if we treat transport cost as a network of relationships between supply and demand, we can evaluate the effects generated on bilateral trade flows. The same idea of network of relationships can be applied to other layers, including economic agreements, financial relationships and cultural ties. Furthermore, the hierarchical relational model described for the trade layer can be replicated for other variables and thus leads to a nested hierarchical system. We represent this idea in Figure 3.3.

In the next section, we will build the model and calibrate it for the steady state that will enable us to test the impact of the possible hazards shown in Figure 3.1b: systematic financial failure, unmanaged migration, illicit trade flow, and cyber attacks or piracy.

Trade

First-layer
determinants

Second-layer
determinants

Figure 3.3   A nested hierarchical multi-layer network. *Source*: Reproduced from Caschili et al. (2015) with permission of Springer

3.3   Model Layers

Various studies have noted tangible and intangible factors that affect international trade (Linders et al., 2005; Anderson and van Wincoop, 2004; Deardorff, 1998). Here, we consider factors as interrelating layers in accordance with the vertical scheme discussed in Section 3.2. We have grouped factors in economic, socio-cultural and physical layers. In the remainder of this section, we provide a description of each layer and the models we have constructed for the estimation process.

3.3.1   Economic Layer

International trade is shaped by the factor productivities of every country, and certainly from this perspective, trade leads to specialisation and interdependence within different national economies. However, as Helpman (2011) observes: 'exports are not valuable *per se*, but rather via the *quid pro quo* of the exchange in which they pay for imports'; in other words, bilateral trade is affected by changes in the levels of various economic elements such as exchange rates (Bergstrand, 1985), border tariffs (Yue et al., 2006) and national wealth (Hufbauer, 1970). In our model, we assume that these economic elements not only interact with the other non-economic considered layers but also in turn have the capacity to shape, re-adapt and create the overall trade pattern. The two relationships are therefore mutual.

3.3.1.1   Bilateral Trade

We use a spatial interaction model framework (Wilson, 1967, 1970) in order to describe the bilateral interactive nature of trade. Let T_{ij} be a measure of the trade flow from country i to country j measured in monetary value (USD). To guarantee the consistency of our model, we impose constraints to ensure that exports (E) and imports (I) are equal to the sum of outward and inward flows of each country:

$$\sum_j T_{ij} = E_i \qquad (3.1)$$

$$\sum_i T_{ij} = I_j \qquad (3.2)$$

Following a classic notation for spatial interaction models, we can write the trade between two countries i and j as follows:

$$T_{ij} = A_i B_j E_i I_j f(\beta, \text{cost}_{ij}) \qquad (3.3)$$

where $f(\beta, \text{cost}_{ij})$ is a general measure of impedance between i and j as a function of the unit cost of shipping, in which cost_{ij} is measured in monetary value per unit of volume per unit of distance. β is a parameter that determines sensitivity to transport costs. A_i and B_j are balancing factors to ensure that Equations (3.2) and (3.3) are satisfied and can be written as

$$A_i = \frac{1}{\sum_j B_j I_j f(\beta, \text{cost}_{ij})} \qquad (3.4)$$

$$B_j = \frac{1}{\sum_i A_i E_i f(\beta, \text{cost}_{ij})} \tag{3.5}$$

We estimate the model in Equation (3.1) from data using the Correlates of War Project's Trade Data (Barbieri et al., 2009). The data set includes annual dyadic and national trade figures. Bilateral trade is obtained from the International Monetary Fund's Direction of Trade Statistics (2012) (see Section 3.5 for details).

We develop the multi-layer nested framework introduced in Section 3.2 by considering physical and cultural layers. We account for the cultural factors as networks, which we mathematically represent as matrices: exchange rate (ER_{ij}), a variable identifying whether two countries share a border (CB_{ij}) and cultural ties (CL_{ij}). The spatial interaction model of Equation (3.3) can now be written as

$$T_{ij} = A_i B_j E_i I_j f(\beta, \text{cost}_{ij}) \text{ER}_{ij} \text{CB}_{ij} \text{CL}_{ij} \tag{3.6}$$

where balancing factors are expressed as

$$A_i = \frac{1}{\sum_j B_j I_j f(\beta, \text{cost}_{ij}) \text{ER}_{ij} \text{CB}_{ij} \text{CL}_{ij}} \tag{3.7}$$

$$B_j = \frac{1}{\sum_i A_i E_i f(\beta, \text{cost}_{ij}) \text{ER}_{ij} \text{CB}_{ij} \text{CL}_{ij}} \tag{3.8}$$

When we examine the schema presented in Figures 3.2 and 3.3, the matrices that are components of Equation (3.6) such as $\{T_{ij}\}$, $\{\text{cost}_{ij}\}$, $\{\text{ER}_{ij}\}$, $\{\text{CB}_{ij}\}$ and $\{\text{CL}_{ij}\}$ can now all be regarded as networks, or layers, of the IMM.

Economic Scale of a Country

For economic scale of a country, we assume here that countries with large net exports of products have equally large domestic production and domestic absorption, consisting of consumption, investment and government spending. We represent this scale by the Gross Domestic Product (GDP). We also consider the GDP per capita (GDPpc) (which is often used to appraise the national wealth). GDP is calculated using the expenditure method as follows:

$$\text{GDP}_i = C_i + \text{INV}_i + G_i + E_i - I_i \tag{3.9}$$

where C_i is spending on personal consumption (durable goods, non-durable goods and services); INV_i is the gross private domestic investment (residential, non-residential and change in business); G_i is the total government spending; and E_i and I_i are, respectively, exports and imports. We can divide GDP into the national component (NC) and net trade component (NetT):

$$\text{GDP}_i = \text{NC}_i + \text{NetT}_i \tag{3.10}$$

$$\text{NC}_i = C_i + \text{INV}_i + G_i \tag{3.11}$$

$$\text{NetT}_i = E_i - I_i \tag{3.12}$$

We estimate NetT$_i$ at each time t by calculating trade flows through Equation (3.6). The national component of GDP is estimated proportional to the population of the country:

$$NC_i(t) = \alpha_i(t) * Pop_i(t) \tag{3.13}$$

$\alpha_i(t)$ is a coefficient that considers the national consumption, investments and government spending per capita. We estimate the value of $\alpha_i(t)$ at each time t as follows:

$$\alpha_i(t) = \alpha_i(t-1) * (1 + \delta_i(t)) \tag{3.14}$$

$\delta_i(t)$ is a stochastic coefficient that is proportional to the annual variation of GDPpc and economic growth rate of country i.

3.3.1.2   Exchange Rate

At present, there is much interest in determining the effect of currency on trade. Indeed, one important reason for forming a currency union is the promotion of trade within the union. With the exception of currency unions, the exchange rate between two countries might favour or impede bilateral trade. We assume that the higher the ratio between exporter country i and importer country j, the higher the trade between the two countries. We construct the exchange rate layer where a link between two countries i and j is equal to the ratio between the two countries' exchange rate (ER) (local currency relative to the US dollar):

$$ER_{ij} = \frac{ER_i}{ER_j} \tag{3.15}$$

3.3.1.3   Transport Cost

Transport cost is one of the major variables affecting bilateral trade. To some extent, import choices are made to minimise transport costs (Hummels, 1999). The literature on international trade has shown that transport costs are mainly influenced by freight rates, economies of scale, infrastructure quality, and importer/exporter location (Immaculada and Celestino, 2005). From the shipper's point of view, there are three categories of costs that determine freight rates: capital costs (fleet acquisition), direct fixed costs (insurance, fleet maintenance, personnel costs, etc.) and variable costs (combustible costs, port costs and type of product traded). Rauch (1999) has shown that differentiated products (i.e. products with a distinct identity and over which the manufacturer has some control regarding price) have higher freight costs due to the search cost buyer–supplier. Against this background, gravity models have been extensively applied to represent bilateral trade (Tinbergen, 1962; Pullianen, 1963; Linneman, 1966; Bergstrand, 1985; Anderson and Van Wincoop, 2004; Baldwin and Taglioni, 2006) and geographical distance has been often used to approximate transport costs. Results of these models have proved that spatial distance is a good proxy for transport costs when trade is studied at the aggregated level, while it is not clear whether this good performance of the proxy remains if trade is analysed at the disaggregated level (Hummels, 1999; Immaculada and Celestino, 2005). In this study, we consider two types of transport costs: geographical distance and a generalised version of transport cost, TC$_{ij}$, between location i and j based on the work of Limao and Venables (2001) as follows:

$$TC_{ij} = f(d_{ij}, CB_{ij}, Isl_i, Isl_j, Y_i, Y_j, Inf_i, Inf_j) \tag{3.16}$$

where d_{ij} is the travel distance between countries i and j, CB_{ij} takes into consideration common borders between countries i and j, Isl accounts for the insularity of a country, Y_i is the per capita income and Inf_i is an indicator of the quality of the infrastructure and the logistics of a country. Because of lack of information we have introduced some simplifications in Equation (3.16): the GDPpc is used as a proxy of the per capita income, and we do not consider insularity in determining transport costs.

3.3.2 Social and Cultural Layer (Socio-cultural)

Population and measures of 'cultural distance' have also been considered as determinants of international trade (Glick and Rose, 2002). In fact, population increases trade and the level of specialisation (Matyas, 1997). Population may have a positive impact on trade flows in the short run, since it may increase the number of the labour force, the level of specialisation and more products to export as a result. However, in the long run, higher population has a tendency to decrease income per capita, making every individual poorer, and therefore it may cause production and exports to decrease. In addition to that, lower income per capita tends to decrease the demand for imports as well.

3.3.2.1 Population

We estimate the variation of population at each time t as the variation due to the internal demographic changes (birth and death rates) and the net migration in the country:

$$Pop_i(t) = Pop_i(t\text{-}1) + births_i(t) - deaths_i(t) + in\text{-}migration_i(t)\text{-}out\text{-}migration_i(t) \quad (3.17)$$

In order to estimate internal demographic change, we consider an average constant value of births and deaths at each time t for each country.

We estimate migration flows between countries as a function of GDPpc of the two countries, the population in the two countries, physical distance and cultural distance:

$$M_{ij} = M_{ij}(GDPpc_i, GDPpc_j, Pop_i, Pop_j, d_{ij}, CL_{ij}) \quad (3.18)$$

We expect that the higher the difference in GDPpc of the two countries, the higher the migration will be from the country with lowest GDPpc to the country with higher GDPpc. Physical distance should negatively influence migration while a cultural linkage should increase migration. Total immigration and emigration of each country are then calculated as follows:

$$in\text{-}migration_i = \sum_j M_{ij} \quad (3.19)$$

$$out\text{-}migration_j = \sum_i M_{ij} \quad (3.20)$$

3.3.2.2 Colonial Link

We use the dummy variable CL_{ij} equal to 1 when there is a colonial history between country i and j, 0 otherwise. The information is obtained from the CEPII data set (Head et al., 2010).

3.3.3   Physical Layer

Infrastructures are the means through which goods are traded and they may have a large effect on trade costs. Travel distance is the first determinant that defines direct monetary outlays for trade – see Section 3.3.1.3, *Transport Costs*.

3.3.3.1   Common Borders

Many studies have confirmed that the influence of distance on trade is non-linear, with trade between bordering countries being significantly greater than countries that are positioned at similar distances, but do not share a border (McCallum, 1995). We take into consideration this determinant by generating a layer in which countries sharing a border are interconnected. We construct the dummy variable CB_{ij} equal to 2 when countries i and j share a border, 1 otherwise.

3.3.3.2   Distance

Transport distance is one of the major variables influencing transport cost because distance represents an impediment to trade. The greater the distance between two countries, the higher is the cost to transport goods, thereby reducing the gains from trade. We approximate the average transport distance between countries by applying the Euclidian distance between national capitals, d_{ij}.

3.3.3.3   Infrastructure Quality

In Section 3.3.1.3, Transport Cost, we have discussed how the quality of infrastructures and the cost and quality of related services are connected to the transport costs. All these factors also influence delivery time. In fact, poor quality infrastructure increases the uncertainty of delivery, which is associated with a higher risk of damage and, therefore, with higher losses and insurance costs. The quality of infrastructure thus largely determines the time required to get products to market and the reliability of delivery. We use the World Bank's Logistic Performance Index to estimate the quality of trade and transport-related infrastructure.

3.4   The Workings of the Model

The IMM is based on the following assumptions:

- immigrant and native are considered as a homogeneous class of workers;
- flexibility in the labour market, which is the flexibility of wages to adjust demand and supply and the capacity to differentiate wages for new labour outcomers in the market;
- intensive margin increase through time (i.e. existing bilateral trading relationships increase through time);
- labour-intensive goods (logistics, manufacturing, agriculture), that is, the costs of labour make up a high percentage of total costs;
- local externalities are constant.

We now show how these assumptions are used in the model.

The model is composed of the layers presented in Section 3.3 and grounded on the assumptions set out above. The IMM operates with a combination of sub-models, which in sequence determine the variation of model parameters through time. The system dynamics are described in a sequence of steps. We begin with two assumptions:

- import, export, population, transport cost, exchange rate, spatial proximity and cultural links are at their initial state; and
- population changes according to death rate, birth rate and net migration of each country.

Given these assumptions, we work with the following algorithm whose structure is shown in Figure 3.4.

1. Start from a fixed configuration on the IMM, nation-by-nation: population ($Pop_i(t)$), GDP ($GDP_i(t)$), transport costs ($Cost_{ij}(t)$) and migration ($M_{ij}(t)$), which are sufficient to calculate $T_{ij}(t)$, the trade. We call this configuration $C(t)$: the initial configuration of the system.
2. Disequilibrium in GDPpc stimulates migration.
3. Migration, birth and death rates change the total population of each country.
4. The change in population generates a modification in the GDP and GDPpc of each country (GDPpc $(t+1)$).
5. The change in the GDPpc generates a modification in transport costs ($Cost_{ij}(t+1)$).
6. The change in the transport costs generates a modification in bilateral trade $T_{ij}(t+1)$.
7. The previous step modifies import and export of each country and thus the GDPs of both.
8. Return to Step 1.

In Figure 3.4, each circle represents a sub-model calibrated at the steady state and provides the forecasts for bilateral trade, GDP and migration, as discussed in Section 3.3. We assume that each country behaves in the same way through time and that no external events affect this configuration. In reality, as we know, internal national and external international events/shocks shape the overall trade pattern of a country. The effect of any shock at any point in time can be introduced into the model by introducing an exogenous change in one or more sub-models for one or more countries. In this way, we can estimate consequences and impacts of an event in the system. In Section 3.7, we test the IMM by simulating the overall effect generated in the system through the imposition of a shock, that is, we introduce a reduction of per capita consumption and investment in the United States and Germany.

3.5   Model Calibration

We have calibrated the IMM at the steady state using data collected from several sources for 40 countries (Tables 3.1 and 3.2). Data at the steady state are for the year 2000. To assure equilibrium in the model, we also introduce a virtual 'country' which accounts for the rest of the world (RoW).

Information on bilateral trade is derived from the Correlates of War Project's Trade, which uses information from the International Monetary Fund's Direction of Trade Statistics (Barbieri and Keshk, 2012; Barbieri et al., 2009). Parts of the information of this study have been extracted from the online World Bank Data Catalog (http://data.worldbank.org/

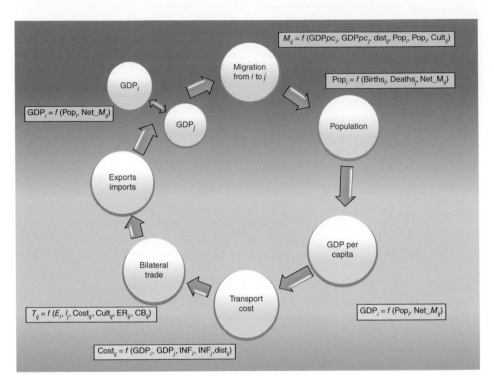

Figure 3.4   Multi-layer interactive process. *Source*: Reproduced from Caschili et al. (2015) with permission of Springer

Table 3.1   IMM variables and sources

Layer type	Variable	Source
Economic	GDP	World Bank Data Catalog[a]
	Trade	Barbieri and Keshk (2012)
	Exchange rate	World Bank Data Catalog[a]
Social and cultural	Population	World Bank Data Catalog[a]
	Birth, death rate	World Bank Data Catalog[a]
	Per capita consumption	Own elaboration
	Migration	Abel et al. (2013)
	Cultural ties	Mayer and Zignago (2011)
Physical	Distance	Mayer and Zignago (2011)
	Physical proximity	Mayer and Zignago (2011)
	Logistic performance index	World Bank Data Catalog[a]

[a] http://data.worldbank.org/data-catalog
Source: Reproduced from Caschili et al. (2015) with permission of Springer.

Table 3.2 List of countries

Geographic area	Countries
Europe	Austria, Belgium, Bulgaria, Cyprus, Czech Republic, Denmark, Estonia, Finland, France, Germany, Greece, Hungary, Ireland, Italy, Latvia, Lithuania, Luxemburg, Malta, Netherlands, Poland, Portugal, Romania, Slovak Republic, Slovenia, Spain, Sweden, United Kingdom
Americas	Brazil, Canada, Mexico, United States
Asia and Pacific	China, India, Indonesia, Japan, Russia, South Korea, Australia, Taiwan, Turkey

Source: Reproduced from Caschili et al. (2015) with permission of Springer.

data-catalog), an open data set providing over 8,000 indicators and covering 214 countries from 1960 to 2011.

Information on migration has been derived from the Abel et al. (2013) data set of estimates of bilateral migration flows between 191 countries. Finally, information on bilateral cultural and physical distance has been extrapolated from the GeoDist project (Mayer and Zignago, 2011). The data set provides several geographical variables, in particular bilateral distances are measured using city-level data. In the remainder of this section, we will present the results for the calibration of the bilateral trade model (see Section 3.3.1) and the bilateral migration model (see Section 3.3.2).

The calibration of the bilateral trade model is based on the following: (i) the impact of distance on trade and generalised trade costs (TC), (ii) the identification of the cost function to be included in the model and (iii) the impact of economic, social and physical determinants on trade. In relation to the first two, we consider the following decay functions:

$$f(\beta, d_{ij}) = \exp(\beta, d_{ij}) \tag{3.21}$$

$$f(\beta, d_{ij}) = d_{ij}^{\beta} \tag{3.22}$$

$$f(\beta, \text{TC}_{ij}) = TC_{ij}^{\beta} \tag{3.23}$$

We have tested both exponential (Equation 3.21) and power decay functions (Equation 3.22) for geographical distance and generalised transport cost (Equation 3.23). The power decay function provides us with better results than the exponential decay. This result is in line with the literature (see inter alia Wilson, 1971) which has found that power decay functions provide better results for long travel distances. Empirical results demonstrate that our model fits the data better if we use geographical distance (Table 3.3), and this result is in line with other studies which have found geographical distance to provide better estimations of bilateral trade for aggregated products (Hummels, 1999; Immaculada and Celestino, 2005).

Finally, economic, cultural and physical layers were included in the model in order to evaluate if they improve its results (Equation 3.6). We have verified that when we introduce Exchange Rate (ER), Colonial Link (CL) and Common Borders (CB) into the model, the statistics always improve. In the simulation presented in Sections 3.6 and 3.7, the trade sub-model is implemented by a power decay function with geographic distance and includes the layers ER, CL and CB.

Table 3.3 Parameter estimates (β coefficient) and Pearson coefficient of correlation (in parenthesis)

	ER	CL	CB	ER, CL	ER, CL, CB
Exponential (d_{ij})	0.000233 (0.507)				
Power decay (d_{ij})	−1.54341 (0.905)	−1.54354 (0.906)	−1.48931 (0.909)	−1.35623 (0.908)	−1.48929 (0.909)	−1.2755 (0.91)
Power decay ($cost_{ij}$)	−2.0096 (0.883)	−2.0112 (0.884)	−1.9843 (0.886)	−2.02495 (0.877)	−1.9856 (0.887)	−1.99499 (0.889)

Source: Reproduced from Caschili et al. (2015) with permission of Springer.

We will next build and calibrate a bilateral migration flows' model using linear regression analysis in the canonical form:

$$Y = \sum_k \beta_k X_k + \varepsilon \tag{3.24}$$

In this calibration, we introduce the following set of hypotheses:

- Migration between two countries is higher from a country with a low GDPpc to a country with higher GDPpc (H1).
- Migration flows are higher between countries with larger populations (H2).
- Distance influences migration negatively (H3).
- Common borders and cultural links increase the chances that one individual moves to another country (H4).

Standard linear regression (Equation 3.24) estimates the dependent variable Y (migration between country i and j) as a linear combination of the independent variables X. The independent variables are the ratio between GDPpc of the two countries (H1), the total population in i and j (H2), the geographic distance d_{ij} (H3), common border CB_{ij} and cultural linkage CL_{ij} (H4). We use a logarithm transformation in order to normalise the variables and a binary form for CB_{ij} and CL_{ij}. Thus, Equation 3.24 assumes the following form:

$$\log(M_{ij}) = \beta_{GDPpc} \frac{\log(GDPpc_j)}{\log(GDPpc_i)} + \beta_{Pop_i} \log(Pop_i) + \beta_{Pop_j} \log(Pop_j) + \beta_{d_{ij}} \log(d_{ij})$$
$$+ \beta_{CB_{ij}} \log(CB_{ij}) + \beta_{CL_{ij}} \log(CL_{ij}) + \varepsilon \tag{3.25}$$

We adjust Equation (3.25) with a lognormal function in order to reduce the model's errors:

$$MT_{ij} = M_{ij} * e^{\frac{\log(M_{ij})}{M_{ij}}} \tag{3.26}$$

All parameters (β) are significant (p-value ≤ 0.001) and their signs are consistent with our hypotheses (Table 3.4). The model approximates the data quite well, having a very high R^2 coefficient of determination ($R^2 = 0.908$).

Table 3.4 Parameter estimates for linear regression on bilateral migration

Model	Unstandardised coefficients		Standardised coefficients	T	Sig.
	B	Standard error	Beta		
log $(GDPpc_j/GDPpc_i)$	-8.242	0.482	-1.311	-17.092	0.000
log (Pop_i)	0.557	0.026	1.511	21.168	0.000
log (Pop_j)	0.695	0.033	1.876	20.786	0.000
log (d_{ij})	-0.904	0.058	-1.169	-15.711	0.000
CL_{ij}	1.420	0.259	0.050	5.491	0.000

$R=0.953$, $R^2=0.908$.
Source: Reproduced from Caschili et al. (2015) with permission of Springer.

3.6 Result 1: Steady State

After having collected data and calibrated the models, we can now focus on the objective of our study, which is to test the IMM in order to verify (i) if the IMM mirrors real-world scenarios and (ii) to trace the effects generated by endogenous or exogenous changes/shocks in the system.

In this part of the analysis, we use a Monte Carlo approach for the simulation. The Monte Carlo strategy allows us to reduce the errors introduced in the model by the stochastic coefficient $\delta_i(t)$ of Equation (3.14). In this way, we can minimise the variance of our final result and thus also its statistical error. We have run the algorithm, described in Section 3.5, over 100 times for each scenario. Each final scenario is obtained by averaging the values of the 100 simulations in each time t.

In Figure 3.5, we depict the four layers of the IMM, in which (a) and (b) denote the trade and migration networks; (c) denotes cultural network (colonial links) and (d) denotes physical linkages (shared borders).

Table 3.5 summarises some statistical information for the trade and migration networks: average values of flows (w), clustering coefficient (c^w and c) and sum of flows through a node (s). For the weighted clustering coefficient c^w we use the equation proposed by Barrat et al. (2004), which measures the probability that two randomly selected nodes connected to a selected node are linked with each other where node cohesiveness is a function of the level of interaction between nodes. In our case, cohesiveness is heterogeneous across both networks: nodes are topologically highly connected (clustering coefficient c 0.88 and 0.96, Table 3.5); when we examine the intensity of interaction among nodes, we can observe that structured clusters emerge. This is due to the fact that although every country has at least one relation with all the other countries, trade and migration flows are heterogeneously distributed among nodes due to the several factors influencing these international relations (Sections 3.3.1 and 3.3.2).

In Figure 3.6a and b, we examine the traffic through each node s by considering the directionality of flows. We plot the cumulative probability distributions of the sum of weighted in-connections s^+ (i.e. total imports and total immigrants) and out-connections s^- (i.e. exports and total emigrants) per country. In the trade network, the distributions of s^+ and s^- are very similar and well approximated by power law decay (we fit the distribution with a

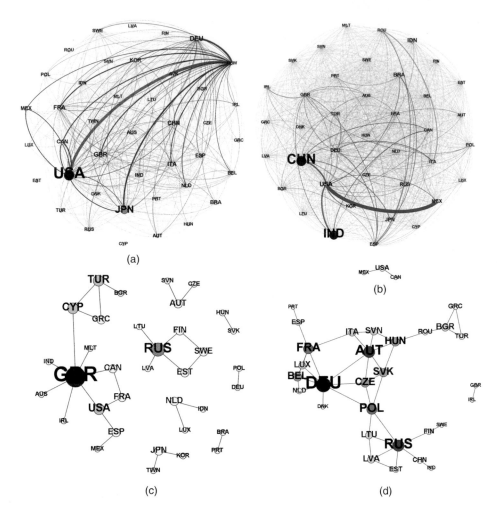

Figure 3.5 Trade network (a), migration network (b), cultural network (c) and network of common borders (d). Link width is proportional to trade and migration flows, while node size and colour are proportional to national GDP in the trade network and to national population in the migration network. Node size and colour for colonial and common border networks are proportional to the number of connections. *Source*: Reproduced from Caschili et al. (2015) with permission of Springer

Table 3.5 Main statistics of trade and migration networks: average link weight $<w>$, weighted and topological clustering $<c^w>$ and C, and average traffic intensity per node $<s>$

	$<w>$	$<c^w>$	C	$<s>$
Trade [$US million]	3,702	0.258	0.88	143,132
Migration [People]	6,298	0.315	0.96	199,167

Source: Reproduced from Caschili et al. (2015) with permission of Springer.

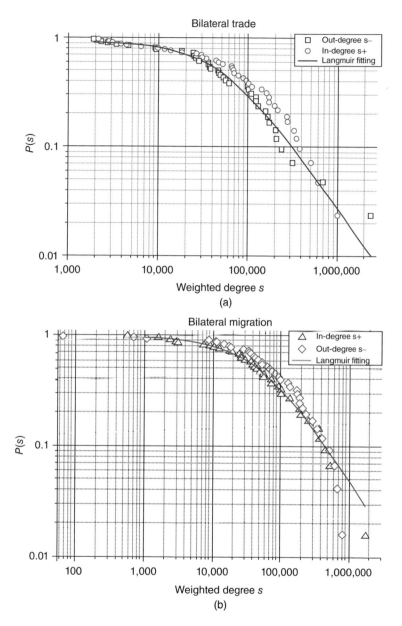

Figure 3.6 Weighted in-degree $s+$ and out-degree s^- distributions for trade (a) and migration (b). *Source*: Caschili et al., 2015. Reproduced by permission of Springer

Langmuir function, exponent $= 1.17$ and $R^2 = 0.987$). The figure emerging from the migration network has similar distributions for immigration (s^-) and emigration (s^+) per country. We fit the cumulative distributions with a Langmuir function, with exponent 0.98 $(R^2 = 0.996)$. Both networks are, however, characterised by pronounced disparities. Power law distributions, short distance between nodes and node clustering (Pastor-Satorras and Vespignani, 2001) are common properties of social networks in which the level of activities of people is very heterogeneous, thus implying highly varying levels of involvement (Muchnik et al., 2013). In other words, a country interacts selectively with other countries. This is one of our main hypotheses tested in this work: we recognise that each country engages with other countries in different ways due to its own spatial, economic and social conditions.

In the remainder of this section, we present the results of the implementation of the algorithm described in Section 3.5. Each sub-model has been calibrated with observed data from the year 2000. We compare the IMM model's results with observed data from 2000 to 2011. We conclude that there is a strong evidence of correspondence between model outputs and the observed data for GDP. The observed visual correlation (Figure 3.7) is also confirmed by the very high Pearson coefficient analysis ($R^2 = 0.988$) in the period 2000–2007. The latter time span is a period of relative stability and growth for the majority of the countries in our sample, hence it is suitable for the validation of the IMM in the steady state. We plot the following model results for three cases: major European economies (Figure 3.7a and b), the Baltic states and Bulgaria (Figure 3.7c and d) and the six major world economies (Figure 3.7e and f).

Apart from the high correlation between model forecasts and observed data, the ranking is consistent for all cases in Figure 3.7 before the 2007–2008 financial downturn. In particular, the model correctly forecasts that the Chinese GDP overtook the Japanese GDP. We can then conclude that the model's validation is positive for the steady state.

3.7   Result 2: Estimation and Propagation of Shocks in the IMM

We have confirmed that the IMM is capable of modelling multi-layered dynamic trade interactions among countries. We aim here to apply the model to shocks and the propagation of exogenous 'events'. Our analysis seeks to identify the countries that can trigger cascading effects and evaluate their magnitude. The IMM comprises several interlinked sub-systems which interact according to the schema presented in Figure 3.4 and Section 3.2. Among these sub-systems, we focus on the trade network in order to identify the countries that are central in the international network and most inclined to diffuse (potentially harmful) effects on the international trade system. We use the eigenvector centrality (Bonacich, 2007) to measure the role of each country in propagating economic impacts on neighbouring countries. Eigenvector analysis provides a measure of the direct influence between a node and its neighbour nodes, which subsequently influences other nodes directly connected to them (who themselves influence still others). The first node is highly influential if it easily propagates 'events' (such as an economic crisis in our case) in the network (Borgatti, 2005). We perform the eigenvector analysis for the trade network for a 10-step simulation as shown in Table 3.6. We can observe that the ranking is consistently stable over the observation period because the model does not introduce dynamic modifications in the topology of the network (i.e. new links are not created nor are existing links removed). Hence in Table 3.6, we report the average value of eigenvector analysis of each country. The zone of highest importance (i.e. central in the network) is North America: the United States is twice as much important than Germany (the most central European country).

We next evaluate the magnitude of the diffusion effect on the IMM layers (vertical propagation) by introducing external shocks for two case studies: the United States, which is the

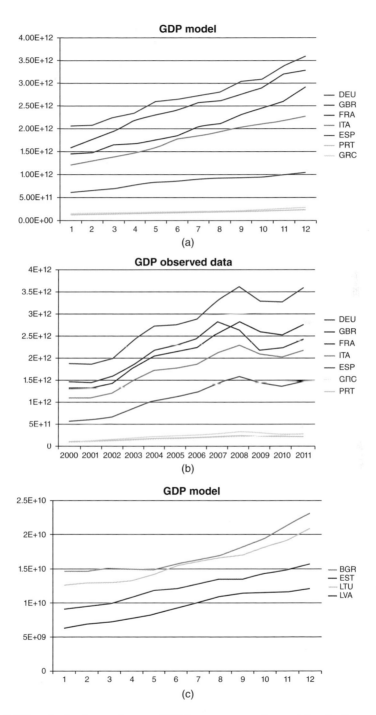

Figure 3.7 IMM results for the steady state. GDP forecast versus observed data for European countries (a and b), Baltic states (c and d) and six major world economies (e and f). *Source*: Reproduced from Caschili et al. (2015) with permission of Springer

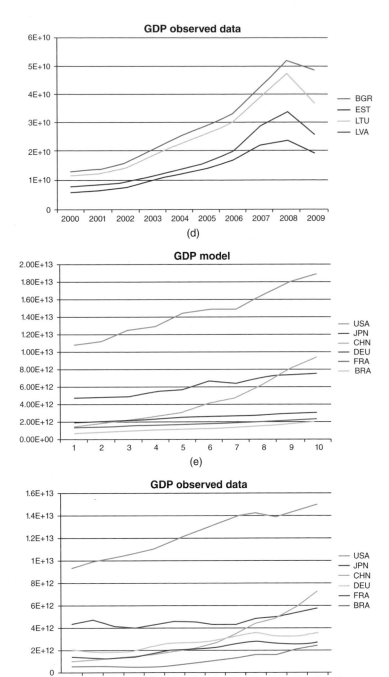

Figure 3.7 (*continued*)

Table 3.6 Country ranking of eigenvector centrality index for the bilateral trade network: the highest is the eigenvector centrality index and the most important is a node

Position	Country	Value	Position	Country	Value
1	USA	0.5723	21	RUS	0.0446
2	CAN	0.4293	22	AUT	0.0360
3	JPN	0.3806	23	DNK	0.0309
4	DEU	0.2541	24	FIN	0.0211
5	CHN	0.2419	25	TUR	0.0220
6	MEX	0.2203	26	POL	0.0198
7	GBR	0.1933	27	CZE	0.0139
8	FRA	0.1651	28	HUN	0.0134
9	KOR	0.1566	29	PRT	0.0122
10	TWN	0.1466	30	LUX	0.0105
11	ITA	0.1351	31	GRC	0.0071
12	NLD	0.0913	32	ROU	0.0044
13	AUS	0.0712	33	SVK	0.0041
14	BEL	0.0694	34	SVN	0.0038
15	IRL	0.0580	35	BGR	0.0020
16	ESP	0.0576	36	MLT	0.0020
17	IDN	0.0520	37	EST	0.0013
18	IND	0.0541	38	LTU	0.0013
19	BRA	0.0505	39	LVA	0.0011
20	SWE	0.0521	40	CYP	0.0008

Source: Reproduced from Caschili et al. (2015) with permission of Springer.

most central country in the network (Figure 3.8a–c), and Germany, which is the most central European country in the network (Figure 3.9a–c).

In the first case, the simulated endogenous shock in the US economy is generated by introducing a decrease in the US GDP of 2% between step 6 and step 9 of the simulation (Figure 3.8). When we compare these results with the steady state, a few countries show a clear dependency from the United States than others. This is the case of Germany, France and Japan. On the other hand, China seems to take advantage of an economic crisis in the United States.

In the second test, we apply a shock in the German GDP. We follow the same strategy used for the United States: a decrease of GDP of 2% between step 6 and step 9 (Figure 3.9). Also in this case, we observe that the imposed endogenous shock in one country has affected other nations, both positively and negatively. The Asian countries (Japan and China) appear to be the most affected by an internal crisis in Germany. In Europe, the economies of Germany and the United Kingdom do not appear to be very connected, as a decrease in the German GDP does not affect the UK economy. The same effect has been recorded in the first test: while the German GDP began to lessen as the US GDP decreased, the UK GDP remained unaffected. Finally, in the Baltic states, Latvia appears to be more affected by an US crisis (Figure 3.8), and Estonia is impacted more by a German economic crisis (Figure 3.9).

We can conclude this section by observing that the model, under the two tests, can potentially analyse shocks and their impacts. Nonetheless, we recognise the need for more extensive

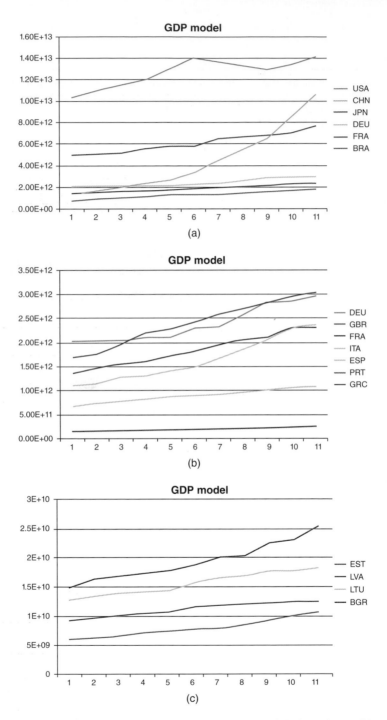

Figure 3.8 Effect of 2% decrease of US GDP over four time steps (6–9) for major world economies (a), European countries (b) and Baltic states (c). *Source*: Reproduced from Caschili et al. (2015) with permission of Springer

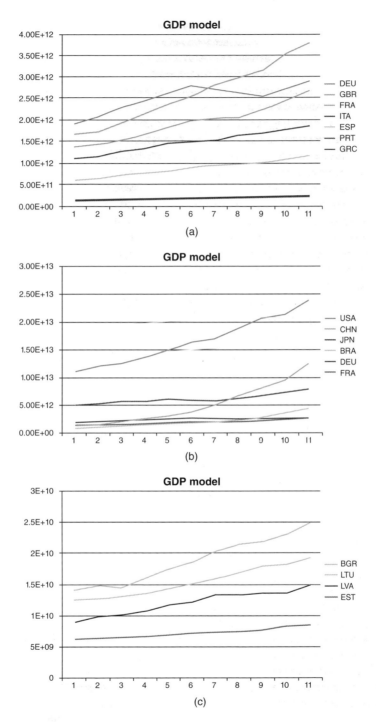

Figure 3.9 Effect of 2% decrease of German GDP over four time steps (6–9) for European countries (a), major world economies (b) and Baltic states (c). *Source*: Reproduced from Caschili et al. (2015) with permission of Springer

analyses in order to verify the robustness of the model in relation to the effects of shocks to the networks. This type of analysis is, however, beyond the scope of the present work but will be the objective of our future research.

3.8   Discussion and Conclusions

We have proposed in this chapter a new model with which to analyse and examine bilateral trade. Underpinning the development of the IMM, we can confirm the presence of interdependence among countries and factors that determine trade and the interactions among the different elements (layers) that define the international trade system. Within each layer we have examined the horizontal connections, and between layers we have modelled the interdependent vertical connections, and in so doing we have been able to represent the inherent complexity of trade networks. The results show that, in the trade market, countries certainly interact with each other but they do not following standard frameworks; instead, their interactions depend on their specific spatial, economic and social conditions.

We have also tested the model under potential possible shocks, such as a decrease in GDP, and we have verified that the model can satisfactorily simulate and identify scenarios of the countries after exogenous events/shocks have taken place.

The IMM can therefore be regarded as a useful tool for decision makers in the global trade market in further understanding the causality and directionality of certain parameters that influence trade. These results are interesting and powerful, particularly in the light of the complexity of international trade processes, and also contribute to the literature since a better understanding of the processes that influence and determine trade can not only shed light on how countries can grow through trade but also help in the design of policies for sustainable economic growth.

References

Aashtiani, H.Z. and Magnanti, T.L. (1981) Equilibria on a congested transportation network. *SIAM Journal on Algebraic Discrete Methods*, **2** (3), 213–226.

Abel, G.J., Bijak, J., Findlay, A.M. et al. (2013) Forecasting environmental migration to the United Kingdom: An exploration using Bayesian models. *Population and Environment*, **35** (2), 183–203.

Akhtar, R. and Santos, J.R. (2013) Risk-based input–output analysis of hurricane impacts on interdependent regional workforce systems. *Natural Hazards*, **65** (1), 391–405.

Anderson, J.E. and Van Wincoop, E. (2004) *Trade Costs*, National Bureau of Economic Research, No. w10480.

Bade, R. and Parkin, M. (2007) *Foundations of Economics*, Pearson Addison Wesley, New Jersey, USA.

Baldwin, R. and Taglioni, D. (2006) *Gravity for Dummies and Dummies for Gravity Equation* NBER working paper no. 12516, National Bureau of Economic Research, Inc.

Barbieri, K. and Keshk, O.M.G. (2012) *Correlates of War Project Trade Data Set Codebook*. Version 3.0. Online: http://correlatesofwar.org, (accessed 05 January 2016).

Barbieri, K., Keshk, O.M.G. and Pollins, B. (2009) TRADING DATA: Evaluating our assumptions and coding rules. *Conflict Management and Peace Science*, **26** (5), 471–491.

Barrat, A., Barthelemy, M., Pastor-Satorras, R. and Vespignani, A. (2004) The architecture of complex weighted networks. *PNAS*, **101**, 3747–3752.

Bergstrand, J.H. (1985) The gravity equation in international trade: some microeconomic foundations and empirical evidence. *The Review of Economics and Statistics*, **67** (3), 474–481.

Bonacich, P. (2007) Some unique properties of eigenvector centrality. *Social Networks*, **29**, 555–564.

Borgatti, S.P. (2005) Centrality and network flow. *Social Networks*, **27** (1), 55–71.

Burger, M.J., Van der Knaap, B. and Wall, R.S. (2014) Polycentricity and the multiplexity of urban networks. *European Planning Studies*, **22** (4), 816–840.

Burns, W.J. and Slovic, P. (2012) Risk perception and behaviors: anticipating and responding to crises. *Risk Analysis*, **32**, 579–582.

Büthe, T. and Milner, H.V. (2008) The politics of Foreign Direct Investment in developing Countries: increasing FDI through international trade agreements. *American Journal of Political Science*, **52**, 741–762.

Caschili S., Medda F., Wilson A. (2015) "An interdependent multi-layer model: resilience of international networks", *Networks and Spatial Economics*, **15**(2), 313–335.

Castet, J.F. and Saleh, J.H. (2013) Interdependent multi-layer networks: modeling and survivability analysis with applications to space-based networks. *PLoS One*, **8** (4), e60402.

Clauset, A., Moore, C. and Newman, M.E. (2008) Hierarchical structure and the prediction of missing links in networks. *Nature*, **453** (7191), 98–101.

Deardorff, A. (1998) Determinants of bilateral trade: Does gravity work in a neoclassical world?, in *The Regionalization of the World Economy* (ed J.A. Frankel), University of Chicago Press, pp. 7–32.

Garelli, S. (2003) *Competitiveness of Nations: The Fundamentals*, IMD World competitiveness yearbook.

Glick, R. and Rose, A. (2002) Does a currency union affect trade? The time series evidence. *European Economic Review*, **46** (6), 1125–1151.

Gunasekaran, A., Patel, C. and Tirtiroglu, E. (2001) Performance measures and metrics in a supply chain environment. *International Journal of Operations & Production Management*, **21** (1–2), 71–87.

Harvie, C. (2004) *East Asian SME Capacity Building, Competitiveness and Market Opportunities in a Global Economy*. Faculty of Commerce-Economics Working Papers, 100.

Head, K., Mayer, T. and Ries, J. (2010) The erosion of colonial trade linkages after independence. *Journal of International Economics*, **81** (1), 1–14.

Helpman, H. (2011) *Understanding Global Trade*, Harvard University Press, London.

Holt, R.P.F., Barkley, J.R. Jr., and Colander, D. (2011) The complexity era in economics. *Review of Political Economy*, **23** (3), 357–369.

Hufbauer, G. (1970) The impact of national characteristics and technology on the commodity composition of trade in manufactured goods, in *The Technology Factor in International Trade* (ed R. Vernon), UMI, pp. 143–232.

Hummels, D. (1999) *Towards a Geography of Trade Costs*, University of Chicago. Mimeographed document.

Immaculada, M.Z. and Celestino, S.B. (2005) Transport costs and trade. Empirical evidence for Latin American imports from the European Union. *The Journal of International Trade & Economic Development: An International and Comparative Review*, **14** (3), 353–371.

Levin, S., Barrett, S., Aniyar, S. et al. (1998) Resilience in natural and socio-economic systems. *Environment and Development Economics*, **3** (2), 222–234.

Limao, N. and Venables, A.J. (2001) Infrastructure, geographical disadvantage, transport costs and trade. *World Bank Economic Review*, **15**, 451–79.

Linders, G.J.M., Slangen, A., de Groot, H.L.F. and Beugelsdijk, S. (2005) *Cultural and Institutional Determinants of Bilateral Trade Flows*. Tinbergen Institute Discussion Paper No. 2005-074/3.

Linneman, H. (1966) *An Economic Study of International Trade Flows*, North-Holland, Amsterdam.

Liu, J., Dietz, T., Carpenter, S. et al. (2007) Complexity of coupled human and natural systems. *Science*, **317** (5844), 1513–1516.

Marincioni, F., Appiotti, F., Pusceddu, A. and Byrne, K. (2013) Enhancing resistance and resilience to disasters with microfinance: Parallels with ecological trophic systems. *International Journal of Disaster Risk Reduction*, **4**, 52–62.

Markose, S.M. (2005) Computability and evolutionary complexity: Markets as Complex Adaptive Systems (CAS). *The Economic Journal*, **115**, F159–F192.

Matyas, L. (1997) Proper econometric specification of the Gravity Model. *World Economy*, **20** (3), 363–368.

Mayer, T. and Zignago, S. (2011) *Notes on CEPII's Distance Measures (GeoDist)*. CEPII Working Paper 2011–2025.

McCallum, J. (1995) National borders matter: Canada-US regional trade patterns. *American Economic Review*, **85** (3), 615–623.

Muchnik, L., Pei, S., Parra, L.C. et al. (2013) Origins of power-law degree distribution in the heterogeneity of human activity in social networks. *Scientific Reports*, **3**. doi: 10.1038/srep01783

Newman, M.E. (2003) The structure and function of complex networks. *SIAM Review*, **45** (2), 167–256.

Pastor-Satorras, R. and Vespignani, A. (2001) Epidemic spreading in scale-free networks. *Physical Review Letters*, **86** (14), 3200–3203.

Pullianen, P. (1963) World trade study: An economic model of the pattern of the commodity flows in international trade 1948–1960. *Ekonomiska Samfundets*, **2**, 78–91.

Rauch, J.E. (1999) Networks versus markets in international trade. *Journal of International Economics*, **48**, 7–35.

Rinaldi, S.M., Peerenboom, J.P. and Kelly, T.K. (2001) Identifying, understanding and analyzing critical infrastructure interdependencies. *Control Systems, IEEE*, **21** (6), 11–25.

Rosser, J.B. (1999) On the complexities of complex economic dynamics. *The Journal of Economic Perspectives*, **13** (4), 169–192.

Strogatz, S.H. (2001) Exploring complex networks. *Nature*, **410** (6825), 268–276.

Tang, C.S. (2006) Robust strategies for mitigating supply chain disruptions. *International Journal of Logistics: Research and Applications*, **9** (1), 33–45.

Teng, S.G. and Jaramillo, H. (2005) A model for evaluation and selection of suppliers in global textile and apparel supply chains. *International Journal of Physical Distribution & Logistics Management*, **35** (7), 503–523.

Tinbergen, J. (1962) *Shaping the World Economy: Suggestions for an International Economic Policy*, Twentieth Century Fund, New York.

Wilson, A.G. (1967) A statistical theory of spatial distribution models. *Transportation Research*, **1**, 253–269.

Wilson, A.G. (1970) *Entropy in Urban and Regional Models*, Pion, London.

Wilson, A.G. (1971) A family of spatial interaction models and associated developments. *Environment and Planning*, **3**, 1–32.

Yue, C., Beghin, J. and Jensen, H.H. (2006) Tariff equivalent of technical barriers to trade with imperfect substitution and trade costs. *American Journal of Agricultural Economics*, **88** (4), 947–960.

4

A Global Inter-country Economic Model Based on Linked Input–Output Models

Robert G. Levy, Thomas P. Oléron Evans and Alan Wilson

4.1 Introduction

The objective of this chapter is to present a global economic model that can be used as the basis for assessing the impacts of future changes in trade, migration, security and development aid. Because these topics are often associated with developing countries, it is important that the model have as few coefficients as possible to enable the future addition of countries who publish little data on their economic structure. The model as it is presented here represents a first 'proof-of-concept' step towards these ambitious goals. At this stage, the focus is on creating a model of global trade which will form a skeleton onto which additional social science models will be added in future work. The economies of individual countries are represented as 35-sector input–output models, each of which is linked through trade flows representing imports and exports. This has recently been made feasible by the publication of the World Input–Output Database (WIOD) (Timmer et al., 2015), a collection of national input–output tables (NIOTs) for 40 (mostly OECD) countries across 17 years from 1995 to 2011. The NIOTs are linked through data from the United Nations, covering trade in both products[1] and services.[2]

While other models of world trade based on the input–output methodology exist, the model presented here is one of the first to integrate the extremely detailed internal economic data provided by WIOD with the enormous wealth of trade data from the UN COMTRADE database, resulting in a picture of global trade that includes an unparalleled amount of empirically

---

[1] comtrade.un.org/db
[2] unstats.un.org/unsd/servicetrade/

Global Dynamics: Approaches from Complexity Science, First Edition. Edited by Alan Wilson.
© 2016 John Wiley & Sons, Ltd. Published 2016 by John Wiley & Sons, Ltd.

derived economic information. The use of such data sources and the simplicity and flexibility of the model proposed will allow for the investigation of a wide range of economic scenarios, including those involving developing countries, in future work.

The remainder of this chapter is structured as follows: Section 4.2 gives an overview of existing work in this area; Section 4.3 gives a description of the present system and outlines how data is used to calibrate the coefficients of the model; The algorithm used to calculate the output of the model is described in Section 4.4 and some preliminary results are given in Section 4.5; Conclusions and potential directions for future research are presented in Section 4.6.

4.2   Existing Global Economic Models

In the mid-1970s, the creator of input–output economics, Wassily Leontief, used the acceptance speech of his Nobel Prize in Economics to announce a very ambitious project to model the global economy:

> Major efforts are underway to construct a data base for a systematic input-output study not of a single national economy but of the world economy viewed as a system composed of many interrelated parts [...] Preliminary plans provide for a description of the world economy in terms of twenty-eight groups of countries, with about forty-five productive sectors for each group.
>
> Leontief (1974)

Duchin (2004) describes how, 20 years on, Leontief's efforts in this area had largely been ignored by economists who describe his departures from the standard, neoclassical modelling in terms of price elasticity and elasticity of substitution as being 'too great to ignore' (see Section 4.3.2 for more on this subject).

In the years since Leontief published his global model, input–output analysis has been largely restricted to regional studies, of which Akita (1993), Khan (1999) and Luo (2013) are examples, and studies related to energy and the environment, such as Leontief (1970), Joshi (1999), van den Bergh (2002) and Hendrickson et al. (2006).

However, much more recently, attention has returned to input–output modelling in a global context more generally. Tukker and Dietzenbacher (2013) describe how several multiregional input-output (MRIO) models have been developed in the very recent literature. These are, along with WIOD which is used in the present model, EORA (Lenzen et al., 2013), EXIOPOL (Tukker et al., 2013) and the more mature GTAP (Walmsley et al., 2012).

Each of these existing models of the global economy extends the idea of the NIOT to an international setting: where in a NIOT the magnitude of every bilateral sector–sector flow is recorded for a given country, these projects record the magnitude of every country/sector–country/sector flow. Thus, for example, the extent to which British agriculture purchases from Belgian manufacturing is recorded in dollar amounts. This results in a large matrix which, through inversion in the normal input–output manner, can be used to predict the impact of a particular exogenous change in demand. This tends to strengthen the case of those economists, mentioned by Duchin (2004), that price elasticity and elasticity of substitution are ignored by input–output economics.

The model presented in this chapter provides a framework in which the power of input–output analysis at the national level can be combined with a more subtle understanding of the dynamics of international trade. The standard MRIO of C countries and S sectors requires $SC \times SC$ coefficients, and a major goal of this model is to minimise the number of coefficients required to model a global system. It also provides modellers with a richer set of coefficients than the standard technical coefficients of MRIOs in order to facilitate the combination of the present model with other models of human systems, such as migration, international security and development aid. Additionally, as mentioned, this reduced set of coefficients will facilitate the addition of countries vital to the study of these systems, but which do not publish data on economic structure. The mathematical estimation of the coefficients for these countries is the subject of upcoming work.

4.3 Description of the Model

4.3.1 Outline

Here, we outline the new global model. In this presentation the model has C countries, the economies of which are divided into S productive sectors.[3] All product flows are given by value, measured in millions of (current price) US dollars.[4] We start with an introduction to input–output tables which, aside from some important notation, can be skipped by the reader familiar with this topic. We then introduce the model of a single country, outline both the normal input–output coefficient sets and introduce the first new set. Finally, we show how these country models are linked with an international trade model and introduce the second new set of coefficients.

4.3.2 Introduction to Input–Output Tables

Input–output is, at its heart, an accounting methodology. The products produced by and imported into a given country in a given year are either: sold as inputs to other sectors ('intermediate supply'); supplied to the 'final demand' of consumers and the government; invested or exported. The total amount imported and produced must equal the amount used, consumed, invested and exported for each sector.

By simple summation, country i's total production of sector s, $x_s^{(i)}$, can be defined as the sum of all intermediate supply plus supply to final demand, investment and export:

$$x_s^{(i)} = \sum_t z_{s,t}^{\dagger(i)} + f_s^{\dagger(i)} + n_s^{\dagger(i)} + e_s^{(i)} \tag{4.1}$$

where $z_{s,t}^{\dagger(i)}$ is the intermediate supply from domestic sector s to sector t in country i, $f_s^{\dagger(i)}$ is the final demand for domestic sector s, $n_s^{\dagger(i)}$ is the investment and $e_s^{(i)}$ are exports.

---

[3] Although each sector produces many distinct products, for the sake of simplicity, these products are considered to be perfect substitutes, allowing the terms 'sector' and 'product' to be used interchangeably throughout this chapter.
[4] This allows us to use the terms 'quantity' and 'value' interchangeably.

Similarly, country i's total import of sector s, $m_s^{(i)}$, is the sum of all intermediate supply of imported products plus demand for and investment of imported products[5]:

$$m_s^{(i)} = \sum_t z_{s,t}^{*(i)} + f_s^{*(i)} + n_s^{*(i)} \tag{4.2}$$

An input–output table, $T^{(i)}$, as described by Miller and Blair (1985), is a particular arrangement of these quantities, which provides a clear and compact summary of the structure of a national economy. Neglecting the (i) superscript for clarity,[6] the NIOT is defined as follows:

$$
\begin{array}{c}
\\
\text{Domestic} \\
\\
\\
\text{Imports} \\
\\
\end{array}
\left\{
\begin{array}{c}
\\ 1 \\ \vdots \\ s \\ 1 \\ \vdots \\ s \\
\end{array}
\right.
\begin{array}{cccccccc}
\text{Sector} & 1 & \cdots & s & \text{F.D.} & \text{Inv} & \text{Exp} & \text{Tot} \\
\end{array}
$$

$$
\left(
\begin{array}{ccccccc}
z_{1,1}^{\dagger} & \cdots & z_{1,s}^{\dagger} & f_1^{\dagger} & n_1^{\dagger} & e_1 & x_1 \\
\vdots & \ddots & \vdots & \vdots & \vdots & \vdots & \vdots \\
z_{s,1}^{\dagger} & \cdots & z_{s,s}^{\dagger} & f_s^{\dagger} & n_s^{\dagger} & e_s & x_s \\
z_{1,1}^{*} & \cdots & z_{1,s}^{*} & f_1^{*} & n_1^{*} & 0 & m_1 \\
\vdots & \ddots & \vdots & \vdots & \vdots & \vdots & \vdots \\
z_{s,1}^{*} & \cdots & z_{s,s}^{*} & f_s^{*} & n_s^{*} & 0 & m_s \\
\end{array}
\right) \tag{4.3}
$$

It will often be convenient to gather those quantities having a single subscript into vectors and those with two subscripts into matrices. We can then characterise a country's economy through the S-vectors $f^{(i)}$, $n^{(i)}$ and $e^{(i)}$ and by the $S \times S$ matrices $Z^{\dagger(i)}$ and $Z^{*(i)}$. In the matrix form, the input–output table may be written as

$$T = \begin{pmatrix} Z^{\dagger} & f^{\dagger} & n^{\dagger} & e & x \\ Z^{*} & f^{*} & n^{*} & 0 & m \end{pmatrix} \tag{4.4}$$

The elements of $Z^{\dagger}$ and Z^{*} provide each sector with a complete 'recipe' for making its output, describing the quantities of each product needed as input, both domestic and imported, to produce total production, x. We are now in a position to introduce our first coefficient set. By dividing each intermediate flow, z, by the total output, x, of the sector using the intermediate, we can arrive at a set of *technical coefficients*, a, which define the input of one sector required per unit output of another. The amount of domestically produced product r required by sector s to produce a single unit of output is thus

$$a_{r,s}^{\dagger} = \frac{z_{r,s}^{\dagger}}{x_s} \tag{4.5}$$

and the equivalent measure for imported r is

$$a_{r,s}^{*} = \frac{z_{r,s}^{*}}{x_s} \tag{4.6}$$

---

[5] Note that, by convention, imported products may not be directly exported. This is a convention we retain throughout.
[6] For the remainder of this chapter, the country superscript will be excluded whenever its presence is clear from the context.

There are therefore $2S \times S$ technical coefficients for each country.[7] These technical coefficients allow the intermediate demand to be calculated for any given set of final demands, investments and exports. The total domestic production of sector s, x_s, is given by

$$x_s = \sum_r a_{s,r}^\dagger x_r + f_s^\dagger + n_s^\dagger + e_s$$

or, in matrix representation, stacking the equations for each sector vertically:

$$x = A^\dagger x + f^\dagger + n^\dagger + e \tag{4.7}$$

Having calculated total domestic production, we can then calculate the import, m, required to satisfy intermediate and final demands as

$$m = A^* x + f^* + n^* \tag{4.8}$$

Thus, the domestic total production, the intermediate demand and the imports may be completely determined from the final demand, the investments, the exports and the technical coefficients.

4.3.3   A Single Country Model

Our country model follows the standard input–output model described in Section 4.3.2 to a large extent. Each country is represented by an input–output table, and many of the coefficients used to describe a country's economy are taken directly from WIOD. Table 4.1 shows a summary of which coefficients are used, unchanged, in our country model.[8]

Table 4.1   Coefficients which define the economy of country i and their source in WIOD national input-output tables

	Description	Source in WIOD
$f_s^{(i)}$	Final demand on sector s	Sum of three final consumption columns: households, government and non-profit organisations
$n_s^{(i)}$	Investment of sector s	Sum of two investment columns labelled 'gross fixed capital formation' and 'changes in inventories and valuables'
$e_s^{(i)}$	Exports of sector s	Column labelled 'Exports'
$z_{s,t}^{\dagger(i)}$	Intermediate demand on sector s by sector t (domestic)	Top-left 35×35 block, labelled 'c1–c35'
$z_{s,t}^{*(i)}$	Intermediate demand on sector s by sector t (imported)	Bottom-left 35×35 block

Note that domestic coefficients are labelled with a dagger superscript (†) and import coefficients with an asterisk (*).
Source: Reproduced with permission from Levy et al. (2014)

---

[7] Note that a single year is assumed throughout this treatment. In time series analyses, there will be $2S \times S$ technical coefficients for every country in every year.
[8] The coefficients relating to final demand and investment are generated by summing across the WIOD columns listed in the table.

Notice that imports, m, are not present in this table. To define m we now introduce the first of two new sets of coefficients.

4.3.3.1 Import Ratios

The above description treats domestic and foreign goods as complements of one another, rather than as substitutes. For example, each aeroplane produced would require some fixed quantity of domestic steel and a (different) fixed quantity of imported steel. Inspired by the description given by Duchin (2004) of Leontief's proposed global model, the present model reverses this assumption and treats foreign and domestic products as perfect complements. This assumption is important for our goal of coefficient parsimony as we shall see.

Leontief assumed that engineers in an importing country do not care where a product originated; they will simply know that domestic production does not meet their demand and instead demand a perfectly substitutable imported product. In a similar spirit, when a product in the present model arrives at the shores of an importing country, it enters a theoretical 'national warehouse' along with domestically produced products, at which point the two become indistinguishable.[9]

We assume that the proportion of domestic to imported goods in this warehouse remains constant. This fraction remains fixed per country and per sector and is called the *import ratio*. It is calculated directly from the country's NIOT as

$$d_s = \frac{m_s}{(x_s - e_s) + m_s} \tag{4.9}$$

where x_s, e_s and m_s are the total production, export and import of sector s calculated via Equations (4.1) and (4.2), respectively. The term $(x_s - e_s)$ represents the products used domestically. It includes all intermediate demand, including that required to fulfil export requirements, as well as direct flows to final demand and investment, but excludes direct flows to export. This is because imports may not directly supply export demand as per Equation (4.4).

The assumption of fixed import ratios halves the number of technical coefficients that must be specified for a given country and, therefore, helps to achieve our goal of model parsimony.

Since we no longer need to track intermediate flows separately for domestic and imported goods, we can create a single matrix of intermediate flows, Z, as follows:

$$Z = Z^\dagger + Z^* \tag{4.10}$$

Now, similar to Equation (4.5), the 'combined' technical coefficients can be calculated by dividing each element of Z by an appropriate element of x. Following Miller and Blair (1985) in defining $\hat{x}$ as a matrix whose diagonal elements are the elements of x, we can calculate the combined technical coefficients as

$$A = Z\hat{x} \tag{4.11}$$

As we have seen, both A and d are calculated directly from data and are then fixed as coefficients of each country model. Once they are calculated, we can use them to calculate, in a

---

[9] Note that this concept is referred to in Miller and Blair (1985) as *import similarity*.

similar style to Equation (4.7), total production x for *any* set of final demands, f, investments, n, and exports, e:

$$x = (I - \hat{d})(Ax + f + n) + e \qquad (4.12)$$

where $\hat{d}$ is an $S \times S$ matrix whose diagonal elements are the import ratios, d_s, and I is the appropriately sized identity matrix. Once x is known, we can calculate imports, m, from Equation (4.9) as

$$m = (I - \hat{d})^{-1}\hat{d}x \qquad (4.13)$$

Finally, every individual inter-sector flow can be recovered using

$$Z = A\hat{x}$$

$$Z^* = \hat{d}Z$$

$$Z^\dagger = (I - \hat{d})Z \qquad (4.14)$$

A further parsimony benefit of the import ratio assumption is that final demand and investment have only S elements to be specified, rather than the $2S$ elements shown in Equation (4.3). This implies that consumers have no preference between domestic and imported products. Rather, the import ratios will set the relationship between demand for domestic goods and demand for imported goods in an identical way to Equation (4.14).

4.3.4 An International Trade Model

The concept of the NIOT is often expanded to an international context using MRIO modelling. In standard MRIO, each sector in each country is explicit about which countries it gets its imports from. This requires each sector to have $S \times C$ technical coefficients, an onerous data requirement and, as we have seen, a challenge to the credibility of the assumptions required.

WIOD presents all these technical coefficients in what it calls world input–output tables (WIOTs), huge MRIO tables for the whole world, but the present model instead takes a second assumption from Leontief via Duchin (2004) which Leontief referred to using the term 'export shares'. It is this assumption which introduces our second new set of coefficients.

4.3.4.1 Import Propensities

Once import demand is fixed by the country model outlined in section 4.3.3, we make the assumption that countries get each of their imported products from their international trading partners in fixed proportions. Thus, country j will always get the same proportion of its total import demand for product s from country i. We refer to these fixed proportions as *import propensities* as they describe a country's propensity to import a given product from each other country. These propensities are calculated directly from empirical observation, using trade flow data provided by the United Nations in their COMTRADE database.[10] This database provides bilateral flows of products at a very fine grain of detail. Using a mapping provided to us by the

---

[10] COMTRADE covers only commodities. An explanation of how import propensities for services sectors are derived is given in the Appendix.

WIOD team, we were able to map each of these product flows to a sector and, by summation, create empirically observed total sector flows. By denoting the flow of sector s from country i to country j as $y_s^{(i,j)}$, we can calculate the associated import propensity directly from data using

$$p_s^{(i,j)} = \frac{y_s^{(i,j)}}{\sum_k y_s^{(k,j)}} \tag{4.15}$$

Given the import requirements of each country from Equation (4.13) and the import propensities from Equation (4.15), the export demand on sector s in country i due to demand from country j can be calculated as

$$e_s^{(i,j)} = p_s^{(i,j)} m_s^{(j)} \tag{4.16}$$

and total export demand in country i is therefore

$$e_s^{(i)} = \sum_k p_s^{(i,k)} m_s^{(k)} \tag{4.17}$$

Equations (4.9)–(4.17) thus describe a system which defines the total productions of all sectors in all countries, the intermediate input–output flows, imports and exports, given a set of technical coefficients, import ratios, import propensities and final demands,[11] all of which can be derived directly from empirically observed data. Following a given set of changes to any of these latter four sets of coefficients, all of the former six categories of flow can be found.

4.3.5 Setting Model Coefficients from Data

As outlined earlier, the model takes four groups of coefficients and produces a complete set of flows within countries (input–output flows) and a complete set of trade flows (imports and exports) between countries. These four groups of coefficients are either calculated from the NIOTs provided by WIOD or from commodity and services trade data provided by the United Nations. Table 4.2 summarises the four groups of coefficients and the way in which they are set from data.

4.4 Solving the Model

4.4.1 The Leontief Equation

To get production requirements from traditional WIOT, it is sufficient to invert the matrix and solve for x directly via Leontief's famous equation:

$$x = (I - A)^{-1} f \tag{4.18}$$

The elegance of Leontief's method also had an additional benefit in that it was computationally efficient. Mathematical methods existed for efficient inversion of matrices which lent the solution an extra appeal in an age of hand-wrought numerical solutions.

---

[11] In the present version of the model, investments, n_s, are set to zero. Data on investments are provided by WIOD, but it is not used in this version.

Table 4.2   The four groups of fixed coefficients in the model and their construction from data. All other flows, including imports and exports, are derived from these coefficients mathematically

Final demand	f_s quantity of a product consumed by the public and the government of a particular country. These are taken directly from the NIOTs using the columns described in Table 4.1.
Technical coefficients	$a_{r,s}$ the quantity of product r required in country i to make a single unit of product s. Calculated from the NIOTs using Equation (4.11).
Import ratios	d_s the proportion of country i's total demand for product s which is supplied by imports. Calculated from the NIOTs using Equation (4.9).
Import propensities	$p_s^{(i,j)}$ the proportion of country j's total of import of product s which comes from country i. Calculated from the UN trade data using Equation (4.15).

Source: Reproduced with permission from Levy et al. (2014).

4.4.2   The Drawbacks of Mathematical Elegance

Behind this elegant and efficient formulation, the assumptions are as follows:

(i) Production functions are linear: to produce double the output, double the amount of every input is required.
(ii) Production possibilities are unlimited: an economy will simply produce whatever is demanded of it.
(iii) Markets always clear: all demand is fulfilled immediately.

The undeniable computational convenience of the traditional method of solving IO models is no longer a sufficient justification for its use if it is accompanied by assumptions which would not otherwise be made.

The model in its current form relaxes none of the three assumptions given earlier, but the separation of international trade from individual domestic economies offers the flexibility to do so in future work, simply by making the desired alterations to the country-level model. Such alterations would be more straightforward to perform and more intuitively comprehensible in this model than in a WIOT, where changing any individual process would require consideration of the entire system and a completely new approach to finding solutions. Recent work on complexity in economics such as Beinhocker (2006) and Ramalingam et al. (2009) supports the thesis that many of the interesting phenomena in macroeconomics happen when systems are out of equilibrium. With this in mind, we now describe a routine for solving the model iteratively, rather than through a single, large matrix inversion.

4.4.3   Algorithm for an Iterative Solution

Figure 4.1 shows a schematic version of the algorithm for calculating imports and exports from the four groups of coefficients.

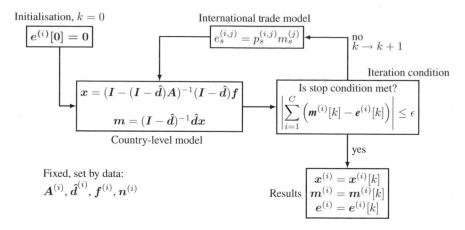

Figure 4.1 The model algorithm: total production, imports and exports are calculated for a given set of fixed coefficients. Note that square brackets have been introduced to designate the iteration variable k, which tracks the quantities that are iteratively recalculated as the algorithm runs. *Source*: Reproduced with permission from Levy et al. (2014)

Initialisation

All export demands are set to 0.

Country-level Model

Inside each country, total demand is calculated using the export demands (which were initially set to zero), final demands, technical coefficients and import ratios, as per Equation (4.12). This step maintains the standard input–output matrix inversion. From total demand, import demand can be calculated via the import ratios as in Equation (4.13).

Iteration Condition

At this point the iteration continues unless the total level of global imports matches, to within some small tolerance $\epsilon > 0$, total exports in each sector.[12] In the first run through the iteration, this stop condition will certainly fail because exports have just been set to zero.

International Trade Model

The import demands calculated in the country-level models can be divided up into export demands using the import propensities by using Equation (4.17). These new export demands replace the export demands from the previous iteration (which were all zero the first time

---

[12] More specifically, as shown in Figure 4.1, we verify that the length of the vector of differences between the global sum of all imports and the global sum of all exports for each sector does not exceed ϵ. This approach ensures that global export and import totals must balance for all sectors simultaneously.

round). With a new set of export demands calculated, the algorithm returns to the country-level model.

The iteration continues in this way, moving between the country-level and the international-level models, until the system of imports and exports converges and the stop condition is met.[13]

Solving the system algorithmically rather than relying on the standard methods of linear algebra allows for a greater flexibility in the way that the model may be extended in future work. For example, provided that care was taken to maintain its stability, the algorithmic approach could be applied even if the assumption of linear production functions were to be relaxed in order to model economies of scale.

4.5   Analysis

4.5.1   Introduction

The model combining input–output country descriptions and the international trade network allows for the measurement of the total input required to make a single unit of a particular sector's product. Since WIOD reports product flows in millions of US dollars ($M), this will also be the unit of all the following analyses. In all that follows, the model was initialised using data from 2010.

4.5.2   Simple Modelling Approaches

4.5.2.1   Economic Significance

A simple way to measure the significance to the global economy (hereafter 'significance') of a particular sector in a given country is to artificially reduce final demand for the products of that country–sector by one unit, to recalculate all flows in the model based on this new final demand and to measure the response of each other sector in each other country in the model.

This takes account of not only the *direct* effects of such a change—a reduction in the output of the sector in question and of those sectors supplying that sector's intermediate demand—but also the full spectrum of *indirect* effects on those sectors which supplied the demand of the sectors who supplied the sector in question, on *their* suppliers and so on. We can formalise the final demand change in sector s in country i as

$$f_s^{(i)} \to f_s'^{(i)} \equiv (f_s^{(i)} - 1) \tag{4.19}$$

The simplest way to define 'response' is as the change of total output of all sectors, r, in all countries, j, caused by the reduction:

$$\Delta_s^{(i)} x_r^{(j)} \equiv (x_r'^{(j)} - x_r^{(j)}) \tag{4.20}$$

where $x_r'^{(j)}$ is the total output of sector r in country j after the change in Equation (4.19) has taken place and the model has been recalculated according to Figure 4.1. $\Delta_s^{(i)} x_r^{(j)}$ is thus the

---

[13] An analysis confirming the convergence of the algorithm is the focus of upcoming work by Dr A. P. Korte of UCL's Centre for Advanced Spatial Analysis.

change in output of sector r in country j induced by the change in final demand for sector s in country i.

The total response across all countries and all sectors induced by sector s in country i is

$$n_s^{(i)} \equiv \Delta x_s^{(i)} = \sum_j \sum_r \Delta_s^{(i)} x_r^{(j)} \tag{4.21}$$

and the average response-inducing power of the sectors in country i is

$$n^{(i)} = \frac{\sum_s n_s^{(i)}}{S} \tag{4.22}$$

where S is, as earlier, the number of sectors in country i's economy.

If a reduction in demand for a sector induces a large reduction in production globally, it follows that sectors with a larger value of n must require, once the entire production network is accounted for, a larger amount of input to produce each unit of their output. n can thus be thought of as a country-level significance measure, with larger values implying an economy which, averaging across the effects of each of its sectors, has a greater significance on production in the rest of the world (RoW).

Table 4.3 shows the 10 most significant modelled countries by this measure. China is the world's most significant economy, a \$1M reduction in final demand leading to an average of \$2.36M reduction in total output worldwide.

This is perhaps a surprising result, as we might have expected the United States to be the world's most significant economy. To investigate further, we split the significance measure, n, into foreign and domestic effects as follows:

$$n_s^{*(i)} = \sum_{j \neq i} \sum_r \Delta_s^{(i)} x_r^{(j)} \tag{4.23}$$

Table 4.3   Response to a \$1M reduction in final demand in terms of the difference induced in the total output, x, of other sectors

(a) most significant countries		(b) China's most significant sectors	
Country	$n^{(i)}$(\$M)	Sector	$n_s^{(China)}$(\$M)
China	2.36	Vehicles	2.47
Czech Republic	1.98	Food	2.20
S. Korea	1.95	Plastics	2.20
Russia	1.94	Metals	2.17
Bulgaria	1.92	Machinery	2.17
Japan	1.90	Wood	2.14
Italy	1.90	Electricals	2.13
Poland	1.89	Construction	2.12
Finland	1.88	Textiles	2.11
Portugal	1.87	Paper	2.10

Only the largest 10 are shown
Source: Reproduced with permission from Levy et al. (2014)

$$\eta^{*(i)} = \frac{\sum\limits_{s} \eta_s^{*(i)}}{s} \tag{4.24}$$

and

$$\eta_s^{\dagger(i)} = \sum_{r} \Delta x_{sr}^{(ii)}$$

$$\eta^{\dagger(i)} = \frac{\sum\limits_{s} \eta_s^{\dagger(i)}}{s} \tag{4.25}$$

Notice that this does indeed split η into precisely two parts as

$$\eta^{(i)} = \eta^{*(i)} + \eta^{\dagger(i)} \tag{4.26}$$

Figure 4.2 shows how the average response across all sectors divides between domestic response, $\eta_s^{\dagger}$, and foreign response,[14] η_s^{*}, for each country. An OLS regression line has been added to the plot. All the countries lying close to the line have a similar total significance, $\eta^{(i)}$. The region below the line is the region of less-than-average significance and that above the line is the region of greater-than-average significance.

China is immediately visible as an outlier. Its economy is around one-third more significant ($\eta = 2.36$) than that of Brazil ($\eta = 1.75$) or India ($\eta = 1.74$), but almost all of this extra significance relates to its impact on its own domestic sectors. Thus, China's significance to the world economy shows itself largely in China itself: it is significant to the extent that it uses disproportionately a large amount of its own productive output in its production technology. This may perhaps be evidence of the much-vaunted end of low labour costs in China (Li et al., 2012), (Economist, 2012), although further research would be needed to verify this.

4.5.2.2 Economic Self-reliance

Countries further to the left of Figure 4.2, such as Brazil, India, Japan, Russia, Turkey and the United States, have a lower global response to a domestic demand shock. We could therefore describe these countries as being more self-reliant than those on the right, such as Luxembourg, Belgium, the Netherlands and Denmark. This self-reliance can be made precise by measuring the ratio of domestic response to foreign response:

$$\phi^{(i)} = \frac{\eta^{*(i)}}{\eta^{\dagger(i)}} \tag{4.27}$$

Figure 4.3 shows the relationship between $\phi^{(i)}$ and population in 2010 (World Bank, 2014).

The positive slope of the regression line (significant at 0.1%) indicates that larger countries are generally more self-reliant. Both Brazil and Australia are more self-reliant than their populations would suggest. Belgium and the Netherlands are less self-reliant in the same sense. An interesting avenue for further research might be to investigate which sectors cause the largest share of these atypical self-reliance measures, particularly that of the largest outlier, Brazil.

---

[14] It should be noted here that the 'rest of the world' (RoW) is not included in this analysis. Thus, 'foreign' here refers only to the countries explicitly modelled. According to Streicher and Stehrer (2012), the 40 WIOD countries cover 85% of the global economy, so the effect of excluding RoW should not be too large.

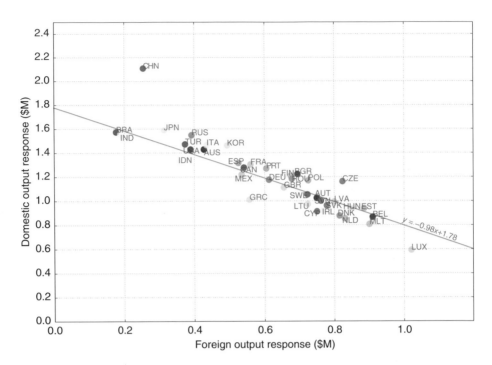

Figure 4.2   The total response of the global economy to a reduction in final demand, averaged across every sector in a country. 'Response' is defined as the total production lost across all sectors. Here, response has been divided into domestic, where the response is measured only in the country whose final demand has been reduced, and foreign, where the response is measured in all other countries only. *Source*: Reproduced with permission from Levy et al. (2014)

4.5.3   A Unified Network Approach

The flow of products between countries can be viewed as a weighted, directed network, where countries are nodes, and the weights of each edge represents the magnitude of the flow between them (Nystuen and Dacey, 1961; Serrano and Boguñá, 2003; Bhattacharya et al., 2008; Baskaran et al., 2011). Similarly, the sector–sector flows constitute an input–output model (Blöchl et al., 2011; Fedriani and Tenorio, 2012). Network science has developed useful ways to analyse the sort of weighted, directed networks that constitute the present model, but a single network representation is required which combines the international and the sub-national networks.

Recall from Section 4.3.3 that products arriving at the shores of an importing country are put into a warehouse along with domestically produced products at which point the two become indistinguishable. If an additional assumption is made that domestic sectors take from this warehouse by means of a random sample, then the fraction of products in each sample from abroad will be identical, as will be the fraction of imported products from each exporter. These fractions will be set by the import ratios and import propensities, respectively.

This additional assumption allows us to specify a complete system of intermediate (input–output) flows, $y_{rs}^{(i,j)}$, from sector r in country i to sector s in country j, thereby

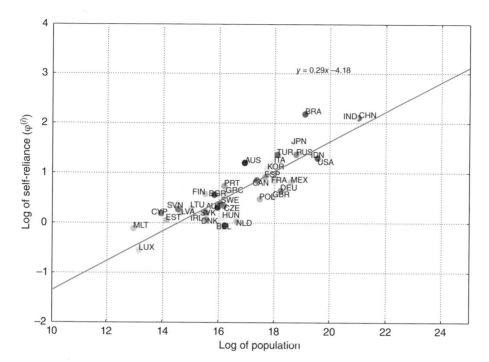

Figure 4.3 Figure showing the positive relationship between economic self-reliance as defined by Equation (1.27) and population. More populous countries tend to be more self-reliant. *Source*: Reproduced with permission from Levy et al. (2014)

effectively reconstructing the MRIO from which we diverted in Section 4.4:

$$y_{rs}^{(i,j)} = p_r^{(i,j)} d_r^{(j)} a_{rs}^{(j)} x_s^{(j)} \tag{4.28}$$

which can be understood as follows: sector s in country j requires an amount $a_{rs}^{(j)} x_s^{(j)}$ of sector r's product to produce its total output; a fraction $d_r^{(j)}$ of this will be supplied by imports; a fraction $p_r^{(i,j)}$ of these imports will come from country i. Note that since countries do not import their own exports, $p_s^{(i,i)} = 0, \forall i$, and the expression holds trivially for $i = j$.

The fractions of each type of demand for s in country j which was exported by country i are given by similar expressions:

$$f_s^{(i,j)} = p_s^{(i,j)} d_s^{(j)} f_s^{(j)}$$

$$e_s^{(i,j)} = p_s^{(i,j)} d_s^{(j)} e_s^{(j)}$$

$$n_s^{(i,j)} = p_s^{(i,j)} d_s^{(j)} n_s^{(j)} \tag{4.29}$$

where, as previously, f is the final demand, e is the export demand and n are investments.

Along with the sector-to-sector flows given by Equation (4.14), this allows for the representation of all the flows in the present model as a single network, to which standard network analysis techniques can then be applied.

This reformulation demonstrates the connection between the model described in this chapter and the established WIOT approach, allowing us to visualise the individual country–sector to

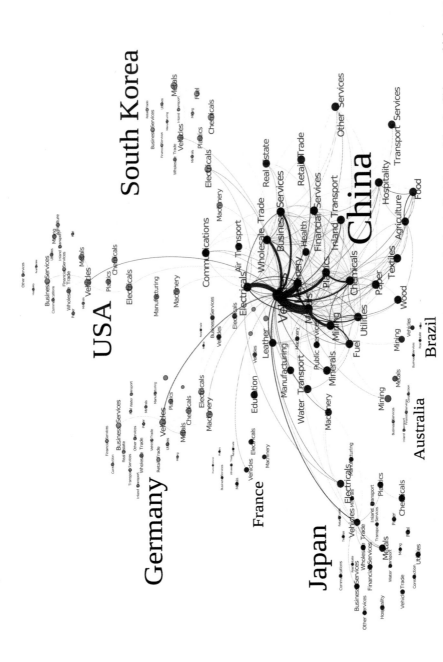

Figure 4.4 A network representation of the seven most-affected countries following a reduction in final demand for the Chinese vehicles sector. Node size is proportional to eigenvector centrality and edge width is proportional to the change in flow. *Source:* Reproduced with permission from Levy et al. (2014)

country–sector flows (or changes in such flows) for any model scenario, such as that calculated in Section 4.5.2 in response to a reduction of demand in China for the vehicles sector by $1M.

The change to every such flow (both between-country trade flows and within-country intermediate flows) in the eight most-affected countries is recorded and visualised in Figure 4.4. Node size is proportional to eigenvector centrality, and the edge weight is proportional to the change induced in the particular flow by the reduction in final demand. This diagram reveals the patterns and shapes of the various economies which would be hard to pick out in the raw flow matrix.

4.5.4 Comparison with a Multi-region Input–Output Model

As discussed in Section 4.2, the modelling approach presented here relaxes the somewhat rigid structure of an MRIO table. In so doing, it also reduces the amount of data required to estimate tables for countries which are not covered by WIOD. But since the present iteration of the model can be made equivalent to a MRIO (as shown in Section 4.5.3), any of the analyses presented here could equally well be performed using the WIOT directly. It is therefore important to verify that the model in its present, most simple, form behaves similarly to an MRIO before any further development is done.

To this end, the results of the experiment of Section 4.5.2, reducing demand for Chinese vehicles, is repeated using an MRIO, and the results are compared in Figure 4.5. The response to a reduction in demand for Chinese vehicles is summed over sectors in each country. The countries are then ranked according to the size of the response. The ranks produced by the MRIO are shown on the left-hand side and those produced by the present model are shown in the right. Visually, the results seem broadly comparable and have a Spearman rank correlation coefficient (Spearman, 1904) of 0.96. This gives some confidence in concluding that this first and simplest iteration of the model behaves in a similar manner to the MRIO it seeks to generalise and extend.

4.6 Conclusions

In this chapter, we have introduced a new model of global trade, combining two distinct modelling approaches for the internal workings of national economies and for international trade. The model incorporates information from two extremely extensive data sets, the UN COMTRADE database and WIOD, giving it a good basis in empirical observation.

The model equations are solved iteratively, thus avoiding the limitations and assumptions of the traditional multi-region input–output approach, lending the model the flexibility to accommodate a wide variety of possible extensions, such as the consideration of non-linear production functions, limited production capacities and non-clearing markets. It also introduces two new sets of coefficients for modellers to work with, which will facilitate its integration with other social science models.

We have also introduced a new measure of the global economic significance of each sector within a national economy, gauging the degree to which that sector drives economic activity across the world. In initial investigations, China was found to be a considerable outlier in terms of its economic significance, driven by high consumption of its own domestically produced products and services. Following this analysis, we introduced a measure of economic

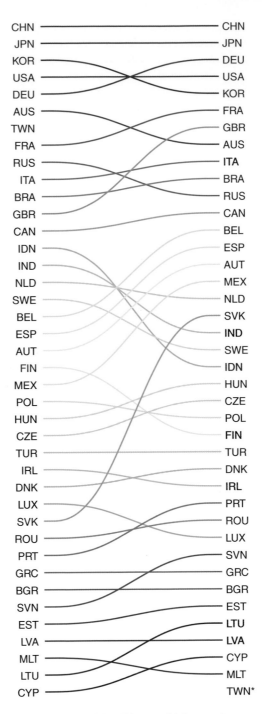

Figure 4.5   A comparison of the results of the Chinese vehicle experiment performed on an MRIO (on the left) and on the present model (on the right). Vertical position represents the rank of the country in terms of how affected it is by a reduction in demand for Chinese vehicles. *Taiwan is ranked last in the present model. This is because the United Nations provides no trade data for the territory. All Taiwan's import ratios are therefore zero (i.e. Taiwan is modelled as being entirely self-sufficient). *Source*: Reproduced with permission from Levy et al. (2014)

self-reliance, which was found to vary inversely with population. These results represent preliminary analyses only, with far more scope for detailed investigation of international trade patterns and alternative global economic scenarios now that the model is operational.

Finally, we have presented a method for transforming the structure of the model into that of a traditional MRIO table, allowing for both traditional MRIO approaches, and the use of the tools and analyses of the network science literature. A very simple example of this approach was presented, with further analyses to follow in future work.

The next step will be to estimate country models for those (mostly non-OECD) countries not in WIOD to facilitate the study of such human systems as migration, international security and development aid. Further work on the model will include the relaxation of its input–output assumptions, beginning with the introduction of limits on production capacity, with the goal of creating a fully dynamic version of the model to analyse the change in trade patterns over time. Upcoming work uses the model to answer specific policy questions such as how to maximise a country's GDP by manipulation of its final demand vector, import ratios and technical coefficients.

Acknowledgements

The authors acknowledge the financial support of the UK Engineering and Physical Sciences Research Council (EPSRC) under the grant ENFOLDing—Explaining, Modelling and Forecasting Global Dynamics, reference EP/H02185X/1.

We would also like to thank Hannah Fry, Rob Downes, Anthony Korte and Peter Baudains for their invaluable input and support during the writing of this chapter.

References

Akita, T. (1993) Interregional interdependence and regional economic growth in Japan: an input-output analysis. *International Regional Science Review*, **16** (3), 231–248.

Baskaran, T., Blöchl, F., Brück, T., and Theis, F.J. (2011) The Heckscher-Ohlin model and the network structure of international trade. *International Review of Economics & Finance*, **20** (2), 135–145.

Beinhocker, E.D. (2006) *Origin of Wealth: Evolution, Complexity, and the Radical Remaking of Economics*, 1st edn, Harvard Business School Press, Cambridge, Mass.

van den Bergh, C.J.M. (2002) *Handbook of Environmental and Resource Economics*, Edward Elgar, Cheltenham.

Bhattacharya, K., Mukherjee, G., Saramäki, J., Kaski, K., and Manna, S.S. (2008) The international trade network: weighted network analysis and modelling. *Journal of Statistical Mechanics: Theory and Experiment*, **2008** (02), P02 002.

Blöchl, F., Theis, F.J., Vega-Redondo, F., and Fisher, E.O. (2011) Vertex centralities in input-output networks reveal the structure of modern economies. *Physical Review E*, **83** (4), 046 127.

Deming, W.E. and Stephan, F.F. (1940) On a least squares adjustment of a sampled frequency table when the expected marginal totals are known. *The Annals of Mathematical Statistics*, **11** (4), 427–444.

Duchin, F. (2004) International trade: evolution in the thought and analysis of Wassily Leontief. Wassily Leontief and Input-Output Economics, pp. 47–64.

Economist, T. (2012) The End of Cheap China. *The Economist*.

Fedriani, E.M. and Tenorio, A.F. (2012) Simplifying the input–output analysis through the use of topological graphs. *Economic Modelling*, **29** (5), 1931–1937.

Hendrickson, C.T., Lave, L.B., and Matthews, H.S. (2006) *Environmental Life Cycle Assessment of Goods and Services: An Input-Output Approach*, RFF Press, Washington, DC.

Joshi, S. (1999) Product environmental life-cycle assessment using input-output techniques. *Journal of Industrial Ecology*, **3** (2-3), 95–120.

Khan, H.A. (1999) Sectoral growth and poverty alleviation: a multiplier decomposition technique applied to South Africa. *World Development*, **27** (3), 521–530.

Lenzen, M., Moran, D., Kanemoto, K., and Geschke, A. (2013) Buxsilding Eora: a global multi-region input-output database at high country and sector resolution. *Economic Systems Research*, **25** (1), 20–49.

Leontief, W. (1970) Environmental repercussions and the economic structure: an input-output approach. *The Review of Economics and Statistics*, **52** (3), 262–271.

Leontief, W. (1974) Structure of the world economy. Outline of a simple input-output formulation. *The Swedish Journal of Economics*, **76** (4), 387–401. ArticleType: research-article / Full publication date: December, 1974 / Copyright © 1974 The Scandinavian Journal of Economics.

Li, H., Li, L., Wu, B., and Xiong, Y. (2012) The end of cheap Chinese labor. *Journal of Economic Perspectives*, **26** (4), 57–74.

Luo, J. (2013) The power-of-pull of economic sectors: a complex network analysis. *Complexity*, **18** (5), 37–47.

Miller, R.E. and Blair, P.D. (1985) *Input-Output Analysis: Foundations and Extensions*, Prentice-Hall, Englewood Cliffs, NJ.

Nystuen, J.D. and Dacey, M.F. (1961) A graph theory interpretation of nodal regions. *Papers of the Regional Science Association*, **7** (1), 29–42.

Ramalingam, B., Jones, H., Reba, T., and Young, J. (2009) Exploring the science of complexity: ideas and implications for development and humanitarian efforts.

Serrano, M.A. and Boguñá, M. (2003) Topology of the world trade web. *Physical Review E*, **68** (1), 015 101.

Spearman, C. (1904) The proof and measurement of association between two things. *The American Journal of Psychology*, **15** (1), 72–101.

Streicher, G. and Stehrer, R. (2012) Whither Panama? Constructing a consistent and balanced world SUT system including international trade and transport margins. *Economic Systems Research*, **27** (2), 213–237.

Timmer, M.P., Dietzenbacher, E., Los, B., Stehrer, R., and de Vries, G.J. (2015) An Illustrated User Guide to the World Input–Output Database: the Case of Global automotive production. *Review of International Economics*, doi: 10.1111/roie.12178.

Tukker, A. and Dietzenbacher, E. (2013) Global multiregional input–output frameworks: an introduction and outlook. *Economic Systems Research*, **25** (1), 1–19.

Tukker, A., de Koning, A., Wood, R., Hawkins, T., Lutter, S., Acosta, J., Rueda Cantuche, J.M., Bouwmeester, M., Oosterhaven, J., Drosdowski, T., and Kuenen, J. (2013) Exiopol—development and illustrative analyses of a detailed global MR EE SUT/IOT. *Economic Systems Research*, **25** (1), 50–70.

Walmsley, T., Aguiar, A., and Narayanan, B. (2012) Introduction to the global trade analysis project and the GTAP data base, *Working Paper 67*, Center for Global Trade Analysis, Purdue University.

World Bank (2014) World Development Indicators 2014, doi: 10.1596/978-1-4648-0163-1.

Appendix

Here, we outline some additional details of the set-up of the model.

A.1   Modelling the 'Rest of the World'

The UN trade data records flows to and from many countries (as well as many non-country 'trade areas') which are not part of WIOD. If these flows were simply ignored, then countries which trade significantly with these areas would be misrepresented in terms of the extent to which they trade with the countries which *are* in the model. To avoid this, the model includes a 'rest of the world' (RoW) entity which receives all exports going to countries not explicitly modelled ('stray' exports) and provides all imports coming from such countries ('stray imports'). The RoW has no technical coefficients and no import ratios.

The RoW is initialised to have a final demand, f_s, equal to the total stray exports in each sector across the whole model.[15] Then, since the RoW has no import ratios, instead of solving Equation (4.13) to calculate import demand, it simply sets

$$m = f \tag{A.30}$$

thus importing a fixed amount of each sector, defined by the initial level of stray exports in the data. The RoW also has a particular way of deriving its total production, x. It has no technical coefficients, so cannot solve Equation (4.12). Rather, it simply sets

$$x = e \tag{A.31}$$

which allows it to produce 'for free' (in the sense that there is no intermediate demand) whatever is required of it from other countries. This also has the effect of decoupling the RoW's import side from its export side.

Other than these aspects, the RoW behaves just like a normal country: it has a set of import propensities defining its imports from each other country, and each other country will have an import propensity relating to trade with the RoW.

A.2   Services Trade Data

Data on trade in goods are based on the very detailed records kept by border agencies for gathering the appropriate taxes. For this reason, the goods trade data can be considered to be of high quality. Since the supply of services is harder to track, the data on trade in service goods[16] are considerably coarser. Specifically, many countries report only the total quantity of services imported and exported, not the origin and destination of any particular bilateral trade flow. In order to estimate the bilateral flows required by Equation (4.15) for the calculation of import propensities, these flow totals must first be divided between trade partners.[17]

---

[15] Flows from countries not in the model to other countries not in the model are ignored.

[16] taken from http://unstats.un.org/unsd/servicetrade/default.aspx.

[17] Where data on a particular flow is reported from both the importer and the exporter, the data from the importer are given preference. Data from the exporter are only used where the importer reports no data.

This is done using a method of iterative proportional fitting introduced by Deming and Stephan (1940). The basic algorithm for balancing n totals into a matrix of $n \times n$ bilateral flows works as follows:

1. Initialise an $n \times n$ matrix with 1s (or any other arbitrary starting value)
2. Multiply each row vector such that the sum of the elements is equal to the required total outflow
3. Multiply each column vector in an identical manner such that they sum to the required total inflow
4. Iterate Steps 2 and 3 until the system converges (in the sense of the sum of the squared difference between each element in the matrix of the current iteration and that of the previous iteration being less than some $\epsilon > 0$)

The procedure is followed S times, producing one flow matrix per sector, with two changes specific to this situation.

A.2.1   Importing Own Exports

The commodities trade data report no flows originating from and arriving to the same country. The import propensities $p_s^{(i,i)}$ are therefore set to zero. To ensure that the outcome of the fitting procedure adheres to this, the initial matrix is modified to have a zero along its diagonal. Since the fitting consists only of multiplications, any zeros in the initial matrix will remain zero in the balanced matrix.

A.2.2   The Rest of the World for Sectors

Since total imports and total exports will inevitably not be equal per sector, a 'rest of world' (RoW) similar to that described in Appendix 1 is added to absorb any excess. If exports exceed imports, the RoW is added as an additional column (importer) and vice versa. The initial matrix is thus either wide or tall, depending on the relevant trade totals.

Part Three

Migration

Part Three

5

Global Migration Modelling: A Review of Key Policy Needs and Research Centres

Adam Dennett and Pablo Mateos

5.1 Introduction

Policies are an important element of international migration research and provide the context – indeed in part the purpose – for model-based research and analysis. Policies may seek to address specific issues such as the integration of migrants, but in most cases they can be boiled down to limiting the flows of some types of migrant or encouraging the flows of others. Judging the impact of both migration flows and migration policy can only be done effectively with good information, but too often data on migration are poor and thus the evidence base lacking. This chapter assesses some of the migration and policy issues, which are currently relevant, before examining where the modelling work being undertaken can help strengthen the evidence base available in order that better policy decisions are made. A directory of international migration research centres and the work they are undertaking is provided in an extensive appendix.

International migration is a perennial research interest to academics across the world. The €28 million NORFACE research programme on migration (http://www.norface-migration. org) is one such reflection of the continuing desire to understand the global patterns, processes and effects of the human population moving around the globe. The inaugural conference for the research programme titled 'Migration: Economic Change, Social Challenge' and held at University College London in April 2011 (http://www.norface-migration.org/sites/ index.php?site=3&page=1) exposed the breadth of interest surrounding different aspects of international migration, with researchers from across the globe presenting research related to a host of topics including migration flows, migration policy, migrants' impact on

Global Dynamics: Approaches from Complexity Science, First Edition. Edited by Alan Wilson.
© 2016 John Wiley & Sons, Ltd. Published 2016 by John Wiley & Sons, Ltd.

labour markets, migrants' impact on sending and receiving communities, migrant networks, second-generation migrants and migrant children, segregation, migrant identities, remittances and migration data. In most cases, interest in these broad areas was locally focused with country-level case studies predominating.

Drawing together a large number of academics who are currently engaged in international migration research, the conference was a snapshot of the current 'state of the art' of many branches in the field and showcased the work advancing current knowledge about international migration to fellow academics. While academics may assess the value of their work in very broad terms – progressing our intellectual understanding, asking new questions and revealing new truths – increasingly the value of research (judged by those who fund it) is measured through its *impact* on a wider society; and so outputs will frequently be geared towards realising this aim (whatever your position on the merits of this approach may be[1]). In the social sciences where much of the work on migration can be found, this impact will invariably be through the influence the research has on forming, shaping or redefining policies that are enacted by governments and other policy makers.

In this chapter, we present a short review of the policies that currently influence and are influenced by international migration. We will endeavour to link policy needs to modelling challenges to provide the context for later chapters. There are a number of research centres that are engaged in research activities related to international migration and international migration policy. The Appendix of this report provides a guide to international migration research centres around the world and the work they carry out, particularly focusing on whether modelling is used to enhance the evidence base from which policy decisions are made.

5.2   Policy and Migration Research

5.2.1   *Key Policy Issues in Contemporary Migration Research*

Studying the various research outputs from the migration research groups outlined in Appendix, 'policy' is a recurrent theme. This is perhaps unsurprising as research will top and tail the theoretical model of the *policy cycle*: this cycle begins with the identification of a problem (research) and progresses through the formulation, adoption and implementation of the policy itself (legislation) before the outcomes of the policy are evaluated (research). Frequently, 'policy briefs' are produced by a number of the research centres. These may be prospective and look to identify particular issues, which may then elicit policy stances by national or international legislatures, or retrospective and review or critique policy decisions which have already been made. Salt and de Bruycker (2011) detail a number of the more recent global policy legislation developments in OECD countries (plus a handful of others), the themes of which are echoed in a number of policy-positioning or policy-reviewing research outputs produced by international migration research centres. Of course, the policy legislation part of the policy cycle does not always follow on from sound research (although undoubtedly if it did, better policy decisions would be made). However, where research may not always come before policy, it almost always will be carried out afterwards by either those who enacted the policy or indeed those who may have opposed the policy in the first place.

---

[1] http://www.guardian.co.uk/science/blog/2012/mar/28/research-council-sacrifices-basic-research-commerce

Since the beginning of the global financial crisis in 2007/2008, countries which have traditionally been destinations for economic migrants have been keener than ever to ensure that labour migration meets labour market needs; and this has been a growing area of policy concern. Salt and de Bruycker (2011) cite a number of examples from Australia, the United Kingdom, Ireland, Russia and Canada, and a full report on this matter for the Home Office was conducted at the UCL Migration Research Unit (Salt et al., 2011). For example, the introduction or expansion of migration points-based systems of entry has been a feature of policy in many of these countries, along with increasingly more stringent selection criteria for migrants and reduced 'shortage-lists' for jobs. In the United Kingdom, it appears that despite the introduction of a points-based system, there is still a gap between current policy and successfully addressing skills shortages through labour immigration. There remains a demand for migrant labour in the United Kingdom, and restricting access will only exacerbate this demand unless there are also changes to the policies and institutions which create the demand in the first place (Ruhs and Anderson, 2012). One particular example of current friction can be found in the elderly care sector – a sector where jobs are typically low paid and low skilled, but the work can be challenging; with an ageing population, demand for these services is clearly growing (Shutes, 2011). In the United Kingdom, recent demand for labour in this area has been filled with migrant workers, although with the recent introduction of the points-based-system which precludes the migration of low-skilled workers, there is a growing mismatch between the supply and demand of these key workers.

Salt and de Bruycker (2011) also highlight the role of family reunion/formation and humanitarian (asylum) policies in broader international migration policy. The pervading sense is that, in relation to family reunification, policies are becoming more restrictive – for example, the tightening of rules governing the level of maintenance support that should be provided by the employed migrant before the rest of their family are allowed to join them from abroad. In Sweden, this now means adequate housing should be provided, whereas in a number of OECD countries, immigrants wishing to bring families into the host country need to ensure (and demonstrate) that these new migrants will not be a financial burden on the destination country. In addition to these restrictions, age restrictions can also operate (stipulating minimum ages for parents) as can pre-arrival integration tests (such as those testing language proficiency). The Netherlands and Denmark require immigrants who are planning to join family members who are already settled, to have knowledge about their societies as well as knowledge of their language, before successful entry applications can be made – similar requirements also exist in France and Germany. One of the most contentious areas of immigration policy in relation to the family and human rights is that of the children of irregular (illegal) migrants. In Europe, the battle between migration and asylum policy, on the one hand, and child protection policy, on the other, has created difficulties in many countries, not least the United Kingdom (Sigona, 2011).

Irregular migration is a contentious policy area in its own right. In a number of countries, border monitoring and enforcement measures have been tightened in locations where irregular migration appears to be increasing. In countries such as Italy, other deterrent policies such penalties for illegal migration have been increased. In Europe as well as in a number of other OECD counties, sanctions against the employers of illegal migrants have been made more severe. Overall, despite the various measures that have been put in place to mitigate the flows of irregular migrants, the impression that Salt and de Bruycker (2011) give is that for a number of countries, dealing with issues such as human trafficking remains a persistent challenge. van Liempt (2007) warns that much policy surrounding irregular migration focuses on the criminal aspects and the last stage of the process – of 'smuggling gangs' and human trafficking across national borders – and too often neglects migrant agency, that is, the proactive decisions (taken with differing motivations at various points along the journey) made by the migrants themselves. The implication is that more attention needs to be paid by the policy makers to factors which – using Lee's (1966) terminology, 'push' migrants out of their countries of origin. These are factors such as poor security or war, economic hardships, social injustices or political oppression.[2] It is important to understand how these factors may assist in the decisions that individuals make, which together with intervening and pull factors in turn lead to the international migration flows.

While many of the ongoing policy debates are concerned specifically with aspects of immigration, the UK government has made 'net migration' a key policy issue, with the Conservative party in 2011 stating publicly that they aim to reduce the current positive net migration balance for the country 'from the hundreds of thousands to the tens of thousands'. While one side of the net migration coin relates to immigration flows, the other relates to emigration, and therefore, by definition, any policies addressed to tackle net migration must necessarily address both sides of the coin. Indeed Salt and de Bruycker (2011) note that a number of countries have enacted policy to encourage emigration – particularly the return of recent immigrants to their home countries. Japan, Spain, the Czech Republic and Bulgaria are all cited as recently implementing such policies. Several examples are also cited in the Scandinavian countries where governments are actively providing funds to facilitate the return migration of asylum seekers who have failed in their applications to remain. In the United Kingdom, part of the persistent levels of positive net migration is due to lower emigration rates of non-nationals and higher immigration levels of nationals (a sort of 'return' to the United Kingdom). To complicate things further, some countries do not count nationals in international migration flows, but with increasing naturalisation rates one could arrive as a non-national and leave as a national being left as a long-term resident in population statistics.

Within Europe, the migration policies of individual countries have been driven by EU legislation. Salt and de Bruycker (2011) outline recent developments which propose new EU-wide policy frameworks in relation to particular areas of current interest:

- The first of these relates to labour migration and would involve the easing of restrictions for third-country nationals working within large multinational companies; this would allow these workers easier access to the EU job market for a defined limited period of time. In addition, seasonal workers will be admitted to countries under EU-wide rules which will both ease admission and limit their exploitation.
- The second relates to combating illegal immigration and sets out a number of policy measures which will assist in the management of external borders of the European Union.
- The third sets out a series of recommendations for the management of legal immigration flows, including improving the annual reporting of trends by national agencies.
- The fourth relates to a series of guidelines to assist countries in setting up integration policies in line with other EU states.

---

[2] See the wealth of resources available at http://www.forcedmigration.org for an overview of countries and regions where these factors influence migration behaviours

- The final development within the European Union concerns stronger multilateral cooperation between EU counties and Africa in relation to international migration and development, and legal and illegal migration flows.

One of the central pillars of EU migration policy has been the Schengen Agreement which followed the Maastricht Treaty in 1991–1993 and has been fully incorporated into EU law with the Amsterdam Treaty since 1999. The free travel area created under this legislation has eased the movement of people between 26 European states, allowing the crossing of borders without the requirement for passport checks to take place. With the right to free movement and barriers to accessing employment for EU citizens in other EU countries removed, national policies have had to adapt to the migration scenarios caused by this EU-wide legislation. The accession of eight former Eastern-bloc and Baltic countries to the European Union in 2004 led to an unprecedented flow of economic migrants from these states to richer Western European countries – flows which whilst in many cases were not permanent moves, but nevertheless imposed significant changes to local social structures in receiving countries, at least for short periods of time. Certainly after the large influx of Polish migrants to the United Kingdom (Düvell and Garapich, 2011) – especially in the market gardening localities around the Wash region in East Anglia, policy makers felt they had to adapt quickly. The Worker Registration Scheme was the resulting attempt by the UK government to temper these flows within existing EU law, although it ceased its operation at the end of April 2011.

The final area of policy activity relates to issues of integration and citizenship, with Salt and de Bruycker (2011) outlining some of the policy decisions taken by some countries in this area. For example, in Denmark, Sweden and Finland, programmes exist to assist new migrants in language training or labour market participation. In other countries, citizenship and integration have been linked explicitly to residence permissions through legislation – some countries have required the signing of agreements or contracts by immigrants which provide assurances of their language competencies or willingness to undertake integration programmes. Countries with such policies include Austria, Denmark, France, Germany, Italy and Norway. There has also been a focus on the integration of established migrants and their children, for example, the recognition of foreign qualifications enabling migrants to access labour markets has now been a common policy objective within a number of European countries. Citizenship can be viewed, in some ways, as the final stop along the route to the integration of migrants. It appears that a common policy within a number of countries (Australia, Canada, the United Kingdom, France, Germany, Denmark, Hungary, the Netherlands) is for citizenship to be contingent on the successful completion of newly introduced 'citizenship tests'. With citizenship seen as the ultimate aspiration of migrants and with it carrying (in many cases) the most rights and benefits, it is perhaps not surprising that withdrawal of citizenship is also a sanction increasingly being employed by countries where immigrants have been seen to harm the national interests of those countries.

The common feature among all of these policy issues is that they are all responses to known or perceived situations. For example, faced with the issue of the oversupply of low-skilled migrant labour, the response is to introduce a points-based system to control these flows. Whereas with the issue of resource pressures caused by families of migrants who cannot engage with the job market due to language, dependency or other issues, the response is then to ensure that the head of family can support dependents and impose language proficiency conditions on their arrival. In order for these policies to be formulated as responses, governments have had to first marshal the available data and use this information to inform an appropriate response. Policies on international migration such as these are reactive, but for *prospective* policy decisions to be made there is a clear need to understand the rapidly evolving dynamics of international migration and attempt to anticipate future trends. As is argued by Penninx (2006), these dynamics are not yet fully understood and this has an impact upon the policy decisions which can be made.

One of the issues within migration research which is affecting the evidence base from which policy makers can draw and therefore limiting the effectiveness of policy – highlighted by Penninx (2006) and also by Kraal (2008) – is the disconnection between different levels of analysis. Much qualitative research is carried out at the level of the individual and rarely links to quantitative research which tends to focus on aggregate populations. For example, how might health and safety issues faced by migrant agricultural workers in Canada (McLaughlin, 2010) be better linked to knowledge of aggregate seasonal flows in order for improved targeting of policy interventions in the agricultural sector? Another example of levels of analysis failing to match up is in geographical space – while we may have some data relating to international migration patterns at a country level, there is often little comparable data at the level of the city or region. The availability of information at multiple levels of analysis is crucial when policy decisions at one level can have very different impacts at another. An instance of this might be where restrictions are placed on the number of unskilled migrants allowed into a country but this has a disproportionate impact on some regions where need for low-skilled migrants is most keenly felt or where social provision has, in the past, been stretched the most.

The *impact* of migration can be seen as a theme underlying all of the policy concerns outlined earlier, and in the United Kingdom this has been expressed explicitly in the recent Migration Advisory Committee report (MAC, 2012) titled 'Analysis of the Impacts of Migration'. The report focuses on a series of areas upon which international migration might have an impact – the impact of migrants on labour markets, specifically native employment; the impact of migrants on the consumption and provision of public services; the impact on transport infrastructure and congestion; the impact on housing provision and housing markets more widely; and societal impacts such as crime, social cohesion and integration.

The authors of the MAC report consider the impacts of extra-European (non-EEA) migration on the United Kingdom and specifically look to quantify, not just the overall net present value (NPV – the single number representing the benefits minus the costs) of a particular migration-related policy, but the NPV for the already resident population. They see this as important as often economic measures such as GDP (which can be used to help compute the NPV) might be calculated per capita where the heads of the immigrant population are included in the denominator along with the previously resident population – the argument being that it is often the case that it is these very immigrants who gain employment and contribute to GDP, but gain most benefit from this additional wealth; and so a more accurate indicator of the value of a policy should only consider the value of migration to those previously living in the country.

Assessing the impact of migration in this way, however, is not straightforward. The authors concede that 'on the basis of current data and knowledge, any attempt to calculate the NPV of migration policies will be subject to considerable uncertainty and likely biases' (MAC, 2012, p. 12). Consequently, they recommend that the quantitative evidence base be strengthened, ideally with data sets including longitudinal information which allow researchers to track behaviours and consumption patterns over a migrant's lifetime; or data containing more detail on the types of migrant moving into different areas, or more geographically detailed information which allows for the impacts to be measured more effectively for smaller areas.

In summary, the unifying themes running through the key policy issues described earlier are impact and evidence. The impact of migrants on the societies they join or the impact of policies on migrant flows, and the evidence for assessing these impacts accurately – this holds true for local, national and international policies.

5.2.2   Linking Policy Issues to Modelling Challenges

Given the policy cycle outlined earlier and the impact agenda, there is a clear potential for effective modelling of migration flows to provide a major contribution in the initial phase of policy development. A model may fill gaps in *current* knowledge – for example, estimating the age profile of recent migrants to London in order that decisions about the amount of money allocated for maternity care in particular boroughs can be made more effectively. A model can also be used to make *predictions* of future populations: it could be formulated to provide intelligence on the likely impact of a specific policy decision – for example, what is likely to happen to migrant flows from *x* to *y* if country *y* decides to remove its border control from *x*? It could also be applied to assess impact in a more general context, perhaps to look at the implications for populations of particular socio-economic conditions resulting from certain overarching policy stances. This is something that the ESPON DEMIFER project did recently using a multi-regional demographic model to look at the likely evolution of populations in Europe under different growth/decline, competitiveness/cohesion policy scenarios (De Beer et al., 2010).

A recent example of a failure in policy preparation linked to inadequate migration flow predictions occurred with the EU enlargement in 2004. Predictions carried out for the UK government estimated that the net migration to the United Kingdom from 8 of the 10 new EU member countries (also called 'accession eight or 'A8') would be between 5,000 and 13,000 migrants per year (Dustmann et al. (2003). Despite these estimates being produced with a number of assumptions and caveats attached, they were broadly accepted before later being widely criticised when in reality the number of accession migrants who came to the United Kingdom exceeded half a million migrants in the first 2 years (BBC, 2006). Clearly when discrepancies between migration estimates and actual figures are so large, there will be a policy vacuum where planned responses will not be able to deal adequately with real events.

In this particular case, the econometric model used to predict migration flows (which was essentially based on relative per-capita income differences and did not take into account the migration policies of the rest of the EU members) proved woefully inadequate. More recent qualitative research (Sriskandarajah et al., 2008) has shown that A8 migrants to the United Kingdom were motivated by more than just income-based factors: changes in origin unemployment, the opportunity to learn English, restrictive policies in adjacent countries and perceived opportunities for enterprise all contributed to a general preference for flows into the United Kingdom. None of these factors were incorporated into the original model devised by Dustmann et al. (2003).

Policy associated with migration responds to lived, perceived or expected impacts for all residents of a jurisdictional entity. The challenge for those engaged in migration modelling is to provide useful intelligence that will assist in better policy decisions being made for particular locations and periods of time. This challenge is of course not an easy one, but it can however be broken down into some specific modelling challenges that can be defined and tackled separately. Overcoming one or more of these sub-challenges in migration modelling will have clear benefits for those involved in migration policy formulation across various countries.

5.2.3 Policy-related Research Questions for Modellers

Consider the following in turn arising from this review.

1. Is it possible to model the bilateral migration flows of individuals at the country level as a function of the various factors at both origin and destination?

   - Spatial interaction models are effective at distributing migrants around a system as long as the number of in and/or out (or net) migrants are known. An improved cost matrix (see later) could assist with more accurate distribution of these migrants, but the generation of migrants at origins and/or destinations can be achieved through regression (or similar) models which use covariates that generate estimates of these figures.

     These covariates might include *inter alia*:

     - Population size and structure: size of the origin and destination populations together with their age and gender structures;
     - Development disparities: wage/income disparities between origin/destination; differences in other development indicators, such as the Human Development Index (HDI), sustainability, transparency/corruption and so on;
     - Other destination attractiveness (pull) factors such as language, historical migration flows, ex-colonial ties, ethnic/ancestry ties, education system, political system, social welfare benefits, perceived enterprise advantages;
     - Other origin repulsiveness (push) factors such as instances of war, famine or natural disasters; political instability; religious intolerance.

   Bilateral flow estimates have recently been produced, and so by fitting explanatory models to these flows, we may learn more about the importance of these and factors influencing each particular origin–destination pair and thus be able to make better future predictions. We have already started to make some inroads into this challenge within ENFOLDING (see Claydon, 2012).

2. Is it possible to define a better proxy for predicting migration flows than physical distance?

   - Distance is not a dynamic variable, but the ways in which people perceive or are affected by distance do change constantly. Borders become more or less permeable; visa restrictions are continuously changing; the costs of travel decrease; and in general, globalisation has made people feel more connected to other parts of the world. Using a time series of migration flows and/or stocks, is it possible to model a '*de facto* distance' as something separate from 'physical distance?'
   - Physical distance is still one of the strongest predictors of international migration behaviour (Cohen et al., 2008), although it is very far from an ideal proxy – for example, Morocco is very much closer to the United Kingdom than Australia, but we would expect significantly more migrants flowing from Australia to the United Kingdom. Using a time series of bilateral flows/stocks estimates, it should be possible to model the *de facto* or '*inferred*' distances experienced by migrants – this could be achieved using the method outlined by Plane (1984). If there are stabilities in these *de facto* distances

over time, then we have the basis for a more accurate proxy in a spatial interaction model which could then be used to create better estimates of bilateral flows. It might also be possible to model external influences on these *de facto* distances. For the ENFOLDing project, it would be interesting to look at how these might be useful in analysing trade flows/comparing trade and migration inferred distances.

3. Can levels of analysis be unified?

- As stated by Kraal (2008), one of the major challenges facing migration research which looks to influence policy is the disconnect between spatial levels of analysis – country, region/city, households or individuals. A significant modelling challenge, therefore, is to try and unify the different levels of analysis so that effective policy decisions can be made local as well as national and supra-national levels. Dennett and Wilson (2011) have already made some inroads with this with a new multilevel spatial interaction model for estimating regional-level flows consistent with country-level information in Europe, but there is scope for experimenting with this methodology in other parts of the world with internal migration data from the IPUMS database and international stock/flow data from the United Nations and other sources such as national censuses.

4. What are the local impacts of inter-regional migration in Europe?

- Building on the work of Dennett and Wilson (2011), there is scope for exploring in more detail the impacts of international migration at sub-national levels. The methodology can be refined to distribute international migrants to cities and regions, and then the potential impacts of these flows can be assessed though their comparison with other variables such as GDP and unemployment. This has particular relevance in the context of the MAC (2012) report where criticism of the evidence base allows accurate assessment of the impact of international migration on, among other things, employment and the labour market in the United Kingdom.[3]

The four modelling challenges outlined earlier should be used as the basis for the ongoing work agenda of the migration work stream. The project is now in possession of data which, while in some cases are provisional (and therefore potentially unreliable), will allow us to begin to explore areas such as bilateral migration flow covariates over time, or the ways in which distance is perceived by international migrants, or indeed whether the decisions made by individuals can be modelled in such a way as to reproduce aggregate behaviours, or whether the multilevel migration modelling framework recently defined by Dennett and Wilson (2011) can be extended globally. Each of these branches of research will lead to tangible outcomes, whether these are new estimates and data sets, or a deeper understanding of the factors which influence migration flows or a re-specification of the global migration system which is not characterised by physical barriers but which is shaped by the repeatedly lived experiences of migrants. All of these could have real influence on the evidence base used by policy makers.

5.2.4   Other International Migration Modelling Research

Despite the large amount of work being carried out on international migration, very little migration modelling is taking place among the various research centres in this field. The exceptions are the Centre for Population Change (CPC) at the University of Southampton (with colleagues elsewhere in the IMEM project), the Wittgenstein Centre in Vienna and CReAM in

---

[3] For an example of analysis which builds on this very recommendation, see Dennett (2014).

UCL (see Appendix for details). Within these three, CReAM is seeking explanations of migration using secondary evidence, while Southampton and the Wittgenstein centre, on the other hand, are engaged in trying to improve our understanding of international migration through increasing the volume of reliable primary data available to researchers. Both are working at the country level – that is, they are attempting to develop models which allow us to reliably estimate the number of migrants flowing between countries for given time periods, although IMEM is focusing on a relatively short European time series whereas the Wittgenstein centre is attempting a 50+ year time series for the whole world and is trying to disaggregate by age, sex and educational achievement. The ENFOLDing project, therefore, is well placed to offer new model-based insights into international migration.

5.3   Conclusion

In the next two chapters, we describe research aimed at the development of model-based solutions to real-world policy problems of the types described earlier. Many of these problems associated with migration have to do with the impact of migrants on the origins they leave as well as destinations they migrate to, and so a great deal of attention is paid by national governments and other policy makers to try and maximise the positive impacts and minimise the negative impacts of these flows in both origin and 'receiving' societies. This review has shown that policies looking to achieve this will be either prospective (anticipating impacts) or retrospective (responding to impacts) but critically in both cases, far too often built atop a rather shallow evidence base.

Academic research on international migration is currently largely concerned with qualitative case studies and policy responses to particular facets of international migration, whether these might be a focus on an individual country or perhaps on a particular classification of migrant. Indeed a journey through the outputs of the research centres described in Appendix will back up this assertion. The poor evidence base from which more extensive quantitative analyses can be conducted is sometimes acknowledged, but rarely tackled directly except for by the few projects engaged in modelling what turns to be rather inadequate data.

Through developing models which can help substitute for deficiencies in data availability and strengthen the evidence base from which effective policy decisions are made, these chapters will make an important contribution to tackling some of the pertinent real-world problems currently presenting themselves.

References

BBC (2006) *Nearly 600,000' New EU Migrants*, BBC, London. http://news.bbc.co.uk/1/hi/uk_politics/5273356.stm (accessed 20 January 2016).
Bijak, J. (2005) *Bayesian Methods in International Migration Forecasting*, Central European Forum for Migration and Population Research, Warsaw. http://www.cefmr.pan.pl/docs/cefmr_wp_2005-06.pdf (accessed 20 January 2016).
Bijak, J. (2006) *Forecasting International Migration: Selected Theories, Models, and Methods*, Central European Forum for Migration and Population Research, Warsaw. http://www.cefmr.pan.pl/docs/cefmr_wp_2006-04.pdf (accessed 20 January 2016).
Black, R., Kniveton, D., Skeldon, R. et al. (2008) *Demographics and Climate Change: Future Trends and their Policy Implications for Migration*, Development Research Centre on Migration, Globalisation and Poverty, Brighton. http://www.migrationdrc.org/publications/working_papers/WP-T27.pdf (accessed 20 January 2016).

Claydon, K. (2012) *A Global Model of Human Migration*, CASA Working Paper 186, UCL, Centre for Advanced Spatial Analysis, London. https://www.bartlett.ucl.ac.uk/casa/publications/working-paper-186.

Cobb-Clark, D.A., Sinning, M. and Stillman, S. (2011) *Migrant Youths' Educational Achievement: The Role of Institutions*, CReAM, UCL, London. http://www.cream-migration.org/publ_uploads/CDP_20_11.pdf (accessed 20 January 2016).

Cohen, J., Roig, M., Reuman, D. and GoGwilt, C. (2008) International migration beyond gravity: A statistical model for use in population projections. *Proceedings of the National Academy of Sciences*, **105** (40), 15269–15274.

De Beer, J., Van der Gaag, N., Van der Erf, R. et al. (2010) *DEMIFER - Demographic and Migratory Flows Affecting European Regions and Cities Applied Research Project 2013/1/3. Final Report*, ESPON and NIDI. http://www.espon.eu/main/Menu_Projects/Menu_AppliedResearch/demifer.html (accessed 20 January 2016).

Dennett, A. and Wilson, A.G. (2011) *A Multi-Level Spatial Interaction Modelling Framework for Estimating Interregional Migration in Europe Working Paper 175*, UCL, Centre for Advanced Spatial Analysis, London. http://www.bartlett.ucl.ac.uk/casa/publications/working-paper-175 (accessed 20 January 2016).

Dennett, A. (2014) Quantifying the Effects of Economic and Labour Market Inequalities on Inter-Regional Migration in Europe – A Policy Perspective. *Applied Spatial Analysis and Policy*. DOI10.1007/s12061-013-9097-4 – http://link.springer.com/article/10.1007/s12061-013-9097-4.

Dustmann, C. and Frattini, T. (2011) *Immigration: The European Experience*, CReAM, UCL, London. http://www.cream-migration.org/publ_uploads/CDP_22_11.pdf (accessed 20 January 2016).

Dustmann, C., Casanova, M., Fertig, M. et al. (2003) *The Impact of EU Enlargement on Migration Flows*, 25/03, Home Office, London. http://webarchive.nationalarchives.gov.uk/20110218135832/http://rds.homeoffice.gov.uk/rds/pdfs2/rdsolr2503.pdf (accessed 20 January 2016).

Düvell, F. and Garapich, M. (2011) *Polish Migration to the UK: Continuities and Discontinuities Working Paper 84*, COMPAS, Oxford. http://www.compas.ox.ac.uk/publications/working-papers/wp-11-84/ (accessed 20 January 2016).

Hynes, W. and Ccrna, L. (2009) *Globalisation Backlash? The Influence of Global Governance in Trade and Immigration*, COMPAS, Oxford. http://www.compas.ox.ac.uk/publications/working-papers/wp-09-74/ (accessed 20 January 2016).

Kraal, K. (2008) *The Future of Migration Research in Europe*, IMISCOE Policy Briefs, Amsterdam. http://library.imiscoe.org/en/record/270915 (accessed 20 January 2016).

Lee, E.S. (1966) A theory of migration. *Demography*, **3** (1), 47–57.

MAC (2012) *Analysis of the Impacts of Migration*, The Home Office, Migration Advisory Committee, London. http://www.ukba.homeoffice.gov.uk/sitecontent/documents/aboutus/workingwithus/mac/27-analysis-migration/ (accessed 20 January 2016).

Martin, P. and Ruhs, M. (2010) *Labor Shortages and US Immigration Reform: Promises and Perils of an Independent Commission*, COMPAS, Oxford. http://www.compas.ox.ac.uk/publications/working-papers/wp-10-81/ (accessed 20 January 2016).

McLaughlin, J. (2010) *Backgrounder on Health and Safety Issues among Migrant Farmworkers in Canada*, International Migration Research Centre, Waterloo. (https://www.wlu.ca/documents/44101/Backgrounder_on_Health_Issues_among_MFWs_in_CanadaIMRC2010.pdf).

Methvin, T.W. (2009) *The New Mexican-Americans: International Retirement Migration and Development in an Expatriate Community in Mexico CMD Working Paper 09-03*, Princeton University. http://www.princeton.edu/cmd/working-papers/papers/wp0903.pdf (accessed 20 January 2016).

Panizzon, M. (2010) *Standing together Apart: Bilateral Migration Agreements and the Temporary Movement of Persons Under "Mode 4" of GATS*, COMPAS, Oxford. http://www.compas.ox.ac.uk/publications/working-papers/wp-10-77/ (accessed 20 January 2016).

Pariyar, M. (2011) *Cast(e) in Bone: The Perpetuation of Social Hierarchy among Nepalis in Britain*, COMPAS, Oxford. http://www.compas.ox.ac.uk/publications/working-papers/wp-11-85/ (accessed 20 January 2016).

Patacchini, E. and Zenou, Y. (2012) *Ethnic Networks and Employment Outcomes*, CReAM, UCL, London. http://www.cream-migration.org/publ_uploads/CDP_02_12.pdf (accessed 20 January 2016).

Penninx, R. (2006) Conclusions and directions for research, in *The Dynamics of International Migration and Settlement in Europe: A State of the Art* (eds R. Penninx, M. Berger and K. Kraal), Amsterdam University Press, Amsterdam.

Plane, D.A. (1984) Migration space: Doubly constrained gravity model mapping of relative interstate separation. *Annals of the Association of American Geographers*, **74** (2), 244–256.

Plewa, P. (2010) *Voluntary Return Programmes: Could they Assuage the effects of the Economic Crisis?*, COMPAS, Oxford. http://www.compas.ox.ac.uk/publications/working-papers/wp-10-75/ (accessed 20 January 2016).

Portes, A. and Shafer, S. (2006) *Revisiting the Enclave Hypothesis: Miami Twenty-Five Years Later* CMD Working Paper 06-10, Princeton University. http://www.princeton.edu/cmd/working-papers/papers/wp0610.pdf (accessed 20 January 2016).

Portes, A., Escobar, C. and Arana, R. (2008) *Divided or Convergent Loyalties? The Political Incorporation Process of Latin American Immigrants in the United States* CMD Working Paper 07-04, Princeton University. http://www.princeton.edu/cmd/working-papers/papers/wp0704.pdf (accessed 20 January 2016).

Raymer, J., de Beer, J. and van der Erf, R. (2011) Putting the pieces of the puzzle together: Age and sex-specific estimates of migration amongst countries in the EU/EFTA, 2002–2007. *European Journal of Population/Revue européenne de Démographie*, **27** (2), 185–215.

Roca iCaparà, N. (2011) *Young Adults of Latin American Origin in London and Oxford: Identities, Discrimination and Social Inclusion*, COMPAS, Oxford. http://www.compas.ox.ac.uk/publications/working-papers/wp-11-93/ (accessed 20 January 2016).

Ruedin, D. (2011) *Conceptualizing the Integration of Immigrants and Other Groups*, COMPAS, Oxford. http://www.compas.ox.ac.uk/publications/working-papers/wp-11-89/ (accessed 20 January 2016).

Ruhs, M. and Anderson, B. (2012) *Responding to Employers: Labour Shortages and Immigration Policy Policy Primer*, The Migration Observatory, Oxford. http://www.migrationobservatory.ox.ac.uk/sites/files/migobs/Labour%20Shortages%20Policy%20Primer_0.pdf (accessed 20 January 2016).

Salt, J. and de Bruycker, P. (2011) Migration policy developments, in *International Migration Outlook 2011: SOPEMI 2011* (ed OECD), OECD Publishing.

Salt, J., Latham, A., Mateos, P. et al. (2011) *UK National Report: Satisfying Labour Demand Through Migration*, Home Office/EMN. http://emn.intrasoft-intl.com/Downloads/download.do;jsessionid=AEF9A64EE84EC118A71644AF18D3D87C?fileID=2136 (accessed 20 January 2016).

Shutes, I. (2011) *Social Care for Older People and Demand for Migrant Workers Policy Primer*, The Migration Observatory, Oxford. http://www.migrationobservatory.ox.ac.uk/sites/files/migobs/Social%20Care%20Policy%20Primer_0.pdf (accessed 20 January 2016).

Sigona, N. (2011) *Irregular Migrant Children and Public Policy Policy Primer*, The Migration Observatory, Oxford. http://www.migrationobservatory.ox.ac.uk/sites/files/migobs/Irregular%20Migrant%20Children%20Policy%20Primer_0.pdf (accessed 20 January 2016).

Skeldon, R. (2007) *On Migration and the Policy Process*, Development Research Centre on Migration, Globalisation and Poverty, Brighton. http://www.migrationdrc.org/publications/working_papers/WP-T20.pdf (accessed 20 January 2016).

Sriskandarajah, D., Latorre, M. and Pollard, N. (2008) *Floodgates or Turnstiles? Post-EU Enlargement Migration Flows to (and from) the UK*, IPPR, London. http://www.ippr.org/publications/55/1637/floodgates-or-turnstilespost-eu-enlargement-migration-flows-to-and-from-the-uk (accessed 20 January 2016).

van Liempt, I. (2007) *Inside Perspectives on the Process of Human Smuggling IMISCOE Policy Briefs*, IMISCOE, Amsterdamhttp://library.imiscoe.org/en/record/234829 (accessed 20 January 2016).

Wessendorf, S. (2011) *Commonplace Diversity and the 'Ethos of Mixing': Perceptions of Difference in a London Neighbourhood*, COMPAS, Oxford. http://www.compas.ox.ac.uk/publications/working-papers/wp-11-91/ (accessed 20 January 2016).

Winters, A.L. (2005) *Developing Country Proposals for the Liberalization of Movements of Natural Service Suppliers*, Development Research Centre on Migration, Globalisation and Poverty, Brighton. http://www.migrationdrc.org/publications/working_papers/WP-T8.pdf (accessed 20 January 2016).

Appendix

A.1 United Kingdom

Centre on Migration, Policy and Society (COMPAS), incorporating The Migration Observatory, University of Oxford (http://www.compas.ox.ac.uk/; http://www.migrationobservatory.ox.ac.uk/)

COMPAS, based at the University of Oxford, is an established international migration research centre in the United Kingdom. *'The mission of COMPAS is to conduct high quality research in order to develop theory and knowledge, inform policy-making and public debate, and engage users of research within the field of migration'.* Producing research output since 2004, COMPAS is now in a phase of work running from 2011 to 2016 which has five workstreams or clusters:

1. Flows and dynamics – exploring global migration flows and the dynamics that drive, facilitate and inhibit migration
2. Labour markets – analysing the socio-economics of international labour migration, particularly the economics and politics of labour shortages and demand
3. Citizenship and belonging – addressing the relationship between mobility, citizenship and the numerous ways in which people 'belong'
4. Urban change and settlement – challenging assumptions around movement and settlement patterns, investigating emergent urbanisms and processes of integration
5. Welfare – addressing the relationship between migration and welfare provision in 'receiving' and 'sending' countries.

Within COMPAS, The Migration Observatory has recently been set up as a conduit for policy-relevant research within the group.

COMPAS publishes an extensive Working Paper series which documents the work across all clusters in the group. The bulk of the work in this series comprises qualitative analyses, with authors focusing recently on areas such as identity and assimilation (Pariyar, 2011; Roca iCaparà, 2011; Ruedin, 2011; Wessendorf, 2011) and Policy (Hynes and Cerna, 2009; Martin and Ruhs, 2010; Panizzon, 2010; Plewa, 2010). Little quantitative analysis appears in the Working Paper series, although The Migration Observatory, with a focus more on UK-based policy, does reproduce data and statistics on international migration trends in relation to the United Kingdom; although much of this is in support of the 'Policy Primer' reports which are the bulk of the output from the sub-group. The group does not engage in any explanatory or predictive modelling.

Centre for Research and Analysis of Migration (CReAM), UCL (http://www.cream-migration.org/).

Another major international migration research centre of the United Kingdom is CReAM, based at University College London. *'CReAM's research focuses on the causes, patterns and consequences of international population mobility and movements affecting the UK and Europe. CReAM aims at informing the public debate on migration in the UK and in Europe by providing new insight, helping to steer the current policy debate in a direction that is based on carefully researched evidence without partisan bias'.* Differing from COMPAS, CReAM also states an emphasis on *quantitative* research, with a research programme divided into four principal strands:

1. Forms of population movement and mobility
2. The non-migrant experience – effects of migration on origin and destination countries
3. The migrant experience – integration, adaptation and exclusion
4. Perception of migrants within receiving countries – identity and aspects of social cohesion.

CReAM has an extensive history of working (discussion) papers dating back to 2004 which make use of quantitative analysis techniques. Papers by Patacchini and Zenou (2012) and Cobb-Clark et al. (2011) use statistical regression-based techniques to examine the effects of ethnic networks on the employment outcomes of migrants and the influence of educational institution arrangements on the attainment of immigrant children, respectively, and are just two examples of the quantitative focus within this large body of work that emanates from the economics tradition within social science. There are, of course, occasional exceptions such as a review paper on immigration in Europe (Dustmann and Frattini, 2011), but the strand tying almost all of the CReAM work together is the empirical analysis of international migration and the associated data using a raft of quantitative techniques. Models are frequently used in these analyses as explanatory models to test assumptions about the relationships between migration and other social or economic variables. They are not used as a method to supplement inadequate data through synthetic estimates.

Centre for Population Change (CPC), Social Sciences: Social Statistics and Demography, S3RI, University of Southampton (http://www.cpc.ac.uk/, http://www.southampton.ac.uk/demography/index.page?, http://www.southampton.ac.uk/s3ri/).

A large amount of work on international migration is carried out under a number of projects run from the various inter-related social sciences and statistics departments/centres based at the University of Southampton. Within the CPC, a work stream exists which focuses on the '*demographic and socio-economic implications of national and transnational migration*'. Contained within this stream are a series of sub-projects focusing on:

1. Migration, mobility and the labour market
2. Migration, mobility and its impact on socio-demographic processes
3. Migration and ageing.

These sub-projects employ a variety of different methodological approaches, both qualitative and quantitative, but tend to concentrate on the effects of migration on aspects of demography and the economy within the United Kingdom.

Another work stream within the CPC project concentrates on '*Modelling population growth and enhancing the evidence base for policy*'. Led by James Raymer, the core aim of this stream is to develop a statistical methodology in order to produce a dynamic population model of the United Kingdom. Linked to this project via James Raymer is another workstream known as the Integrated Modelling of European Migration (IMEM) project, funded by NORFACE programme (http://www.norface.org/migration12.html). The aim of IMEM is to use statistical techniques to develop a consistent, harmonised time series of inter-country flows across Europe. These modelled migration flows are designed to improve the migration evidence base where empirical data collected via censuses and surveys are inconsistent across countries and, in some cases, unreliable.

IMEM's work in modelling and estimating international migration flows uses Bayesian techniques to incorporate auxiliary (expert) information, which assesses the quality of empirical data, into estimates which also fill gaps in intra-European migration flow matrices. The work follows a previous project called MIMOSA (Migration Modelling for Statistical Analysis – http://mimosa.gedap.be/), which also attempted to develop a consistent time series of inter-country flows in Europe (Raymer et al., 2011), although did not deal with uncertainties in the recorded data in the way that IMEM is currently attempting to.

Sussex Centre for Migration Research (SCMR) – http://www.sussex.ac.uk/migration/.

The Sussex Centre for Migration Research has been at the vanguard of international migration research in the United Kingdom since 1997. The home for (and contributing to) dozens of research projects over the last decade or so, the SCMR, has produced a considerable volume of output in this time. Similar to COMPAS and CReAM, much of this output is documented in an extensive Working Paper series – a series which reflects the broad interests which the centre has supported, but which also exhibits themes reminiscent of those popular within COMPAS and CReAM: migration and identity, migration and economic impact, remittance flows and patterns of migration.

Among the large number of research projects supported wholly or partially by the SCMR, one of the largest was the Development Research Centre on Migration, Globalisation and Poverty (http://www.migrationdrc.org/). The project ran from 2003 to 2010 and operated with a mission statement to '*to promote new policy approaches that will help to maximize the potential benefits of migration for poor people, whilst minimizing its risks and costs*'. Much of the output, therefore, had a distinct policy focus exemplified by papers by Winters (2005), Skeldon (2007) and Black et al. (2008).

The use of models does not appear explicitly in much of the body of work presented by the Migration DRC; however, some model assumptions are employed in the Global Migrant Origin Database sub-project (http://www.migrationdrc.org/research/typesofmigration/global_migrant_origin_database.html), which carried out the collation of global migrant stock data with some subsequent estimation to produce a complete matrix of bilateral migrant stocks. Models used in this project were relatively elementary, using, for example, average rates of attrition or the propensity for a country to send migrants abroad to update stock estimates for years where data did not exist.

Migration Research Unit (MRU) – University College London (UCL) (http://www.geog.ucl.ac.uk/mru/).

The MRU is another centre of international migration research based at UCL. With links to CReAM through the 'UCL Global Migration Network', the MRU was founded in 1988 and has similar objectives to CReAM and COMPAS based around policy-relevant research with a particular focus on migration statistics, high-skilled migration and irregular migration in the United Kingdom and Europe. With a long history of research output, the MRU focuses particularly on empirical analyses to inform policy and a growing interest in qualitative analysis of the implications of migration at local level. Modelling related to migration is not a current concern.

Centre for Interaction Data Estimation and Research (CIDER) – University of Leeds (http://cider.census.ac.uk/).

CIDER is principally concerned with supplying census-based interaction data (commuting, internal and international migration) to academic users within the United Kingdom. While synthetic data creation through various modelling techniques has been a feature of CIDER's work since its inception in 2005, this work has concentrated on the modelling of internal migration flows in the United Kingdom rather than international flows.

A.2 Rest of Europe

Wittgenstein Centre for Demography and Global Human Capital – Vienna Institute of Demography/International Institute for Applied Systems Analysis, Austria (http://www.oeaw.ac.at/wic/; http://www.oeaw.ac.at/vid/; http://www.iiasa.ac.at/).

The Wittgenstein Centre for Demography and Global Human Capital is a research collaboration among the Vienna Institute of Demography, the International Institute for Applied Systems Analysis and the Vienna University of Economics and Business. The overarching goal of the centre is '*to better understand the role of human capital—the human resource base in terms of the number of people and their changing structure by age, gender, place of residence, level of education, health status, cognitive skills, and participation in the "production" of human wellbeing'*.

Many research teams within the centre are engaged in a variety of data and modelling intensive activities related to global human issues. The 'Migration and Education' team focuses on the following:

1. Estimating annual bilateral international migration flows for the globe between 1960 and 2010
2. Modelling the age–sex–education structure of international migration flows
3. Determining the spatial structure of international migration and how it has changed between 1960 and 2010
4. Developing a set of alternative scenarios describing the alternative futures of migration from 2010 to 2050
5. Using case studies to determine the impact of climate change on international migration flows in Asia, focusing especially on educational selectivity.

Preliminary estimates of a time series of bilateral, country-level, migration flows have already been achieved, although have yet to be published.

Central European Forum for Migration and Population Research (CEFMR), Poland (http://www.cefmr.pan.pl/).

Founded in 2002 as a research partnership among the Swiss Foundation for Population, Migration and Environment (PME); the Institute of Geography and Spatial Organization of the Polish Academy of Sciences; and the International Organization for Migration, CEFMR continues to pursue a research programme which '*specialises in multidisciplinary research on international migration in Central Europe. It conducts research in demography, population statistics, modelling of migration and population, geography, migration policies, sociology and economics'*.

CEFMR has a notable Working Paper series which reveals a heavily quantitative approach in much of the work which has been carried out, for example, papers by Bijak (2005, 2006)(now at the Southampton CPC) reviewing and advocating a variety of modelling approaches (but particularly Bayesian statistical methods) for forecasting international migration flows.

Members of CEFMR have been and still are engaged in a number of research projects with an international migration focus. Perhaps the largest recent project has been the demographic

and migratory flows affecting European regions and cities (DEMIFER) project (http://www
.espon.eu/main/Menu_Projects/Menu_AppliedResearch/demifer.html). CEFMR researchers
were responsible for the multi-level, multi-regional cohort component projection model
known as MULTIPOLES (http://www.espon.eu/export/sites/default/Documents/Projects/
AppliedResearch/DEMIFER/FinalReport/DEMIFER_Deliverable_D4_final.pdf) which projected, for a range of growth scenarios, demographic components (including in, out and net
migration) for over 250 European regions up until 2051.

The European Migration Network (EMN), Belgium (http://emn.intrasoft-intl.com/html/
index.html).

The objective of the EMN is to provide *'up-to-date, objective, reliable and comparable
information on migration and asylum'* in the European Union, in order to support policy making in these areas. The EMN was launched in 2003 as a pilot project, but has since been
established by the European Council and given legal status in 2008. EU member states supply
the EMN with information through a network of national 'contact points'. The main outputs
from the EMN are a series of annual policy reports which relate to political and legislative
developments associated with migration and asylum, as well as an accompanying annual series
on migration statistics. Various ad hoc studies are also produced on a range of migration and
asylum-related topics as and when specific requests are made by EU partner countries. For
example, recent reports have been produced on the impact of immigration on European Societies, irregular migrants living in the European Union, return migration, temporary and circular
migration and satisfying labour demand in EU member states through migration.

Despite an extensive programme of work, the EMN states that it *'does not normally engage
in primary research'* and instead focuses on collecting, analysing and redistributing information on migration which is already available. As such, it hosts an 'information exchange
system' which facilitates the sharing of migration information between EU partner countries.

IMISCOE – International Migration, Integration and Social Cohesion in Europe, Netherlands (http://www.imiscoe.org/).

IMISCOE is a network of researchers and research centres across Europe. The organisation
has a mission organised around a series of ongoing research questions: *'What are the causes
and nature of current migration processes? How can migration flows be managed and influenced? How can societies maintain social cohesion and societal viability? How can scientists,
policy makers and practitioners in the field exchange knowledge and experience?'*

IMISCOE is coordinated by the Institute for Migration and Ethnic Studies (IMES) based
at the University of Amsterdam, but has an additional 27 institutional partner members from
Universities across Europe. These include the following:

1. CEDEM (Centre d'Études de l'Ethnicité et des Migrations), University of Liège, Belgium
2. CEG (Centro de Estudos Geográficos), University of Lisbon, Portugal
3. CEIFO (Centre for Research in International Migration and Ethnic Relations, University
   of Stockholm, Sweden
4. CEMIS (Centre for Migration and Intercultural Studies), University of Antwerp, Belgium
5. CEREN (Centre for Research on Ethnic Relations and Nationalism), University of
   Helsinki, Finland
6. CES (Centro de Estudos Sociais), Universidade de Coimbra, Portugal
7. CES-NOVA, New University of Lisbon, Portugal
8. CESS (Centre for Economic and Social Studies), Tirana, Albania

9. CMR (Centre of Migration Research), Warsaw University, Poland
10. DEUSTO (Research Unit on Migration, Management of Diversity and Social Cohesion), University of Deusto, Bilbao, Spain
11. efms (European Forum for Migration Studies), University of Bamberg, Germany
12. EUI (European University Institute), Florence, Italy
13. EUR (Erasmus University of Rotterdam), Netherlands
14. FIERI (The Forum Internazionale ed Europeo di Ricerche sull'Immigrazione), Italy
15. GRITIM (Interdisciplinary research group in immigration), Universitat Pompeu Fabra, Spain
16. ICMPD (International Center for Migration Policy Development), Austria.
17. IEM (Instituto Universitario de Estudios sobre Migraciones de la Universidad Pontificia Comillas de Madrid), Spain
18. IMIS (Institute for Migration Research and Intercultural Studies), University of Osnabrück, Germany
19. INED (Institut National d'Études Démographiques), France
20. ISR (Institute for Urban and Regional Research), Austrian Academy of Sciences
21. MIGRINTER (Migrations Internationales, Espaces et Sociétés), University of Poitiers, France
22. MIM (Malmö Institute for Studies of Migration, Diversity and Welfare), Malmö University, Sweden
23. MiReKoc, Koç University, Istanbul, Turkey
24. NIDI (Netherlands Interdisciplinary Demographic Institute), The Hague, Netherlands
25. NOVA (Norwegian Social Research), Oslo, Norway
26. SCMR (Sussex Centre for Migration Research), United Kingdom
27. SFM (Swiss Forum for Migration and Population Studies, the University of Neuchâtel, Switzerland.

IMISCOE members are organised into a series of Research Groups:

1. Ageing migrants: demography, agency and welfare
2. Common European Economic Space and Migration Initiative
3. Diaspora and Development
4. Employment and Migrant Legality in Contemporary Europe
5. MIGCITPOL (Migration, Citizenship and Political Participation – former cluster B3)
6. Standing Committee on International Migration and its Regulation
7. Popular arts, diversity and cultural policies in post-migration urban settings (POPADIVCIT)
8. Research-Policy Dialogues on Migration and Integration in Europe
9. The Multilevel Governance of Immigrant and Immigration Policies
10. The Social Nexus between Irregular Migration, the Informal Economy and Political Control
11. TRANSMIG.

IMISCOE continues to produce a very large volume of work. The IMISCOE-AUP series is a collection of peer-reviewed books on themes ranging from transnationalism, to European Policy, through to local case studies. The series of almost 50 titles are publicly available through the open-access portal, OAPEN (http://oapen.org/search?title=&creator=& seriestitle=imiscoe&subject=&isbn=&year=&year-max=&smode=advanced). In addition to the AUP series, a Working Paper series is also available on the IMISCOE website (http:// www.imiscoe.org/index.php?option=com_content&view=category&layout=blog&id=13& Itemid=21). Quantitative analysis is not uncommon in these outputs, but is used among various other analysis techniques. Explanatory or predictive modelling, however, is not something which features in these publications. Related to the books and working papers are a series of 'Policy Briefs'. These are shorter documents which focus on a series of specific policy questions or statements – for example, 'How can we turn Europe's increasing cultural diversity into an economic and social asset?'

Most of the research partners within IMISCOE are based within universities, although there are examples of institutions without direct University affiliation, such as the International Centre for Migration Policy Development or the Institut National d'Études Démographiques (INED) in France.

The International Centre for Migration Policy Development (ICMPD), Belgium (http://www.icmpd.org/).

The ICMPD was founded in 1993 to provide expertise and services at a time when countries within Europe were beginning to cooperate on issues related to migration and asylum. The ICMPD specifically looks to promote '*comprehensive and sustainable migration policies*' and act as a mechanism for the exchange of services between governments and organisations.

The ICMPD has a dedicated research unit which explores international migration trends, patterns and policies, with a particular focus on seven distinct themes:

1. Illegal Migration & Return
2. Trafficking in Human Beings
3. Border Management & Visa
4. Asylum
5. Migration & Development
6. Legal Migration & Integration
7. Multi-Thematic Research.

The ICMPD publishes details of much of its work – working papers, reports and policy briefs – through its separate research website: http://research.icmpd.org. In many cases, however, the research outputs are held separately often on the websites of the projects which the ICMPD has been involved with. One of the projects which has had more of a quantitative-/data-driven focus is PROMINSTAT (Promoting Comparative Quantitative Research in the Field of Migration and Integration in Europe) (http://www.prominstat.eu/).

PROMINSTAT ran between 2007 and 2010 and was a programme funded by the European Commission with an aim to contribute to a better understanding of migration and associated social issues within Europe. This contribution was defined through the main project exercise – an exhaustive documentation of statistical data and metadata on migration held by 29 countries (EU27 + Norway and Switzerland). The project produced three main outputs:

1. A series of country reports on the various national migration data collection systems

2. An online, searchable, meta-database containing information on all quantitative data sets featuring migration-related data in the 29 study countries
3. Thematic working papers on data collection.

PROMINSTAT was coordinated by the International Centre for Migration Policy Development (ICMPD), although featured contributions from 18 partner institutions. It built on work which was completed in earlier ICMPD projects, particularly COMPSTAT (http://research.icmpd.org/ 1243.html?&F=ylohmjdtl) and THESIM (http://www.uclouvain .be/en-7823.html), which were projects designed to address the lack of comparable data on the social and economic integration of migrants in the European Union and the reliability of international migration data, respectively.

Migration Policy Centre (http://www.migrationpolicycentre.eu/).

Founded in January 2012, the Migration Policy Centre (MPC) is based at the European University Institute (EUI) in Florence. The broad aim of the centre is to 'conduct research on global migration to serve migration governance needs at the European level. This means developing, implementing and monitoring migration-related policies to assess their impact on the economy and on society more generally'.

The MPC is focusing its attentions on two recent global developments which have the potential to shape migration flows and migration policies in the near future: the global economic crisis and particularly its impact in Europe and the 'Arab Spring', which has led to major political changes in the Arab world, especially in North Africa.

Assisting the work in these two areas, the MPC will oversee three regional 'migration observatories' which will monitor migration into Europe from three main regions of the world. The first of these, CARIM-South, was set up before the inception of the MPC in 2004 with a remit to 'document and to analyse migration in seventeen countries of the Southern and Eastern Mediterranean and of Sub-Saharan Africa'. Two new observatories were set up in 2011: CARIM-East looking at the flows associated with seven countries to the East of the European Union; and CARIM-India focusing on the exchanges of migrants between the European Union and India.

Despite its young age, the MPC is already a home to a large volume of work on migration. Currently, much of the research consists of empirical studies related to the regional observatories, although a section for policy briefs has been set aside on the new website.

A.3   Rest of the World

Queensland Centre for Population Research (QCPR), Australia (http://www.gpem.uq.edu.au/ qcpr).

The QCPR, based at the University of Queensland in Australia, is one of the leading centres of population research in the southern hemisphere. With a focus on understanding demographic processes and population dynamics, particularly in Australia, the QCPR is also concerned with some global patterns and processes, particularly in relation to internal migration.

The current 'Internal Migration Around the GlobE' (IMAGE) project, in collaboration with the University of Leeds in the United Kingdom, aims to build an understanding of global *internal* migration patterns through developing a database of flows for more than 150 countries, with associated indicators and comparison metrics.

Institute for the Study of International Migration – Georgetown University, United States (http://www12.georgetown.edu/sfs/isim/).

The Institute for the Study of International Migration is concerned with analysis of international migration issues affecting the United States and other countries 'including various bilateral, regional, and multilateral approaches to migration and refugee policy. Understanding forced migration and responses to humanitarian emergencies is another important area of policy research'.

Current research projects cover three main substantive areas:

1. Immigration and Integration of Immigrants
2. Migration and Development
3. Refugee and Humanitarian Emergencies.

The institute has published a range of reports on areas such as transatlantic migration, migration and climate change, impacts immigrants on health sector employment and refugee flows. Migration modelling does not feature in current or previous research projects within the centre.

International Migration Research Centre (IMRC) – Wilfrid Laurier University, Canada (https://www.wlu.ca/homepage.php?grp_id=2599).

The IMRC has a mission statement to 'foster research in the areas of new policy development and alternative models and practices of managing both temporary and permanent forms of international migration'.

As with a number of other centres of international migration research, the work in the centre comprises a number specific case studies (migrant farm workers in Canada, temporary worker programmes in Canada, migration of nurses, effects of the global financial crisis on remittances, etc.). In the case of the IMRC, a theme which emerges is that of temporary and agricultural migrants with a specific focus on Canada. Again, while quantitative elements exist within the analysis, migration modelling of any sort is not prevalent.

The Centre for Migration and Development – CMD (http://www.princeton.edu/cmd/).

The CMD, established in 1998, at Princeton University has a particular interest in immigrant communities living in the developed world and in the growth and development prospects of sending nations. A Working Paper series documents various pieces of work undertaken to this end, and a data archive makes accessible a series of data sets which have been used in CMD research projects. These datasets include the following:

- The Adaptation Process of Cuban and Haitian Refugees (CHR)
- The Comparative Immigrant Entrepreneurship Project (CIEP)
- The Children of Immigrants Longitudinal Study (CILS)
- The Comparative Immigrant Organizations Project (CIOP)
- Cuban and Mexican Immigrants in the US (CMI)
- Caribbean Urbanization in the years of the Crisis (CUIC)
- Latin American Migration Project (LAMP)
- Mexican Migration Project (MMP)
- New Immigrant Survey (NIS)
- Project on Ethnicity and Race in Latin America (PERLA)

The working papers are very much focused on individual issues – ethnic enclaves in Miami (Portes and Shafer, 2006), the political incorporation of Latin American migrants into the

United States (Portes et al., 2008) or American retirement in New Mexico (Methvin, 2009). As is common with much of the research carried out by other institutions focusing on international migration, methods are mixed with some analysis of quantitative data but qualitative techniques such as interviews are also common place. Modelling does feature in some of the work, but again these will be explanatory regression models (or similar) used to analyse data which are readily available, rather than models which are used to supplement inadequate data holdings or predict future patterns.

6

Estimating Inter-regional Migration in Europe

Adam Dennett and Alan Wilson

6.1 Introduction

In this chapter, we show how methods of biproportional fitting – assembled through the use of entropy-maximising methods – can be used to generate estimates of missing data, and particularly flows, from partially complete sets of data. This enables us to generate inter-regional migration flows within Europe.

Understanding migration is one of the enduring challenges facing geographers and demographers worldwide. The challenge persists, thanks to the range of territories and geographical scales of interest, the difficulty in dealing with inconsistent definitions of migrants and migration events, the variable (and often poor) quality of data and the large and sometimes complex array of tools available. While an understanding of migration patterns and processes at the global scale presents possibly the largest challenge, in Europe we still know far less about the movements of people within the Union than may be expected given the continued desire for knowledge about population change and the amount of demographic data made available from member countries (Poulain et al., 2006). Acknowledging this, a number of recent projects have made attempts to address some of the limitations of (intra-) European migration data. Against a background of varying migrant definitions, inconsistent data relating to the same flows collected for origins and destinations, and incomplete matrices, the MIMOSA (Modelling migration and migrant populations) project (Raymer and Abel, 2008), produced a series of inter-country migration estimates for years between 2002 and 2006 through harmonising available data and using a multiplicative modelling framework to model flows between countries. Following on from this, the IMEM (Integrated Modelling of European Migration) project (van der Erf et al. – http://www.nidi.nl/Pages/NID/24/842.TGFuZz1FTkc.html) is currently looking to improve upon the methodology employed in MIMOSA through a Bayesian statistical approach. Further work has also been carried out by Abel (2010) who used a negative binomial regression (spatial interaction) model to estimate inter-country flows using a suite of predictor variables.

All of these projects have limited their scope to inter-country flows, but within Europe much of the focus of the EU commission is on regional policy (http://ec.europa.eu/regional_policy/index_en.cfm) which is intended to address the quite marked socio-economic disparities which persist between smaller zones within the Union. A recent project which had a partial focus on migration at the regional (Nomenclature of Territorial Units for Statistics level 2 – NUTS2) level in the European Union was the DEMIFER project (De Beer et al., 2010). One of the outputs from this project is a set of regional population projections for four different growth/cohesion scenarios which include a model of regional in- and out-migration based upon annual transition rates (Kupiszewska and Kupiszewski, 2010). While in- and out-migration rates tell us something about migration at the regional level within Europe, they reveal little about the interaction between regions and the hotspots of population exchange which occur within the Union helping drive the dynamism and evolution of local population structures. Indeed our knowledge of these exchanges across the whole Union is poor.

Within the United Kingdom, migration policy is rarely far from the headlines, although as Cangiano (2011) points out, there has been a certain disconnect between immigration policy and wider acknowledgement of demographic issues such as the ageing of the population. Compounding these macro policy problems, there is a local dimension to demographic issues and a current knowledge gap in relation to local immigration concentrations and emigration flows. The UK government has a limited capacity to predict or control the flows of EU nationals into the country and then where in the country they go once they have arrived; conversely knowledge of areas which are likely to experience increased pressures due to migration is vital for effective policy decisions to be made. Where these issues exist in the United Kingdom, we can be sure that similar issues are experienced in other EU member states.

Therefore, in this chapter, we propose a methodology for estimating the inter-regional flows which pose these particular local policy problems within Europe. The work builds on previous research which has made use of variations on the entropy maximising spatial interaction models (SIMs) first introduced by Wilson (1970, 1971) and used in migration research (He and Pooler, 2003; Plane, 1982; Stillwell, 1978). A new Multi-Level Spatial Interaction Model (MLSIM) is proposed which incorporates data at both country and regional levels in Europe to produce estimates of the inter-regional inter-country flows consistent with known information at these different levels. The heart of the method is biproportional fitting.

6.2   The Spatial System and the Modelling Challenge

2006 is the year for which the maximum amount of migration data at all levels is available, and so we use this as our temporal base. The spatial system of 287 NUTS2 regions nested within 31 countries (EU 27 + Norway, Iceland and Switzerland – which will be referred to as the 'EU system' in this chapter subsequently) is shown in Figure 6.1. Migration data for some of the flows occurring are available. These data, along with cells representing missing data, can be visualised as an origin/destination matrix as shown for a sample of countries in Figure 6.2. The grey cells in Figure 6.2 represent inter-regional intra-country (internal migration) migration flow counts which are available for most counties in the system. Flows

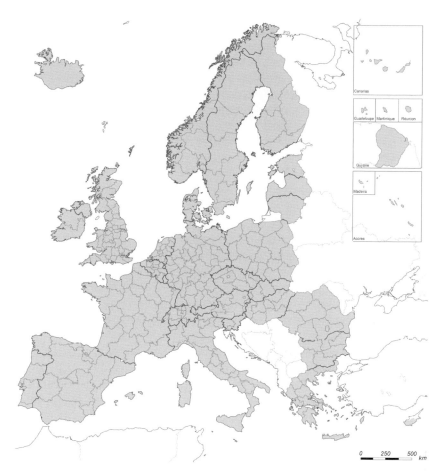

Figure 6.1 The 287 NUTS2 regions of EU 27 + 3 counties. *Source*: Reproduced with permission from Dennett and Wilson (2013)

within NUTS2 regions (the white cells on the diagonal) are not included in this analysis. The internal migration data were collated for use in the ESPON-funded DEMIFER project (http://www.espon.eu/main/Menu_Projects/Menu_AppliedResearch/demifer.html), although in almost all cases, these data are freely available from the Eurostat statistics database (often referred to as 'New Cronos' – http://epp.eurostat.ec.europa.eu/portal/page/portal/statistics/ search_database). Internal migration data for two countries – France and Germany – are not available on this database and were procured separately for DEMIFER from national statistical agencies. It should be noted, although, that while technically European NUTS2 zones, the French overseas departments of Guadeloupe, Martinique, Reunion and French Guiana are not included. The coloured cells represent inter-country flows. Consistent estimates of international (intra-Europe) origin/destination flows have been created for the 31 countries for our year of interest by Raymer and colleagues for the MIMOSA project (Raymer and Abel, 2008).

		Country1			Country2			Country3		
		zone1	zone2	zone3	zone1	zone2	zone3	zone1	zone2	zone3
Country1	zone1	0	1131	1887		7211			4856	
	zone2	1633	0	14055						
	zone3	2301	20164	0						
Country2	zone1		9885		0	1608	328		8190	
	zone2				1252	0	1081			
	zone3				346	1332	0			
Country3	zone1		4992			4773		0	630	106
	zone2							546	0	569
	zone3							112	587	0

Figure 6.2   Example migration data availability within Europe

Missing data in this EU system matrix are the inter-country, inter-regional flows – for example, the flows from the three zones in Country 1 to the three zones in Country 3 which sum to the 4,856 migrants we know flowed between Country 1 and Country 3 in Figure 6.2. The modelling challenge, therefore, is to estimate this missing data in the matrix making use of information available at both the country and regional levels. The ultimate goal is to produce a full set of inter-regional estimates which make the most use of all available flow information at all levels within the system. Therefore, it will be necessary to understand the full range of the models which can be built from the elements of the migration system. In defining a suite of models, it will become apparent that some are more likely to produce better results than others in different data scenarios – the model which produces the best results in this current data scenario may not be feasible to use where less data exist, and so other less-optimum models in the family might produce the next best estimates given different data availability.

One question that arises from this challenge in the current context is whether it is feasible to treat this 287 zone EU system as a whole when it is the convention to make a distinction between 'internal migration' flows and 'international migration' flows. It could be argued that where national borders are real barriers to travel then two systems should be defined, however, in a post-Schengen Europe (Convey and Kupiszewski, 1995; Kraler et al., 2006) national boundaries are not the rigid constructs (both metaphorically and physically) they once were, with flows of migrants between member countries now (in principle) as easy as flows within them. Indeed it is not uncommon for another type of human flow – daily commutes – to occur between countries such as Denmark and Sweden or Luxembourg and Belgium (Mathä and Wintr, 2009). With this being the case, we might expect internal migration and international migration in these areas of Europe to be virtually interchangeable in terms of, for example, the motivations for moves or the limiting factors such as distance which curtail flows. Whether this is actually the case will be explored although the modelling experiments with different models in the family are detailed later in the paper.

6.3   Biproportional Fitting Modelling Methodology

To achieve the task set out in Section 6.1, we will make use of a variation on the doubly constrained entropy maximising SIM (Wilson, 1970, 1971). SIMs are particularly appropriate in the context of migration where empirical studies and model experiments have demonstrated that the propensity to migrate decreases with distance (Boyle et al., 1998; Flowerdew, 2010; Fotheringham et al., 2004; He and Pooler, 2003; Singleton et al., 2010; Stillwell, 1978; Taylor,

1983). Indeed, Olsson (1970, p. 223) notes that '*Under the umbrella of spatial interaction and distance decay, it has been possible to accommodate most model work in transportation, migration, commuting and diffusion*'.

If T is the number of migrant *transitions*, (Rees, 1977), let capital letters such as I and J denote countries and let lower case letters such as i and j denote NUTS2 regions within a country. Then let T^{IJ} be the number of migrants from country I to country J in some time period, say t to $t + 1$ (which we will leave implicit for ease of notation). Then we can denote by T^{IJ}_{ij} the number of migrants from region i in I to region j in J. For convenience we denote all the migration flows by T, but the different subscripts and superscripts indicate the different geographical levels in the system. This notation implies that we number the NUTS2 zones from $1, \ldots, n$ for country I rather than numbering them consecutively for the whole system.

The available data described in Figure 6.2 can then be shown as in Figure 6.3. We have inter-regional, intra-country data for each country – T^{IJ}_{ij} where $I = J$. These internal migration flows could also be described with the notation T^{II}_{ij} to distinguish them from inter-country inter-regional flows. Intra-regional flows – T^{II}_{ii} – are not available. At the country level, inter-country flows T^{IJ} are available.

The row and column totals are known for the T^{II}_{ij} elements, that is, at the NUTS2 level, and also for the T^{IJ} inter-country levels. Let these be M^I_i and N^J_j and O^I and D^J, respectively, so that:

$$M^I_i = \sum_j T^{II}_{ij} = \sum_j T^{IJ}_{ij}, I = J \tag{6.1}$$

$$N^J_j = \sum_i T^{II}_{ij} = \sum_i T^{IJ}_{ij}, I = J \tag{6.2}$$

$$O^I = \sum_J T^{IJ}, J \neq I \tag{6.3}$$

$$D^J = \sum_I T^{IJ}, I \neq J \tag{6.4}$$

These row and column totals are depicted in expanded versions of Figures 6.2 and 6.3, as shown in Figure 6.4a and b. Note that the O^I and D^J totals do not include intra-country data contained in the M^I_i and N^J_j totals – consistent with the common practice of not including intra-country flows in international migration analysis. Internal migration data are assumed to

		Country1			Country2			Country3		
		zone1	zone2	zone3	zone1	zone2	zone3	zone1	zone2	zone3
Country1	zone1	T^{11}_{11}	T^{11}_{12}	T^{11}_{13}						
	zone2	T^{11}_{21}	T^{11}_{22}	T^{11}_{23}		T^{12}			T^{13}	
	zone3	T^{11}_{31}	T^{11}_{32}	T^{11}_{33}						
Country2	zone1				T^{22}_{11}	T^{22}_{12}	T^{22}_{13}			
	zone2		T^{21}		T^{22}_{21}	T^{22}_{22}	T^{22}_{23}		T^{23}	
	zone3				T^{22}_{31}	T^{22}_{32}	T^{22}_{33}			
Country3	zone1							T^{33}_{11}	T^{33}_{12}	T^{33}_{13}
	zone2		T^{31}			T^{32}		T^{33}_{21}	T^{33}_{22}	T^{33}_{23}
	zone3							T^{33}_{31}	T^{33}_{32}	T^{33}_{33}

Figure 6.3 Sample system in Figure 6.2 using defined notation

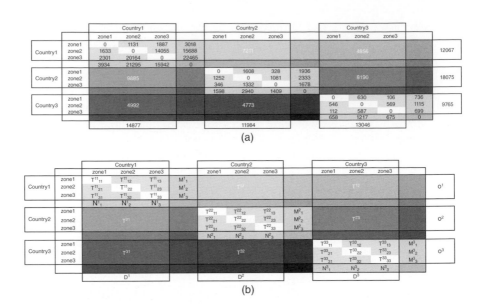

Figure 6.4 Expanded sample system with margins and sub-margins

be consistent such that:

$$\sum_i M_i^I = \sum_j N_j^J = \sum_{ij} T_{ij}^{II} \tag{6.5}$$

The sample data shown in Figure 6.4a and b represent the information we currently have about our system of interest. The formulation thus far implies that we are not seeking to model flows at the NUTS2 level within each country I (we have these data) and to and from other countries, $J, J \neq I$. The ultimate modelling goal, however, is to estimate these inter-country regional level flows, effectively filling all T_{ij}^{IJ} interior cells in the matrix.

In order to model these NUTS2 level flows between countries, we introduce another element of notation: T_{iJ}^I and T_{Ij}^J are, respectively, the out-migration flows from NUTS2 i in country I to country $J (\neq I)$ and the in-migration flows to NUTS2 j in country $J (\neq I)$ from country I. T_{iJ}^I and T_{Ij}^J can be viewed as table sub-margins and are equivalent to M_i^I and N_j^J (where the country subscripts are dropped as flows are internal) so that

$$T_{iJ}^I = \sum_{j \in J} T_{ij}^{IJ} \tag{6.6}$$

$$T_{Ij}^J = \sum_{i \in I} T_{ij}^{IJ} \tag{6.7}$$

Then T^{IJ} in (6.3) and (6.4), for $I \neq J$, would be given by

$$T^{IJ} = \sum_{i \in I} T_{iJ}^I = \sum_{j \in J} T_{Ij}^J = \sum_{i \in I} \sum_{j \in J} T_{ij}^{IJ} \tag{6.8}$$

These sub-margin elements are shown in Figure 6.5a and b. In addition to these new sub-margins, two new row and column margins can also be calculated. O_i^I and D_j^J are directly

Figure 6.5 (a)

Country	zone	Country1 zone1	zone2	zone3		Country2 zone1	zone2	zone3		Country3 zone1	zone2	zone3				
Country1	zone1	0	1131	1887	3018				529				356	885	3903	
	zone2	1633	0	14055	15688		7211		2748		4856		1850	4598	20286	12067
	zone3	2301	20164	0	22465				3935				2650	6584	29049	
	Σ	3934	21295	15942	41171	1181	2455	3575	7211	2513	1604	738	4856			
Country2	zone1				3218	0	1608	328	1936				2666	5884	7820	
	zone2		9885		3878	1252	0	1081	2333		8190		3213	7091	9424	18075
	zone3				2789	346	1332	0	1678				2311	5100	6778	
	Σ	818	3598	5470	9885	1598	2940	1409	5947	4239	2706	1245	8190			
Country3	zone1				1441				1378	0	630	106	736	2177	3554	
	zone2		4992		2183		4773		2087	546	0	569	1115	3298	5385	9765
	zone3				1368				1308	112	587	0	699	2067	3376	
	Σ	413	1817	2762	4992	782	1625	2366	4773	658	1217	675	2550			
		1230	5415	8232		2380	4565	3775		4897	3923	1920				
		5164	26710	24174		3561	7020	7350		7410	5527	2659				
		14877				11984				13046						

(a)

Figure 6.5 (b)

Country	zone	Country1 z1	z2	z3		Country2 z1	z2	z3		Country3 z1	z2	z3				
Country1	zone1	T^{11}_{11}	T^{11}_{12}	T^{11}_{13}	M^1_1				T^1_{12}				T^1_{13}	O^1_1	P^1_1	
	zone2	T^{11}_{21}	T^{11}_{22}	T^{11}_{23}	M^1_2		T^{12}		T^1_{22}		T^{13}		T^1_{23}	O^1_2	P^1_2	o^1
	zone3	T^{11}_{31}	T^{11}_{32}	T^{11}_{33}	M^1_3				T^1_{32}				T^1_{33}	O^1_3	P^1_3	
	Σ	N^1_1	N^1_2	N^1_3												
Country2	zone1					T^{22}_{11}	T^{22}_{12}	T^{22}_{13}	M^2_1				T^2_{13}	O^2_1	P^2_1	
	zone2		T^{21}		T^2_{21}	T^{22}_{21}	T^{22}_{22}	T^{22}_{23}	M^2_2		T^{23}		T^2_{23}	O^2_2	P^2_2	o^2
	zone3					T^{22}_{31}	T^{22}_{32}	T^{22}_{33}	M^2_3				T^2_{33}	O^2_3	P^2_3	
	Σ	T^2_{11}	T^2_{21}	T^2_{31}		N^2_1	N^2_2	N^2_3								
Country3	zone1				T^3_{11}				T^3_{12}	T^{33}_{11}	T^{33}_{12}	T^{33}_{13}	M^3_1	O^3_1	P^3_1	
	zone2		T^{31}		T^3_{21}		T^{32}		T^3_{22}	T^{33}_{21}	T^{33}_{22}	T^{33}_{23}	M^3_2	O^3_2	P^3_2	o^3
	zone3				T^3_{31}				T^3_{32}	T^{33}_{31}	T^{33}_{32}	T^{33}_{33}	M^3_3	O^3_3	P^3_3	
	Σ									N^3_1	N^3_2	N^3_3				
		D^1_1	D^1_2	D^1_3		D^2_1	D^2_2	D^2_3		D^3_1	D^3_2	D^3_3				
		Q^1_1	Q^1_2	Q^1_3		Q^2_1	Q^2_2	Q^2_3		Q^3_1	Q^3_2	Q^3_3				
		D^1				D^2				D^3						

(b)

Figure 6.5   Sample system including all sub-margin and margin elements

related to O^I and D^J in that:

$$O^I_i = \sum_J T^I_{iJ} \tag{6.9}$$

$$O^I = \sum_{i \in I} O^I_i \tag{6.10}$$

$$D^J_j = \sum_I T^J_{Ij} \tag{6.11}$$

$$D^J = \sum_{j \in J} D^J_j \tag{6.12}$$

A final set of margins can be calculated for all interior cells in the matrix where

$$P^I_i = O^I_i + M^I_i \tag{6.13}$$

$$Q^J_j = D^J_j + N^J_j \tag{6.14}$$

With a complete system description, we can then consider the variety of models which can be built. Equations (6.1)–(6.4), (6.6), (6.7), (6.9), (6.11), (6.13) and (6.14) can provide the core constraint equations for a suite of entropy maximising models, which can be used to estimate various elements and aggregations of the T^{IJ}_{ij} flows in the multi-level system matrix. We might describe this as a family of MLSIMs, with the model possibilities being the following:

(i) Model the NUTS2 flows within each country separately – that is, model T_{ij}^{II} (in which case I simply functions as a label for each country model). Equations (6.1) and (6.2) would be the accounting/constraint equations.

(ii) Model the inter-country flows, T^{IJ}, separately. Equations (6.3) and (6.4) would be the accounting equations.

(iii) Model asymmetric NUTS2 flows i and j in and out of each I and ($\neq I$), T_{iJ}^{I} and T_{Ij}^{J}. Three versions of the asymmetric model can be formulated.

    (a) Equations (6.9) and (6.3) would hold as accounting/constraint equations for Equation (6.6) and Equations (6.11) and (6.4) would be the constraints for Equation (6.7).

    (b) Known T^{IJ} flows with Equation (6.9) would hold as constraints for Equation (6.6) and known T^{IJ} flows with Equation (6.11) would hold as constraints for Equation (6.7).

    (c) It would also be possible to use Equations (6.13) and (6.3) as the constraints for Equation (6.6) and Equations (6.14) and (6.4) as the constraints for Equation (6.7). This model is almost identical to (a), although in this case we would also be modelling M_i^I as T_{iJ}^I and N_j^J as T_{Ij}^J.

(iv) Model T_{ij}^{IJ} for each country separately using sub-margins (6.6) and (6.7) as constraints.

(v) Model T_{ij}^{IJ} where $I \neq J$ with Equations (6.9) and (6.11) as constraints.

(vi) Model the full array of NUTS2 regions, T_{ij}^{IJ} using Equations (6.13) and (6.14) as the accounting constraints.

If the accounting equations (6.1)–(6.4) are deployed as in Models (i) and (ii), this leads to the construction of doubly constrained models for which the main task would be to identify impedance functions, associated generalised costs c_{ij}, and the model parameter values. In migration, research cost is often the physical distance between places: the propensity to migrate decreases with distance and thus the cost of travel can be inferred to increase. Empirical studies have shown that this distance decay in migration propensity will often follow either a negative exponential or inverse power law (Stillwell, 1978). In SIMs, this is represented by a parameter β, (normally negative), which can be calibrated endogenously if data exist. In the equations which follow, we write the distance decay function f, as exponential $- f(c_{ij}) = e^{\beta c_{ij}}$ – although it would be just as appropriate to write it as a power law $- f(c_{ij}) = c_{ij}^{\beta}$.

6.3.1 Model (i)

Model (i) is the most straightforward and would produce

$$T_{ij}^{II} = A_i^I B_j^I M_i^I N_j^I e^{\beta^I c_{ij}^I} \tag{6.15}$$

$$A_i^I = \frac{1}{\sum_j B_j^I N_j^I e^{\beta^I c_{ij}^I}} \tag{6.16}$$

$$B_j^I = \frac{1}{\sum_i A_i^I M_i^I e^{\beta^I c_{ij}^I}} \tag{6.17}$$

where the generalised distance decay parameter β can be calibrated endogenously using T_{ij}^{II} data. An alternative version of this model could calculate origin or destination-specific β parameters:

$$T_{ij}^{II} = A_i^I B_j^I M_i^I N_j^I e^{\beta_i^I c_{ij}^I} \tag{6.18}$$

$$T_{ij}^{II} = A_i^I B_j^I M_i^I N_j^I e^{\beta_j^I c_{ij}^I} \tag{6.19}$$

6.3.2 Model (ii)

The inter-country Model (ii) would be

$$T^{IJ} = A^I B^J O^I D^J e^{\mu C^{IJ}} \tag{6.20}$$

where balancing factors are calculated with equivalent equations to (6.16) and (6.17).

6.3.3 Model (iii)

The asymmetric models in Model (iiia) would take the form

$$T_{iJ}^I = A_i^I B^J O_i^I D^J e^{\beta_i^I c_{iJ}^I} \tag{6.21}$$

$$T_{Ij}^J = A^I B_j^J O^I D_j^J e^{\beta_j^J c_{Ij}^J} \tag{6.22}$$

With the balancing factors for (6.21):

$$A_i^I = \frac{1}{\sum_J B^J D^J e^{\beta_i^I c_{iJ}^I}} \tag{6.23}$$

$$B^J = \frac{1}{\sum_i A_i^I O_i^I e^{\beta_i^I c_{ij}^I}} \tag{6.24}$$

and the balancing factors for (6.22):

$$A^I = \frac{1}{\sum_j B_j^J D_j^J e^{\beta_j^J c_{iJ}^J}} \tag{6.25}$$

$$B_j^J = \frac{1}{\sum_I A^I O^I e^{\beta_j^J c_{Ij}^J}} \tag{6.26}$$

Equations (6.21) and (6.22) can be visualised easily by collapsing the matrices in Figure 6.5a and b into just the relevant margins and sub-margins (Figures 6.4a and b, 6.5a and b). These margins then become, effectively, the i, j values in a standard two-dimensional matrix.

It is important to note that while in the examples in Figures 6.6a and 6.7a, corresponding country to country sums are equal – for example $\sum_i T_{iJ}^I = \sum_j T_{Ij}^J$ – as they should be, in Model (iiia) the modelled values will not correspond in this way, due to the constraints used. To exemplify, consider Figures 6.8 and 6.9. The marginal values in these figures are almost identical

T^1_{ij}		Country1	Country2	Country3		
Country1	zone1		529	356	885	
	zone2		2748	1850	4598	12067
	zone3		3935	2650	6584	
Country2	zone1	3218		2666	5884	
	zone2	3878		3213	7091	18075
	zone3	2789		2311	5100	
Country3	zone1	1441	1378		2818	
	zone2	2183	2087		4270	9765
	zone3	1368	1308		2677	
		14877	11984	13046	39907	

(a)

T^1_{ij}		Country1	Country2	Country3		
Country1	zone1	T^1_{11}	T^1_{12}	T^1_{13}	O^1_1	
	zone2	T^1_{21}	T^1_{22}	T^1_{23}	O^1_2	O^1
	zone3	T^1_{31}	T^1_{32}	T^1_{33}	O^1_3	
Country2	zone1	T^2_{11}	T^2_{12}	T^2_{13}	O^2_1	
	zone2	T^2_{21}	T^2_{22}	T^2_{23}	O^2_2	O^2
	zone3	T^2_{31}	T^2_{32}	T^2_{33}	O^2_3	
Country3	zone1	T^3_{11}	T^3_{12}	T^3_{13}	O^3_1	
	zone2	T^3_{21}	T^3_{22}	T^3_{23}	O^3_2	O^3
	zone3	T^3_{31}	T^3_{32}	T^3_{33}	O^3_3	
		D^1	D^2	D^3		

(b)

Figure 6.6   Collapsed matrix showing only region-to-country sub-margins depicted in Figure 6.5

to those in Figures 6.6a and 6.7a (only two migrants are misplaced in Figure 6.8). The interior T^I_{iJ} and T^J_{Ij} values are quite different. In these modelled matrices, $\sum_i T^I_{iJ} \neq \sum_j T^J_{Ij}$. For example, the total flows from Country 1 to Country 2 in Figure 6.8 are 6,915, whereas the total flows from Country1 to Country 2 in Figure 6.9 are 7,776. The reason for this is that the T^I_{iJ} and T^J_{Ij} flows are only constrained to the marginal totals – either O^I_i and D^J or O^I and D^J_j, respectively. In these models, T^I_{iJ} and T^J_{Ij} have multiple equilibria, only a small number of which result in $\sum_i T^I_{iJ} = \sum_j T^J_{Ij}$. This has implications for Model (iv) in our suite of models.

T^1_{ij}	Country1 zone1	zone2	zone3	Country2 zone1	zone2	zone3	Country3 zone1	zone2	zone3	
Country1				1181	2455	3575	2513	1604	738	12067
Country2	818	3598	5470				4239	2706	1245	18075
Country3	413	1817	2762	782	1625	2366				9765
	1230	5415	8232	1963	4080	5941	6752	4310	1984	39907
		14877			11984			13046		

(a)

T^1_{ij}	Country1 zone1	zone2	zone3	Country2 zone1	zone2	zone3	Country3 zone1	zone2	zone3	
Country1	T^1_{11}	T^1_{12}	T^1_{13}	T^2_{11}	T^2_{12}	T^2_{13}	T^3_{11}	T^3_{12}	T^3_{13}	O^1
Country2	T^1_{21}	T^1_{22}	T^1_{23}	T^2_{21}	T^2_{22}	T^2_{23}	T^3_{21}	T^3_{22}	T^3_{23}	O^2
Country3	T^1_{31}	T^1_{32}	T^1_{33}	T^2_{31}	T^2_{32}	T^2_{33}	T^3_{31}	T^3_{32}	T^3_{33}	O^3
	D^1_1	D^1_2	D^1_3	D^2_1	D^2_2	D^2_3	D^3_1	D^3_2	D^3_3	
		D^1			D^2			D^3		

(b)

Figure 6.7 Collapsed matrix showing only country-to-region sub-margins depicted in Figure 6.5

T^1_{ij}		Country1	Country2	Country3		
Country1	zone1	0	682	203	885	
	zone2	0	2441	2157	4598	12067
	zone3	0	3792	2702	6584	
Country2	zone1	2940	0	2944	5884	
	zone2	4461	0	2630	7091	18075
	zone3	2780	0	2320	5100	
Country3	zone1	1173	1645	0	2818	
	zone2	2086	2184	0	4270	9765
	zone3	1437	1240	0	2677	
		14877	11984	13046	39907	

(a)

Figure 6.8 T^I_{iJ} values modelled using the entropy-maximising model in (6.21)

T^1_{ij}	Country1 zone1	zone2	zone3	Country2 zone1	zone2	zone3	Country3 zone1	zone2	zone3	
Country1	0	0	0	760	2968	4048	910	2459	922	12067
Country2	610	3820	4890	0	0	0	5842	1851	1062	18075
Country3	620	1595	3342	1203	1112	1893	0	0	0	9765
	1230	5415	8232	1963	4080	5941	6752	4310	1984	39907
		14877			11984			13046		

Figure 6.9 T^J_{Ij} values modelled using the entropy-maximising model in (6.22)

6.3.4   Model (iv)

Model (iv) takes T_{iJ}^I and T_{Ij}^J as constraints, with the doubly constrained version of the model defined as

$$T_{ij}^{IJ} = A_{iJ}^I B_{Ij}^J T_{iJ}^I T_{Ij}^J e^{\beta_i^I c_{ij}^I}$$

(6.27)

$$T_{ij}^{IJ} = A_{iJ}^I B_{Ij}^J T_{iJ}^I T_{Ij}^J e^{\beta_j^J c_{ij}^J}$$

(6.28)

With the balancing factors for (6.27):

$$A_{iJ}^I = \frac{1}{\sum_{j \in J} B_{Ij}^J T_{Ij}^J e^{\beta_i^I c_{ij}^I}}$$

(6.29)

$$B_{Ij}^J = \frac{1}{\sum_{i \in I} A_{iJ}^I T_{iJ}^I e^{\beta_i^I c_{ij}^I}}$$

(6.30)

If $\sum_i T_{iJ}^I = \sum_j T_{Ij}^J$, then it is possible to solve Equations (6.27) and (6.28) – the iterative procedure which calculates that the A_{iJ}^I and B_{Ij}^J balancing factors are able to converge when $\sum_i T_{iJ}^I$ and its corresponding sub-margin $\sum_j T_{Ij}^J$ are the same value. If T_{iJ}^I and T_{Ij}^J values are estimated using the entropy-maximising procedure described in Equations (6.21) and (6.22), then $\sum_i T_{iJ}^I \neq \sum_j T_{Ij}^J$, meaning that the iterative balancing factor routine will not converge and Equations (6.27) and (6.28) cannot be solved.

One solution to this issue is to estimate T_{iJ}^I and T_{Ij}^J using a method other than the entropy-maximising model described. As already noted, T_{iJ}^I and T_{Ij}^J are equivalent to M_i^I and N_j^J. In this system, we already know the values of M_i^I and N_j^J from the T_{ij}^{II} internal migration data available. Given this information, the following equations can be used to estimate T_{iJ}^I and T_{Ij}^J:

$$T_{iJ}^I = \left(\frac{M_i^I}{T^{II}} \right) T^{IJ}$$

(6.31)

$$T_{Ij}^J = \left(\frac{N_j^J}{T^{II}} \right) T^{IJ}$$

(6.32)

where these T_{iJ}^I and T_{Ij}^J estimates are constrained to the corresponding T^{IJ} values, $\sum_i T_{iJ}^I = \sum_j T_{Ij}^J$, and thus it is possible to solve Equations (6.27) and (6.28).

There is, however, an entropy-maximising solution to this issue as well. In Model (iiib), the constraints used to estimate T_{iJ}^I and T_{Ij}^J are not the matrix margins as shown in Figures 6.6 and 6.7. By using these margins in Model (iiia), we are not taking advantage of all known information in the system. As T^{IJ} flows are known, a combination of matrix margins and known interior T^{IJ} values can be used as constraints, thus the equations for T_{iJ}^I and T_{Ij}^J become

$$T_{iJ}^I = A_i^I X^{IJ} O_i^I T^{IJ} e^{\beta_i^I c_{iJ}^I}$$

(6.33)

$$T_{Ij}^J = Y^{IJ} B_j^J T^{IJ} D_j^J e^{\beta_j^J c_{Ij}^J}$$

(6.34)

with the balancing factors for (6.33) calculated:

$$A_i^I = \frac{1}{\sum_J X^{IJ} T^{IJ} e^{\beta_i^I c_{iJ}^I}} \tag{6.35}$$

$$X^{IJ} = \frac{1}{\sum_i A_i^I O_i^I e^{\beta_i^I c_{iJ}^I}} \tag{6.36}$$

and the balancing factors for (6.34):

$$Y^{IJ} = \frac{1}{\sum_j B_j^J D_j^J e^{\beta_j^J c_{IJ}^J}} \tag{6.37}$$

$$B_j^J = \frac{1}{\sum_J Y^{IJ} T^{IJ} e^{\beta_j^J c_{Ij}^J}} \tag{6.38}$$

In constraining T_{iJ}^I and T_{Ij}^J to T^{IJ} flows, $\sum_i T_{iJ}^I = \sum_j T_{Ij}^J$. This means that when Equations (6.33) and (6.34) are used as inputs into (6.27) and (6.28) in Model (iv), the balancing factors will always converge and the equations can be solved. Model (iv) represents the T_{ij}^{IJ} estimates which will adhere most closely to the known information about the system, and as such might be described as the *optimum* model for the EU system in this study.

6.3.5 Model (v)

If Model (iv) is the optimum model, then Models (v) and (vi) which produce alternative T_{ij}^{IJ} estimates using less information might be described as being *suboptimal*. Model (v) will only produce T_{ij}^{IJ} estimates where T^{IJ}, $i \neq j$. This model can be written as

$$T_{ij}^{IJ} = A_i^I B_j^J O_i^I D_j^J e^{\beta_i^I c_{ij}^I} \tag{6.39}$$

where

$$A_i^I = \frac{1}{\sum_j B_j^J D_j^J e^{\beta_i^I c_{ij}^I}} \tag{6.40}$$

$$B_j^J = \frac{1}{\sum_i A_i^I O_i^I e^{\beta_i^I c_{ij}^I}} \tag{6.41}$$

In this model, O_i^I and D_j^J can be estimated in exactly the same way as T_{iJ}^I and T_{Ij}^J in Equations (6.31) and (6.32) so

$$O_i^I = \left(\frac{M_i^I}{T^{II}}\right) O^I \tag{6.42}$$

$$D_j^J = \left(\frac{N_j^J}{T^{II}}\right) D^J \tag{6.43}$$

The T_{ij}^{IJ} estimates in Model (v) will not adhere as closely to known T^{IJ} values as those in Model (iv), as the constraints are the outer margins on the expanded matrix shown in Figure 6.5.

6.3.6   Model (vi)

Finally, Model (vi) models the whole T_{ij}^{IJ} matrix, including T_{ij}^{II} flows. This model (with an origin-specific distance decay parameter) takes the form

$$T_{ij}^{IJ} = A_i^I B_j^J P_i^I Q_j^J e^{\beta_i^I c_{ij}^I} \sqrt{b^2 - 4ac} \tag{6.44}$$

where

$$A_i^I = \frac{1}{\sum_j B_j^J Q_j^J e^{\beta_i^I c_{ij}^I}} \tag{6.45}$$

$$B_j^J = \frac{1}{\sum_i A_i^I P_i^I e^{\beta_i^I c_{ij}^I}} \tag{6.46}$$

with the P_i^I and Q_j^J constraints calculated as in Equations (6.13) and (6.14).

This new family of doubly constrained MLSIMs allows estimates of a full matrix of 287×287 flows within the defined European system to be made. While Model (iv) defined in Equations (6.33) and (6.34) will produce estimates which are forced to adhere most closely to the known information in the system, other models in the family, which by definition will produce results constrained to less information, will allow us to examine features of the European migration system which do not fit our model assumptions. In doing this we might, for example, be able to identify areas where it would be prudent to adjust the cost proxy in order to distribute migrant flows more effectively within the system without the 'helping hand' that constraints give, or indeed answer the question posed in the introduction to this chapter relating to whether it is feasible to treat the European system as, effectively, an internal migration system where national boundaries have little influence on migration flows. First, however, a number of technical challenges relating to the implementation of the models need to be overcome.

6.4   Model Parameter Calibration

All of the models described in the MLSIM family make use of a calibrated distance decay parameter (or parameters), but in making use of such a parameter, a number of problems present themselves. Firstly, calibration can only be carried out using known data within the system – therefore, the β parameter(s) will have to be calibrated using either T_{ij}^{II} flows of T^{IJ} flows. This means that, potentially, these parameters may not be completely appropriate for T_{ij}^{IJ} flows. In the absence of other means of estimating appropriate parameters, however, it could be argued this is the best option available at this time, and so it is the option we will have to take.

Accepting that available observed data will be used to calibrate the best-fit parameter(s), the next issue relates to the method used to carry out the calculation. Distance decay parameters in SIMs have historically been calibrated using maximum-likelihood techniques employed in computer algorithms – these commonly use iterative procedures to search for the 'best-fit' between the estimates created by the model and the sample data. As an aside, while standard iterative procedures are most frequently used in this type of modelling, it should be noted that a significant amount of work has been carried out by Openshaw and colleagues on the calibration of SIMs using genetic algorithms (Diplock and Openshaw, 1996; Openshaw, 1998): an approach perhaps operationalised most recently by Harland (2008) – we will not explore these methods here, but will use a conventional iterative approach. Batty and Mackie (1972) discuss a range of maximum-likelihood calibration methods, but the Newton–Raphson search algorithm has been shown to perform better than most and has been adopted in both the SIMODEL computer program developed by Williams and Fotheringham (1984) and the IMP program developed by Stillwell (1978); both Fortran programs using the search routine to find the parameter estimates which minimise the divergence between the mean value of the total distance travelled in the observed and modelled flow matrices – an approach also used by Pooler (1994). Thanks to its successful implementation in SIMs for migration analysis, the Newton–Raphson algorithm is the one that we choose to use here.

Initially two versions of the doubly constrained model were run to calibrate a best-fit general distance decay β parameter for the whole system. The results of these models are shown in Table 6.1 and are contrasted with a more basic singly constrained model for comparison. Here, a selection of goodness-of-fit (GOF) statistics are displayed – the coefficient of determination (R^2), the square root of the mean squared error (SRMSE), the sum of the squared deviations and the percentage of misallocated flows – although they all display very similar findings. It is clear that the doubly constrained model with the inverse power function applied to the distance matrix produces the best fit to the original data, with an R^2 of some 87%. This compares to an R^2 of 72% for the negative exponential function and 62% for the reference production constrained model.

The question that follows is: should this overall distance decay parameter be used as the distance decay input to the estimation model? If this parameter is representative of the whole system, then it could be argued that it could. To test this, a T_{ij}^{II} model with an inverse power distance decay function (akin to that in the second row of Table 6.1) was run separately for each of the 21 countries in the system comprised of more than a single zone in order to calibrate a series of β^I parameters. The results of these experiments are shown in Table 6.2.

In this instance, we chose the inverse power distance decay function as it was the best-performing function in the T_{ij}^{II} experiment. Serendipitously, the power function is scale independent whereas the exponential function is not (Fotheringham and O'Kelly, 1989), meaning we are able to directly compare the β^I parameters directly. In Table 6.2, we use the R^2 value as our measure of goodness of fit. We are aware that there has been some

Table 6.1 Goodness-of-fit statistics for T_{ij}^{II} model experiments

Model equation	β	R^2	SRMSE	Sum Sq Dev	% Misallocated
$T_{ij}^{II} = A_i^I B_j^J M_i^I N_j^J e^{\beta c_{ij}^I}$	−4.2986	0.718	39.393	10,456,839,051	21.554
$T_{ij}^{II} = A_i^I B_j^J M_i^I N_j^J c_{ij}^{\beta}$	−0.9136	0.865	27.992	5,280,098,085	17.008
$T_{ij}^{II} = A_i^I M_i^I c_{ij}^{\beta}$	−1.2201	0.623	45.457	13,886,764,628	28.131

Table 6.2 Goodness-of-fit statistics for inter-regional migration data modelled with a doubly constrained model with a power distance decay β parameter

Country code	Country	R^2	β^l (power function)
FI	Finland	0.996	−0.754
SE	Sweden	0.974	−0.771
AT	Austria	0.972	−0.747
HU	Hungary	0.963	−0.567
SK	Slovakia	0.948	−0.773
NL	Netherlands	0.936	−1.279
DK	Denmark	0.930	−0.969
NO	Norway	0.919	−0.814
BG	Bulgaria	0.901	−0.825
CZ	Czech Republic	0.889	−0.807
UK	United Kingdom	0.884	−0.927
PL	Poland	0.877	−1.068
CH	Switzerland	0.788	−0.867
BE	Belgium	0.772	−1.049
RO	Romania	0.745	−0.763
DE	Germany	0.715	−0.760
IT	Italy	0.699	−0.718
ES	Spain	0.621	0.154
FR	France	0.549	1.093

debate over which is the most appropriate metric to use (Knudsen and Fotheringham, 1986); however, R^2 is commonly used and for comparative proposes, the choice of statistic has little relevance to the outcome. A number of points can be made about the results displayed in Table 6.2. Firstly, the countries are ranked according to their goodness of fit and we can observe that around half of the list have R^2 values over 90%, with Finland, Sweden and Austria ranked the highest – Finland with an exceptionally high R^2. It is clear, however, that there is a considerable variation in the β^l parameters for each country. This would suggest that it may not be ideal to use the generalised β parameter to model flows for the whole EU system. Furthermore, the reliability of some of the β^l parameters can be called into question with particularly low R^2 values for Spain and France – countries which exhibit positive β^l parameter values. The exact way in which these parameters can be understood has been questioned (Fotheringham, 1981); however, one interpretation is that the value can be read behaviourally and the number is an index of the deterrent to migration, with high negative values representing distance being a strong deterrent to migration and low negative values inferring that distance is a weak deterrent. Positive values in this context would indicate that distance is an attraction to interaction – that is, the further away origins and destinations, the more likely migration is to occur. Clearly this is unlikely to be the case across the whole of Spain and France.

Given this evidence, generalised distance decay parameters are currently poor candidates for inputs into an estimation model for the whole of Europe. A potential solution, therefore, would be to use distance decay parameters which are specific to each NUTS2 zone – a technique first outlined by Stillwell (1978). This returns us to Model (i) and Equations (6.18) and (6.19).

Table 6.3 Goodness-of-fit statistics for Model (i) with β_i^I and β_j^J parameters

Model equation	R^2	SRMSE	Sum Sq Dev	% Misallocated
$T_{ij}^{II} = A_i^I B_j^J M_i^I N_j^J e^{\beta_i^I c_{ij}^J}$	0.928	19.802	2,642,462,153	12.284
$T_{ij}^{II} = A_i^I B_j^J M_i^I N_j^J e^{\beta_j^J c_{ij}^J}$	0.931	19.582	2,583,959,209	12.163

The GOF statistics for Model (i) – taken for all internal migration flows in the system rather than for each separate country) – are shown in Table 6.3. Evidently, these models provide much better fits than the generalised parameter models, with R^2 values around 93%. A geography to these distance decay parameters can be observed, with the frictional effects of distance operating very differently for in- and out-migration flows across the EU system, as is shown in Figures 6.10 and 6.11. It should be noted that the nature of the algorithm used to carry out this calibration means that where it is not possible to calculate a zone-specific distance decay parameter (e.g. in those countries where T_{ij}^{II} data do not exist such as Greece), a generalised distance decay parameter which is calculated for the whole system prior to zone-specific calibration is allocated. Given the results of these experiments, it is these origin and destination-specific parameters calibrated on internal migration data which will be used as distance decay inputs into our later estimation models.

6.5   Model Experiments

The first step is the estimation of margin constraints. In the section of the MLSIM family of models outlined in Section 6.3, which used to estimate T_{ij}^{IJ} flows, all require some inputs which are not available directly from the data to hand. In addition to the distance decay parameters that will be calibrated only on internal migration data, Models (iiia), (iiib), (iv), (v) and (vi) make use (directly and indirectly) of O_i^I and D_j^J margins. Consequently, sub-models are required to make estimates of these data. When $\sum_{i \in I} O_i^I = O^I$ and $\sum_{j \in J} D_j^J = D^J$, it follows that it should be feasible to estimate the NUTS2-level O_i^I and D_j^J margins from the country-level O^I and D^J margins, given the appropriate ratio values. But which are the appropriate ratios to use?

As information at the internal migration T_{ij}^{II} level is complete, it might be possible to use the distribution of internal migrants to estimate the distribution of international migrants such that

$$O_i^I = \left(\frac{M_i^I}{O^I} \right) O^I \tag{6.47}$$

$$D_j^J = \left(\frac{N_j^J}{D^J} \right) D^J \tag{6.48}$$

The assumption here is that the distribution of internal in- and out-migrants within countries is the same as the distribution of immigrants and emigrants moving between countries. But can internal migrant distributions be used to estimate distributions of international migrants within

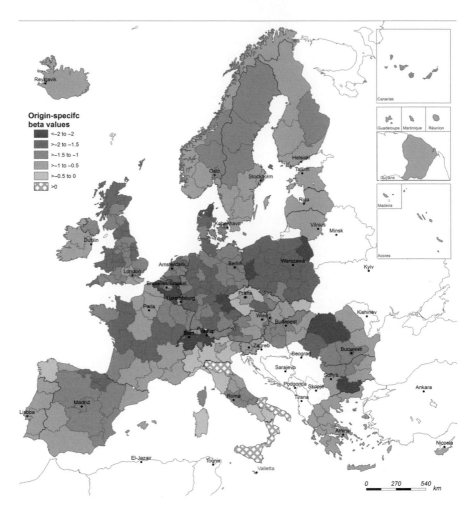

Figure 6.10 β_i^I values calibrated on inter-regional, intra-country migration data, 2006

countries accurately? We might expect, for example, capital cities to dominate these distributions with larger urban areas also providing significant origins and destinations at both levels. Is this the case in reality? Figure 6.12 shows the comparable distributions of internal and international migration for a selection of European countries at NUTS2 level (all countries where comparable data exist at this level), taken from Census data from the 2000–2001 census round and compiled by Eurostat. Broadly speaking, there are positive correlations between internal and international migration distributions, although there are some noticeable differences in the correlation coefficients denoted by the R^2 values (and the scatter plots). For most countries in the selection, R^2 values are over 80%, indicating that internal migration distributions are reasonably good predictors of international migration distributions. For some countries, however, this predictive relationship is weak. Poland, for example, has an R^2 value of only 17%, with the Czech Republic (23%) and Switzerland (28%) not faring much better. The reasons for the

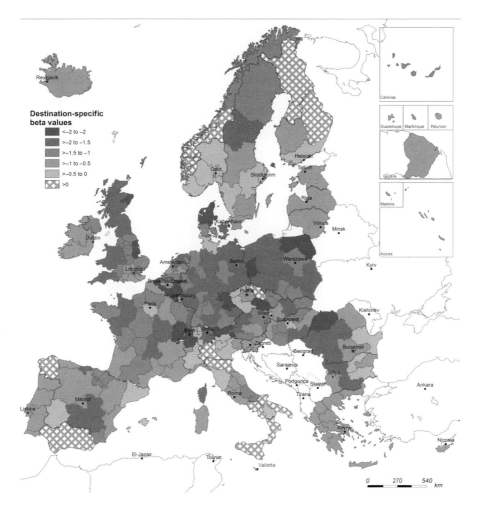

Figure 6.11 β_j^J values calibrated on inter-regional, intra-country migration data, 2006

lack of correlation in these countries are difficult to ascribe, but differences in the perceived attractiveness of particular destinations to internal and international migrants will affect the correlations. Studying Figure 6.12, the scatter plots show that there is very little pattern in the association between internal in-migration and immigration in Poland, although examining Switzerland and the Czech Republic, it appears that were it not for one or two outliers in the scatter plots, the correlation would be far stronger. Through mapping the differences between internal and international migrant distributions, it is possible to interrogate these and other outliers a little further.

Figure 6.13 maps the distribution of the differences between the regional shares of internal and international (in-)migration across NUTS2 zones in Europe (where data are available). A number of points should be made about this map. Firstly, all yellow zones signify less than a 1% deviation between the distribution of internal and international migrants – these zones

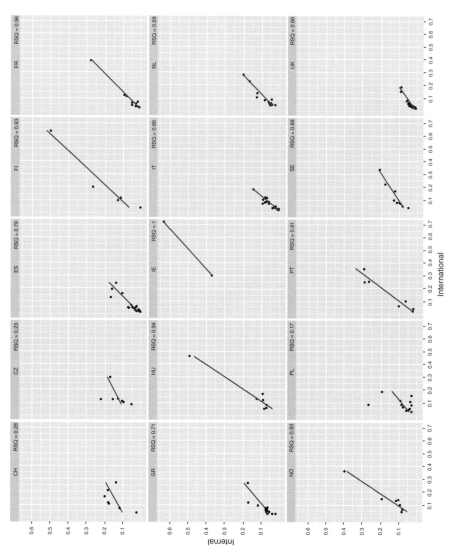

Figure 6.12 Correlation between internal ('Place of residence changed outside the NUTS3 area') and international ('Place of residence changed from outside the declaring country') migrant distributions for NUTS2 regions, selected EU countries, 2001. *Source:* Raymer and Abel (2008)

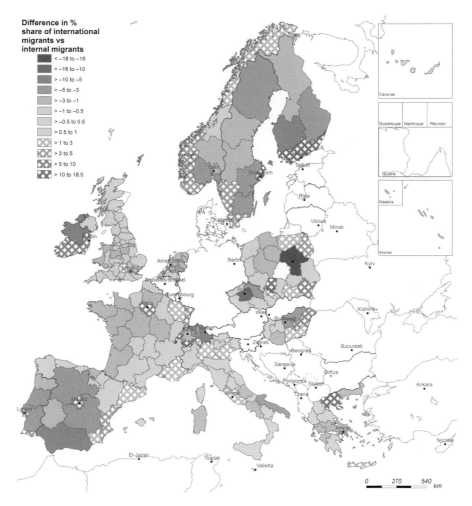

Figure 6.13 Distribution of NUTS 2 regions where shares of internal and international in-migrants differ, selected EU countries, 2001

include much of the United Kingdom and large parts of France, Italy, the Czech Republic, Poland and Greece. In these areas, internal migration distributions can be seen to be good predictors of international migration distributions. Secondly, zones in light orange and light green show only up to a 3% deviation – these include most of the rest of France, a number of regions in Scandinavia, the Netherlands, Poland, the United Kingdom, Italy and Greece. Perhaps the most important point of note, however, which becomes very apparent when examining Figure 6.13, is that there appears to be a 'capital city effect'. The regions containing London, Paris, Madrid, Rome, Amsterdam, Stockholm, Helsinki, Prague, Lisbon and Dublin all exhibit a noticeably higher (average 8.7%) proportion of the national share of international immigrants compared to the national share of internal in-migrants. Some capital cities go against this pattern, although Bern can probably be discounted as in terms of city status, Zurich (which matches this trend) could be argued to be a city of more importance within Switzerland.

Oslo, Athens and Budapest have lower proportions of international immigrants than internal in-migrants, but the city region where a very large trend in the opposite direct occurs is Warsaw in Poland. Here, the proportion of internal migrants to Warsaw is over 18% higher than the proportion of international migrants. The fact that Warsaw is an attractive destination for internal migrants would not be surprising, but why it accounts for a much larger proportion of these migrants compared to international migrants is unclear without further investigation of the particular motivations of migrants in Poland.

Based on this, it could be argued that if this capital city effect could be accounted for consistently, and the proportions of migrants associated with other regions in the country adjusted accordingly, then internal migration distributions could be used to make international migration D_j^J margin estimates relatively reliable, assuming that these associations hold over time.

Incidentally, the time dimension provides us with another option for modelling the sub-national distributions of international migrants. Where decennial census (or other periodic) data can provide sub-national immigrant distributions, if country-level immigrant data are available, sub-national distributions can be estimated with the formula:

$$D_j^{Jt+n} = \left(\frac{D_j^{Jt}}{D^{Jt}} \right) D^{Jt+n} \tag{6.49}$$

Even if more up-to-date national data are not available, an assumption could be made that these ratios hold over time so that

$$D_j^{Jt+n} = \left(\frac{D_j^{Jt}}{D^{Jt}} \right) D^{Jt} \tag{6.50}$$

Returning to Equations (6.47) and (6.48), unfortunately the nature of the data collated by Eurostat means that it is not possible to assess whether emigrant distributions also follow the distributions of internal out-migrants (these data are census/population register data relating to resident populations in recording countries and therefore cannot contain emigrant data). Given the high degree of association between internal migration in- and out-migration distributions (Figure 6.14), it might be reasonable to use international immigrant distributions to estimate international emigrant distributions, but the capital city effect would need to be explored before this could be done with confidence. Here our concern is to present a general methodology for estimating the full EU matrix of NUTS2 flows and so we will not dwell on this element of the estimation process at this stage, although it should be stressed that the estimation of O_i^J and D_j^J marginal values will have an important bearing on reliability of the final modelled outputs.

As a consequence of the data to hand and the investigations of internal/international migration associations, at this stage internal migration distributions will be used to estimate O_i^J and D_j^J marginal values for the model as in Equations (6.47) and (6.48), but we recognise that this is an area of the methodology which could be improved in the future.

6.6   Results

In a file containing the full suite of Models (iv), (v) and (vi), T_{ij}^{IJ} and T_{ij}^{II} estimates are publicly available for anyone wishing to make use of the data through the following link:

http://dl.dropbox.com/u/8649795/Multilevel_SIM_Results.xlsx.

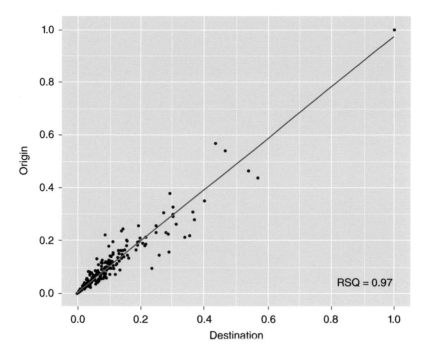

Figure 6.14   Correlation between the NUTS2 regional share of internal in- and out-migration flows across EU countries, 2006

Model (iv) takes in T_{iJ}^I and T_{Ij}^J inputs from Model (iiib) and can be viewed as the *optimum* model as any outputs will be constrained to known T^{IJ} flows and estimated O_i^I and D_j^J margins (where T_{iJ}^I and T_{Ij}^J estimates used these constraints). Models (v) and (vi), in contrast, are *suboptimal* as estimates will not be constrained to T^{IJ} flows, only O_i^I and D_j^J, or P_i^I and Q_j^J margins. Running suboptimal models is an important part of the model-building process as they allow us to explore the reliability of some of the general model assumptions; however here, we present just the results from the optimum Model (iv) of the family of MLSIMs.

MLSIMs offer the opportunity to examine inter-regional flows between all countries in our chosen EU system – examining all flows or even all significant flows would be an extensive task; therefore, we will take the United Kingdom as exemplification. Figure 6.15 depicts all flows over 200 persons entering UK regions from other EU regions, and it is clear that particular origin and destination combinations predominate. Firstly, the importance of London and the South-East corner of the United Kingdom is very apparent – nearly all flows are concentrated in this area, with only a small number entering regions containing other large cities such as Manchester and Birmingham. A large number of these flows originate in Polish regions and many terminate in London. Interestingly, flows from Poland into East Anglia, which have gained much media attention in the United Kingdom are picked up by the model, despite other explanatory factors such as increased job opportunities in the agricultural sector not taken into consideration by the model. One small caveat in relation to these flows can be made referring back to our observations about the poor relationship between internal and international

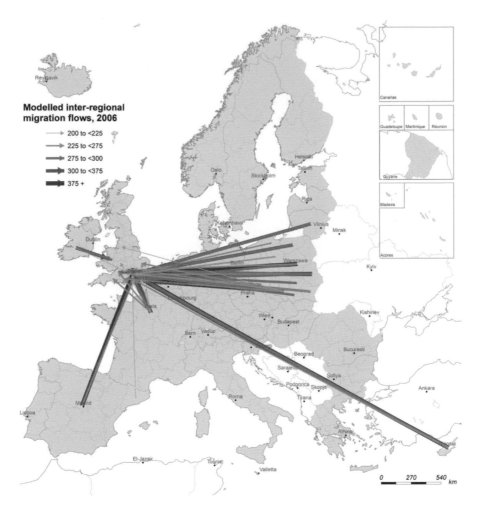

Figure 6.15   Flows greater than 200 migrants entering UK regions from other EU system regions, 2006

migration distributions in Poland made in Section 5.1. Where the relationship between these flow distributions is poor in Poland, some of the precise flow volumes originating from these Polish NUTS2 regions should be treated with caution. Where these relationships are stronger in France and Spain, the large flows from other major capital cities such as Paris and Madrid can be viewed more reliably, indeed given the 'capital city effect' also noticed, these flows may even be larger in reality. High-volume flows are also noticeable from Cyprus, although these may well be associated with the movement of armed forces.

Examining the flows out of UK regions to the rest of the EU system (Figure 6.16), the South-East – and especially London – predominates as with immigration. Destinations for migrants leaving the United Kingdom are quite different to the origins for those arriving in 2006. The large volumes of migration (we may assume related to retirement) can be observed flowing into Spanish regions – regions including the largest cities of Madrid, Barcelona and

Figure 6.16   Flows greater than 200 migrants leaving UK regions for other EU system regions, 2006

Valencia, as well as the Costa del Sol. Large flows can also be observed from London and other regions of the United Kingdom into Ireland – this is partially a function of Ireland consisting of only two regions and so these flows appear more concentrated, although the close ties between all countries of the United Kingdom and Ireland mean that these flows are entirely expected.

6.7   Conclusions and Comments on the New Framework for Estimating Inter-regional, Inter-country Migration Flows in Europe

In this chapter, we have introduced a new family of models for estimating inter-regional migration flows in Europe. Our guiding principle was a simple one – to make use of the maximum amount of available data (embodied in the constraints imposed within the model and the parameters used to influence the patterns) to produce estimates of the maximum likelihood given the information available.

The estimates produced by Model (iv) represent the current 'best-guess' given the data to hand. They embody all known information about flows into and out of countries, the behaviour of internal migrants within their home countries and the relationships between the destination preference of internal and international migrants. There are, of course, a number of areas where these estimates could be improved. Firstly, the country-level international migration data constraints are themselves estimates. The data used were taken from the MIMOSA project (Raymer and Abel, 2008) – data which the authors recognise the limitations of and which will soon be superseded by improved estimates from the IMEM project mentioned in the introduction. When these model inputs can be improved, then there will be a knock-on improvement to our own estimates. We have already acknowledged that there are issues with the methodology we employed to estimate the O_i^I and D_j^J matrix margins which formed constraints either directly or indirectly for all models. As outlined, in these estimates, we have simply taken the national distributions of internal migrants to distribute international migrants. While there are high correlations between these distributions for in-migration, demonstrated across Europe from Census and register data, a 'capital city effect' persists where these destinations can attract up to 10% more migrants internationally than internally. Furthermore, we have been unable to ascertain whether a similar situation exists for out-migration flows. Finally, in using distance decay parameters calibrated with internal migration data, we could be introducing error where internal migration flows, even in an open border Europe, act very differently to international flows. Experimentation with suboptimal models which are not reported in this chapter suggest that this might be the case, with country border effects far stronger than the in- and out-migration constrained models estimate.

Model (vi) constrains inter-regional estimates to known (estimated) inter-country flows allowing us to explore the likely inter-regional international flows within Europe. This is an important development as for the first time we are able to examine, at a much higher resolution than previously possible, pressure points within the migration system. Not shown in this chapter, but evident in the results which are available through the web link given at the beginning of Section 6.6, are the regions in central and southern Spain which are likely destinations for the large influx of migrants from Romania, along with the areas of Romania which are equally as affected (if not more) socially, demographically and economically by these large flows of people. In the United Kingdom, we have shown the localised concentrations of migrant flows particularly into London, the South-East and East Anglia, and especially from regions in Poland.

While even in this optimum model, there are improvements that can be made. Now the modelling framework is in place, when improved inputs can be supplied to the model, then improved outputs can be very easily achieved. In this chapter we have concentrated on 2006, but data (albeit less comprehensive) for other years exist, and so a natural extension to this work would be to explore the temporal dynamics of particular sets of inter-regional flows in the system. Furthermore, in this analysis, we have chosen Europe to exemplify our models, but clearly we need not be limited to Europe – the model can easily be applied to estimate sub-national flows in a global context, opening up exciting possibilities for a more complete global sub-national understanding of migration. Implementation of a wider spatial system and broader temporal base means that the model framework introduced in this chapter should provide a useful tool for policy decisions related to demographic trends both in Europe and further afield.

References

Abel, G.J. (2010) Estimation of international migration flow tables in Europe. *Journal of the Royal Statistical Society: Series A (Statistics in Society)*, **173** (4), 797–825.

Batty, M. and Mackie, S. (1972) The calibration of gravity, entropy, and related models of spatial interaction. *Environment and Planning*, **4** (2), 205–233.

Boyle, P.J., Flowerdew, R. and Shen, J. (1998) Modelling inter-ward migration in Hereford and Worcester: The importance of housing growth and tenure. *Regional Studies*, **32** (2), 113–132.

Cangiano, A. (2011) *Demographic Objectives in Migration Policy-Making, The Migration Observatory*, University of Oxford, Oxfordhttp://migrationobservatory.ox.ac.uk/sites/files/migobs/Demographic%20Objectives%20Policy%20Primer.pdf (accessed 12 January 2016).

Convey, A. and Kupiszewski, M. (1995) Keeping up with Schengen: Migration and policy in the European Union. *International Migration Review*, **29** (4), 939–963.

De Beer, J., Van der Gaag, N., Van der Erf, R., Bauer, R., Fassmann, H., Kupiszewska, D., Kupiszewski, M., Rees, P., Boden, P., Dennett, A., Stillwell, J., De Jong, A., Ter Veer, M., Roto, J., Van Well, L., Heins, F., Bonifazi, C. and Gesano, G. (2010) *DEMIFER - Demographic and Migratory Flows affecting European Regions and Cities*, Applied Research Project 2013/1/3. Final Report, ESPON and NIDI. http://www.espon.eu/main/Menu_Projects/Menu_AppliedResearch/demifer.html (accessed 12 January 2016).

Dennett A. and Wilson A. (2013) A multi-level spatial interaction modelling framework for estimating inter-regional migration in Europe. *Environment and Planning A* **45**, 1491–1507. http://www.envplan.com/abstract.cgi?id=a45398

Diplock, G. and Openshaw, S. (1996) Using simple genetic algorithms to Calibrate Spatial Interaction Models. *Geographical Analysis*, **28** (3), 262–279.

Flowerdew, R. (2010) Modelling migration with poisson regression, in *Technologies for Migration and Commuting Analysis: Spatial Interaction Data Applications* (eds J. Stillwell, O. Duke-Williams and A. Dennett), IGI Global.

Fotheringham, A.S. (1981) Spatial structure and distance-decay parameters. *Annals of the Association of American Geographers*, **71** (3), 425–436.

Fotheringham, A.S., Rees, P., Champion, T. et al. (2004) The development of a migration model for England and Wales: Overview and modelling out-migration. *Environment and Planning A*, **36** (9), 1633–1672.

Fotheringham, A.S. and O'Kelly, M.E. (1989) *Spatial Interaction Models: Formulations and Applications*, Kluwer Academic Publishers.

Harland, K. (2008) *Journey to Learn: Geographical Mobility and Education Provision*, University of Leeds.

He, J. and Pooler, J. (2003) Modeling China's province-to-province migration flows using spatial interaction model with additional variables. *Geographical Research Forum*, **23**, 30–55.

Knudsen, D.C. and Fotheringham, A.S. (1986) Matrix comparison, goodness-of-fit, and spatial interaction modeling. *International Regional Science Review*, **10** (2), 127–147.

Kraler, A., Jandl, M. and Hofmann, M. (2006) The evolution of EU migration policy and implications for data collection, in *Towards the Harmonisation of European Statistics on International Migration (THESIM)* (eds N. Poulain, N. Perrin and A. Singleton), Université Catholique de Louvain–Presses Universitaires de Louvain, Louvain-la-Neuve, pp. 35–75.

Kupiszewska, D. and Kupiszewski, M. (2010) *Deliverable 4 – Multilevel Scenario Model, DEMIFER – Demographic and Migratory Flows Affecting European Regions and Cities*, ESPON & CEFMR, Warsaw. http://www.espon.eu/export/sites/default/Documents/Projects/AppliedResearch/DEMIFER/FinalReport/DEMIFER_Deliverable_D4_final.pdf (accessed 12 January 2016).

Mathä, T. and Wintr, L. (2009) Commuting flows across bordering regions: A note. *Applied Economics Letters*, **16** (7), 735–738.

Olsson, G. (1970) Explanation, prediction, and meaning variance: An assessment of distance interaction models. *Economic Geography*, **46**, 223–233.

Openshaw, S. (1998) Neural network, genetic, and fuzzy logic models of spatial interaction. *Environment and Planning A*, **30** (10), 1857–1872.

Plane, D.A. (1982) An information theoretic approach to the estimation of migration flows. *Journal of Regional Science*, **22** (4), 441–456.

Pooler, J. (1994) An extended family of spatial interaction models. *Progress in Human Geography*, **18** (1), 17–39.

Poulain, M., Perrin, N. and Singleton, A. (eds) (2006) *THESIM: Towards Harmonised European Statistics on International Migration*, Presses universitaires de Louvain, Louvain-la-Neuve.

Raymer, J. and Abel, G. (2008) *The MIMOSA model for estimating international migration flows in the European Union, Joint UNECE/Eurostat Work Session on Migration Statistics*, UNECE/Eurostat, Geneva, (Working paper 8). http://www.unece.org/stats/documents/ece/ces/ge.10/2008/wp.8.e.pdf (accessed 12 January 2016).

Rees, P. (1977) The measurement of migration from census and other sources. *Environment and Planning A*, **9**, 257–280.

Singleton, A., Wilson, A. and O'Brien, O. (2010) Geodemographics and spatial interaction: An integrated model for higher education. *Journal of Geographical Systems*, **14**, 1–19.

Stillwell, J. (1978) Interzonal migration: Some historical tests of spatial-interaction models. *Environment and Planning A*, **10**, 1187–1200.

Taylor, P.J. (1983) *Distance Decay in Spatial Interactions, CATMOD, 2*, Geo Books, Norwich.

Williams, P.A. and Fotheringham, A.S. (1984) *The Calibration of Spatial Interaction Models by Maximum Likelihood Estimation with Program SIMODEL*, Dept. of Geography, Indiana University.

Wilson, A. (1970) Entropy in urban and regional modelling, in *Monographs in Spatial and Environmental Systems Analysis* (eds R.J. Chorley and D.W. Harvey), Pion, London.

Wilson, A. (1971) A family of spatial interaction models, and associated developments. *Environment and Planning A*, **3**, 1–32.

7

Estimating an Annual Time Series of Global Migration Flows – An Alternative Methodology for Using Migrant Stock Data

Adam Dennett

7.1 Introduction

Globally, international migration flow data are poor. Putting aside issues of data quality and definitional nuances for one moment and concentrating just on coverage, it is fair to say that little is known of the annual flows of people between the circa 250 countries which make up the political map of the world. If we imagine a large two-dimensional data matrix with each of the world's counties heading the rows (as origins) and columns (as destinations), with the interior cells of the matrix containing counts of migrants moving between each country over the course of a year, then for any random year, the chances that there are non-zero data filling even a small proportion of the cells will be low.

Even if the matrix happens to be representing flows in the first or second year of a decade when, curiously, many of the countries that still carry out a full population census do so almost in synchronisation with each other, coverage would be patchy. The matrix may be slightly more populated with data than in other years, but it would very likely be characterised by vertical lines of data with far fewer horizontal lines – this is because Censuses are quite good at counting people into countries, but rather poor at counting people out (it is hard to count someone after they have left).

Some sections of the matrix, even in a non-census year, would be rather more populated with data than others. In the section containing the countries of Europe, it would be relatively complete across a number of years. National statistical agencies in EU countries have been bound

by legislation for sometime now to supply the European statistical agency, Eurostat, with estimates of immigration. Many of the previous empty cells in this section of the matrix have
been filled by various academic projects (Abel, 2010; Raymer and Abel, (2008a,b); de Beer
et al., 2010; Raymer et al., 2013; Wiśniowski et al., 2013). These projects have been successful in both harmonising immigration and emigration estimates between countries (estimates
which, among other things, vary due to definitional differences and differences in the methods
or systems used to record these moves) and estimating flows between countries where data are
incomplete.

Flows into or out of some of the more economically developed countries in the world (some
of the matrix margins rather than the internal cells) will also be a little more detailed. Members
of Organisation for Economic Co-operation and Development (OECD), for example, supply
the organisation with information on total inflows and/or outflows of migrants and have been
doing so for a decade or so.

Some sections of the matrix would be particularly sparse and not just because there are some
combinations of countries between which people do not often migrate (e.g. those with small
populations), but because detailed censuses, population registers, re-purposed administrative
data or border crossing surveys are simply not available to provide estimates of the true flows
that take place. Compounding this lack of statistical infrastructure, parts of the world such as
those recently ravaged by war or civil unrest experience a frequent ebb and flow of refugees
and more permanent migrants across porous borders which are exceedingly hard to measure
accurately.

Observing this situation from afar, one may be forgiven for thinking that the poor state of
international migration flow data must be indicative of a lack of importance – surely governments across the world, if they were really interested in these flows, would put in place better
systems for capturing population movements? But can this really be the case? Among countries in Europe, North America and Australasia, issues such as immigration are rarely far from
the top of the political agenda, indeed a recurrent news item in the United Kingdom will be
the latest cry from a different Member of Parliament, pressure group or think tank that the
migration statistics in the country are 'not fit-for-purpose'.[1] While the concerns voiced aloud
by politicians tend to be responses to voter anxieties about jobs, or posturing in relation to
those who are seen as deserving (or not) of the social benefits, some of the less politically
motivated worries about immigration can be traced back to the pragmatic concerns of plan-
ners who, understandably, want good information about the populations residing in different
parts of a country so that service provision can be met effectively. While births and deaths can
be recorded quite accurately, it is the volatility of migration and the quality of the statistics
that can cause the largest headaches (UKSA, 2009).

In this context, it is perhaps surprising that very little work has been carried out to try
and populate the sparse matrix of global migration flows, although the parochial nature of
policy-making means that, in the main, what is happening in the rest of the world is of little
immediate importance to those making national decisions. Even if local statistics are tackled,
the collective will from politicians to address the global lack of data appears to be, at best,

---

[1] http://www.parliament.uk/business/committees/committees-a-z/commons-select/public-administration-select-
committee/news/migration-statistics-report-published/,   http://www.leftfootforward.org/2012/11/migration-statis-
tics-uk-border-agency-still-not-fit-for-purpose/,   http://www.ft.com/cms/s/0/2be2f4a2-2524-11dd-a14a-000077b07-
658.html#axzz2bkRIYhUO

weak. Outside of official statistics and an apparent lack of governmental interest in global migration flows, academic interest has been not much more buoyant. Notable exceptions to this are the recent efforts by Abel (2013a,b) who has attempted to generate a series of 5-year migration flow estimates for the whole world by comparing changes in bilateral (origin/destination) migration stock data. Migrant stocks can be viewed as running totals of migration flows and as such are directly related. These stocks are recorded as counts of people who are either born in another country or citizens of another country; the latter can change with naturalisation (or less frequently denaturalisation) and so country of birth is often the more straightforward measure to use – indeed stocks are counts of foreign-born people in about 80% of the countries that collect these data.[2] Changes in these stocks (after accounting for deaths – and births in the case of foreign national stocks) are due to migration flows. Therefore, by comparing the stocks of migrants born in different countries at 5-year intervals, Abel (2013a) is able to generate estimates of the migration flows between these countries for 5-year time periods.

Abel's method is innovative and makes intelligent use of stock data which are published by the World Bank. In order to validate the effectiveness of the model employed, his model estimates are compared with the published net migration data. These comparisons show encouraging correlations, but as vastly different in- and out-migration data can produce the same net flow, there may be problems with validating against such statistics. As such it is possible, even with good net migration correlations, that poor flow estimates are generated. The reason Abel chooses to use net migration to validate his model is because other data are not available for the whole system – hence the existence of the problem in the first place. But flow data *do* exist for sections of the global matrix so it is possible to carry out a partial validation on the gross flows rather than the net – this is exemplified below.

The best inter-country migration estimates available at this time come from the IMEM project[3] (Raymer et al., 2013; Wiśniowski et al., 2013). This work is an evolution from the earlier work of the MIMOSA[4] project, which produced similar, but less accurate, estimates of the same intra-European flows. These annual flow estimates are for the years 2002–2008 which, unfortunately, cannot be aggregated to cover either the 2000–2005 or 2005–2010 periods Abel uses. However, whilst a little crude, dividing Abel's estimates by 5 gives an approximation of the estimated single-year flow volume which can be compared with each of the IMEM migration years. Now the IMEM data are not without issues and so any disagreement should not be taken as definitive evidence of a problem or otherwise, but they certainly may give an indication of possible complications in Abel's estimation process.

Table 7.1 shows that the estimates generated using Abel's method are some way away from the IMEM estimates for the 29 countries in the European system. Total annual flow volumes (around 400,000 between 2002 and 2008) are a long way from the figures of between 1.2 and 1.9 million estimated by IMEM. More importantly, the correlations represented by overall coefficient of determination (R^2) values for the 841 origin/destination interactions are low (2002 and 2007 chosen as close to the mean year for each 5-year period). The R^2 value gives an

---

[2] http://www.migrationobservatory.ox.ac.uk/data-and-resources/data-sources-and-limitations/unpd-international-migrant-stock-data.
[3] http://www.cpc.ac.uk/research_programme/?link=IMEM_project.php.
[4] http://mimosa.gedap.be/.

Table 7.1 Comparison of Abel's 5-year migration flow estimates with IMEM data

| Year | IMEM data | Abel's 5-year estimates/5 | | | |
| | | Total | | R^2 | |
	Total migrants	2000–2005	2005–2010	2000–2005	2005–2010
2002	1,246,762	407,792		0.24	
2003	1,257,237	407,792			
2004	1,453,802	407,792			
2005	1,494,601	407,792	419,727		
2006	1,639,059		419,727		
2007	1,849,528		419,727		0.15
2008	1,860,165		419,727		

indication of how close Abel's estimates replicate the overall structure of the IMEM data – the variations in the large and small flow volumes between different origins and destinations – and it shows that they are not very close.

It is clear then that there is scope to improve upon the estimates of global migration flows that are presently available in terms of both their temporal resolution (annual estimates would be preferable to 5-year transitions) and their distribution. Taking the latter, we can observe that while Abel makes use of bilateral migrant stock data collated by the World Bank/United Nations, his method employs an inverse distance function in order to distribute flows estimated from differencing stocks and disregards perhaps the most valuable information contained within: the large hints towards likely origin/destination interactions contained within the accumulation of historical flows. Using distance to distribute migration flows within a spatial interaction model is, of course, well established (Flowerdew and Lovett, 1988; Willekens, 1999; Stillwell, 1978; Fotheringham et al., 2004), but distance is only ever an imperfect proxy for the observed distribution of flows. Where an empirical record of historical flows exists, as it does with the bilateral stock data, then it should be possible to achieve far more accurate estimates using this as a distribution function. In his introduction, Abel discounts the use of stock data due to the comment from Massey et al. (1999) that migrant stock data can 'yield a misleading portrait of the current migration system' (Abel, 2013a, p. 507) as migration flows are far more volatile. But by not using the considerable amount of evidence from historical flows that are contained in stocks as to the potential preferences of migrants, then we are really ignoring some absolutely vital data. Distances between countries do not change at a perceptible rate (certainly not over the century or so at most that modern migration scholars are likely to be interested in) and so are even less flexible as a distribution term than stocks. Massey's assertion holds even less water when we look at the evidence. Studying Figure 7.1, there is a clear (log) linear relationship between bilateral stocks (as a proportion of all stocks) and bilateral flows (as a proportion of all flows) in the European system (where we have data to carry out such a comparison). As such, it is reasonable to assume a similar proportional relationship exists between stocks and flows in the rest of the global system, and this relationship can be employed to achieve more accurate estimates of global flows.

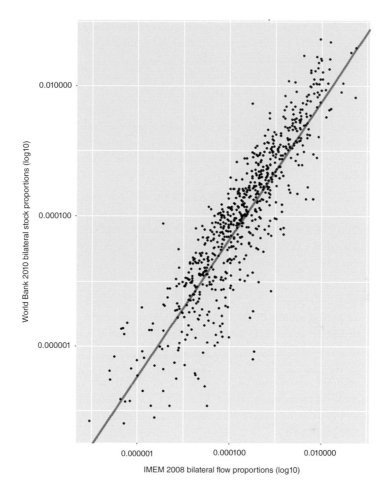

Figure 7.1 Comparison between bilateral stock and flow proportions for European countries $R^2 = 0.794$

7.2 Methodology

7.2.1 Introduction

The following section will detail a new methodology for taking advantage of the structure inherent in migration stock data to produce an alternative set of global estimates. The method is very straightforward and makes a number of crude assumptions, but despite this, the outcomes are still an improvement on current best estimates.

7.2.2 Calculating Migration Probabilities

The World Bank publishes bilateral migrant stock data for the years 1960, 1970, 1980, 1990, 2000, 2010[5] – these are data collated from censuses and population registers and represent

---

[5] http://data.worldbank.org/data-catalog/global-bilateral-migration-database.

total stocks at these years, primarily for foreign-born migrants (Özden et al., (2011)). The first stage of the new estimation process is to calculate bilateral migration rates or probabilities relative to the total stock population:

$$PS^t_{ij} = \frac{S^t_{ij}}{S^t} \tag{7.1}$$

where PS^t_{ij} is the probability of being born in country i and currently residing in country j at a particular time t, which is the ratio of the stock of migrants born in country i and currently residing in country j, at this time S^t_{ij}, to the total global stock S of migrants residing in a different country to their birth:

$$S^t = \sum_i \sum_j S^t_{ij} \tag{7.2}$$

where the probability of a migration flow between country i and country j in any given year, PF^t_{ij} can be calculated in a very similar way such that

$$PF^t_{ij} = \frac{M^t_{ij}}{M^t} \tag{7.3}$$

where M^t_{ij} is the observed migration flow (transition) between origin country i and destination country j in year t and M^t is the total of all migration flows in the system at this time:

$$M^t = \sum_i \sum_j M^t_{ij} \tag{7.4}$$

If we assume that the relationship in Figure 7.1 holds for the whole global system, then we can say that it is approximately the same as the cumulative historical flow probability captured by bilateral stock relationship PS^t_{ij}

$$PF^t_{ij} \approx PS^t_{ij} \tag{7.5}$$

Therefore, if we are able to calculate or estimate the total number of migrants moving around the world in a given year, then it is elementary to estimate bilateral migration flows, $\hat{M}^t_{ij}$, from the stock probabilities:

$$\hat{M}^t_{ij} = M^t PS^t_{ij} \tag{7.6}$$

7.2.3   Calculating Total Migrants in the Global System

Equation (7.6), in effect, is equivalent to the unconstrained (or total constrained) spatial interaction model proposed by Wilson (1971), but it does of course leave the non-trivial problem of estimating M^t. One obvious solution would be to collate data on immigration or emigration. The widely available net migration statistics for countries would suggest that these data are in existence, as conventionally net migration data are calculated as immigration minus emigration. Consider Table 7.2 that represents a sample migration system:

In this system with total immigration (D_j), emigration (O_i) and represented as the inner set of matrix margins (superscript for time, t, is implicit), net migration (N_i and N_j) is calculated such that in this closed system:

$$\sum_i N_i = \sum_j N_j = 0 \tag{7.7}$$

Table 7.2 Sample migration flows between countries in a three-country system

M_{ij}	Country 1	Country 2	Country 3	O_i	N_i
Country 1	0	120	300	420	−295
Country 2	50	0	92	142	178
Country 3	75	200	0	275	117
D_j	125	320	392	837	
N_j	295	−178	−117		

where

$$N_i = D_j - O_i \tag{7.8}$$

$$N_j = O_i - D_j \tag{7.9}$$

and which also means that

$$\sum_{i>0} N_i = \left| \sum_{i<0} N_i \right| \tag{7.10}$$

where we have a full and complete data set:

$$D_j = \sum_i M_{ij} \tag{7.11}$$

$$O_i = \sum_j M_{ij} \tag{7.12}$$

As such

$$\sum_i O_i = \sum_j D_j \tag{7.13}$$

In demographic accounting, the net migration balance is conventionally calculated as for N_i (in-migration minus out-migration) rather than N_j. However, where data on in- or out-migration are unavailable or unreliable, net migration is also calculated as the residual difference between the overall population growth rate and the rate of natural increase (births minus deaths) (UN, 2013). In the latest revision of the UN global net migration estimates,[6] of the 201 countries for which net migration data are available, 75% (148) have their statistics calculated in this way.

Now this, of course, poses a problem if we are hoping to use immigration or emigration statistics to calculate the total number of migrants – the relative abundance of net migration statistics belies this lack of immigration or emigration information. As is evident in Table 7.2, the positive or negative net balance of migrants will always be some way short of the total migrants – although the extent of the undercount is unknown. The impression from Table 7.2 is that the positive or negative net balance (295 migrants) is quite a long way short of the total migrants in the system (837). But this stylised system has migrants flowing in both directions

---

[6] http://esa.un.org/wpp/excel-data/migration.htm.

from each country in a relatively balanced way – in reality, for most bilateral country inter-actions, net migration balances will be asymmetric, especially where flows are between the Global South and countries in the more economically developed regions of the world.

So the question is: how asymmetric are the flows? Are they asymmetric enough for it to be reasonable to use net balance data as a proxy for total migration flows in our system? We know that the net balance will always be an underestimate of the true number of migrants, but it is, at least, an empirical observation. If the underestimation is too severe, then it will not be practical to use. Figure 7.2 gives us some clues as to the level of underestimation by comparing the positive net migration balance between 1965 and 2010 with the total migration estimated by Abel as the difference in foreign-born migration stocks (adjusted for deaths) between two successive time periods.

Figure 7.2 shows the relationship between the UN net migration data and the total migrant estimates derived from Abel's (2013b) flow estimates. There is clearly an association over

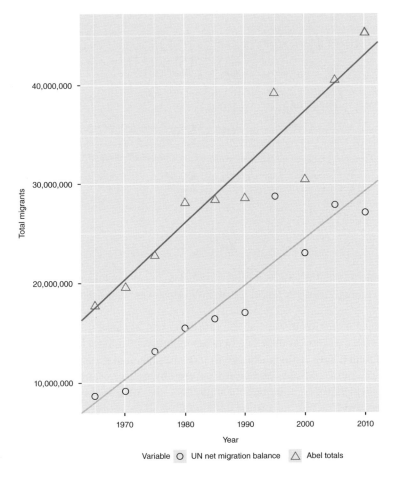

Figure 7.2 Comparison between total global migration flows (5-year periods) as estimated by Abel (2013b) and UN net migration positive balance (5-year periods), 1965–2010

time between the two sets of statistics, which is broadly linear. Both totals increase at roughly the same rate over the period, and there are also similarities in the flattening out in terms of growth between 1980 and 1990, as well as a similar outlier in 1995. The difference between the two total estimates remains at around 10 million migrants between 1965 and 2005, with the absolute difference only really departing from this gap in 2010 when the difference is closer to 20 million. One interesting outcome of plotting these two data sets against each other is that we can get a sense of the asymmetry of net migration balances over time. The closer the two statistics for a given year, the more asymmetric the flows in the system (as there is less cancelling of immigration with emigration) – the further apart, the less asymmetric (i.e. people tend to be migrating to and fro from countries in both directions). As we can see in Figure 7.2, the gap is widening slightly over time, signifying that people are less likely to flow just in one direction. Given the increase in globalisation and cyclical migration over this period, along with the overall increase in flows, this is probably what we would expect.

7.2.4 Generating a Consistent Time Series of Migration Probabilities

The next stage in the estimation process is to use the estimates of M^t to generate sets of bilateral flow estimates, $\hat{M}^t_{ij}$, using the corresponding stock probabilities, PS^t_{ij}. To do this for the years where data exist is very straightforward, although some added complications for time periods emerge. As previously outlined, UN (World Bank) bilateral migration stock data (and therefore probabilities) exist for six single years, on the decade, from 1960 to 2010. These single years are represented with the superscript index, t. Total migrant data estimated from either net migration balances or Abel's (2013b) data set exist for 5-year periods, T, where $t \in T$ and

$$M^T = \sum_{t \in T} M^t \qquad (7.14)$$

which means that as it stands:

$$\hat{M}^t_{ij} \neq M^T PS^t_{ij} \qquad (7.15)$$

Therefore, in order to proceed, we need to harmonise the time periods for either the stock probabilities or the total migration estimates, so that we arrive at either the estimate shown in Equation (7.6) or an alternative:

$$\hat{M}^T_{ij} = M^T PS^T_{ij} \qquad (7.16)$$

Either way, we are presented with two options: estimate PS^T_{ij} from PS^t_{ij} or estimate M^t from M^T – both of which have options which will influence the final flow estimates. We will begin with the estimation of M^t from M^T (consider Figure 7.3). Figure 7.3 shows the data points for the total flow estimates, M^T, from Abel's data, where $T = \{1965, 1970, \ldots, 2010\}$. Our first task is to decide whether M^T is the true value of M^T or just an approximation. In Figure 7.3, we can observe from the plotted points that while in general the value of M^T increases over time, there is noticeable fluctuation in the values. These might be real fluctuations, but perhaps more probably they correspond to errors in data collection. If we believe that the trend is more probable than the observed data, then new values of M^T could be calculated. The most obvious way to do this, as shown by the straight purple line, is to fit a regression line through the points, minimising the sum of the squared differences between the line and the residual data points.

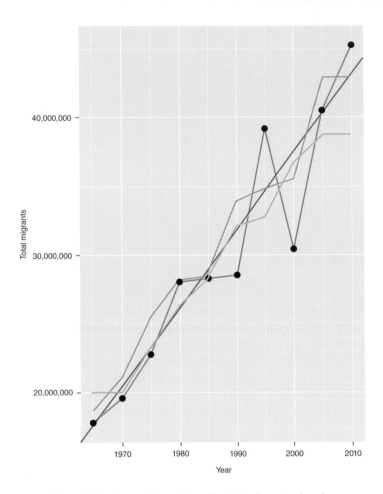

Figure 7.3 Smoothing options for total migrant estimation

Alternatively, we can retain some of the fluctuation in the original values of M^T by using some kind of moving average. The pink and blue lines in Figure 7.3 demonstrate a two-case and three-case moving average, respectively. Clearly, the choice of starting values for M^T will influence any subsequent calculation of M^t.

Once we have settled on a set of values for $M^T, M^{T+1}, \ldots, M^{T+n}$, then assuming a linear trend between any of the values of M^T, we are able to estimate M^t. Perhaps the most straightforward (although crude) option is to assume that the regression line is the best estimate of all values of M^T. When M^T is the sum of all flows during a 5-year period, then

$$\overline{M^{t \in T}} = \frac{M^T}{5} \tag{7.17}$$

Using the values shown in Figure 7.3 (the purple linear regression line), for $T = 2010$, $M^T = 43,057,000.\therefore \overline{M^{t \in T}} = 8,611,400$. For $T = 1965$, $\overline{M^{t \in T}} = 3,520,100$. As we are assuming the same linear relationship between all values of M^t, then by simply calculating the slope and intercept values, we are able to estimate total migrants for any year in the range. As $M^{t \in T}$ is the arithmetic mean of the 5 previous years summing to M^T, then where $T = 2,010$, $t = 2,007.5$.

7.2.5   Producing Annual Bilateral Estimates

We now have a series of total flow estimates for all years between 1965 and 2010 and bilateral flow probabilities, PS_{ij}^t, for 6 years, t: 1960, 1970, 1980, 1990, 2000 and 2010. Using Equation (7.6), it is straightforward to apply these probabilities to estimate bilateral flows for each of these 6 years. Interpolating the probabilities for intervening years between each known decade means that a full set of estimates can be produced for all intervening years.

7.3   Results and Validation

7.3.1   Introduction

The aim of this estimation exercise was to generate a set of global bilateral flow estimates with values more accurate than others presently available. Having generated a new time series of estimates, we are now able to assess their validity (as far as is possible). In addition to the IMEM data for Europe mentioned earlier, some empirical flow data exist for a small set of global countries and are collected by the United Nations.[7] Kim and Cohen (2010) collate some of these data for analysis in their paper and it is these data we make use of here. Kim and Cohen's outflow data are recorded flows from 13 developed world countries (Australia, Belgium, Croatia, Denmark, Finland, Germany, Hungary, Iceland, Italy, New Zealand, Norway, Sweden and the United Kingdom) between 1950 and 2007. Their inflow data are to these same countries, plus Canada, France, Spain and the United States over the same time period (although for both inflows and outflows, not all years are available for every origin/destination pair).

7.3.2   IMEM comparison

As noted earlier, the comparison with IMEM data is not ideal as we are having to divide Abel's estimates by 5 to get an approximation of the 1-year migration estimates. If the comparisons are close, then this will be more of a problem than if there are significant differences between the estimates. Table 7.3 presents three key statistics for the years 2002 and 2007: the total estimated migrants for the 841 interactions; the coefficient of determination (R^2) for all flows compared with the equivalent IMEM data – which gives an indication of the overall correlation; and the root mean squared error – which gives an indication of the accuracy of the estimates.

Immediately apparent is that both estimation methodologies underestimate flows in both years quite considerably, but the new methodology places more migrants in the system and gets a little closer to the IMEM estimate. In terms of the structure of the migration system, the new estimates perform much better, with R^2 values of 0.44 and 0.51 – significantly better than the 0.24 and 0.15 shown in Abel's data – normalising the skewed distribution with a $\log_{10}$

---

[7] http://www.un.org/en/development/desa/population/migration/data/empirical2/index.shtml.

Table 7.3 Goodness-of-fit statistics for new estimates and Abel's estimates when compared with flows in Europe in 2002 and 2008 from the IMEM project

Year	Goodness of fit	IMEM	New estimates	Abel estimates
2002	Total	1,246,762	673,056	407,792
	R^2		0.44	0.24
	R^2 log(data)		0.77	0.40
	RMSE		3679	4533
2007	Total	1,849,528	676,791	419,727
	R^2		0.51	0.15
	R^2 log(data)		0.75	0.40
	RMSE		5923	7557

transformation does not change this relationship, although does improve the R^2 for both sets of estimates. This is perhaps not surprising, given the aggregate structure of the migration systems implicit in the flow probabilities used to generate these estimates. Finally, the root mean squared error shows that the new estimates are also more accurate than Abel's with much lower error.

Overall, it is clear that the new method produces better results for Europe, but exploring the data, it is interesting to see where the errors are being made with both estimates – doing so confirms the strengths and weaknesses of both methods. For example, in 2002, the IMEM estimate for flows from Ireland into Britain is 15,230, whereas the Abel estimate is 56 and the new estimate some 27,549. In terms of the magnitude of the error, both are a long way away, but the overestimate of the new method is indicative of the historical stocks of Irish migrants in Britain. Abel's huge underestimate exposes the problems in dealing with foreign born/versus foreign national stock definitions and the complex patterns of national allegiances among people living in Northern Ireland and the Irish Republic that are very hard to capture with the data to hand. Other similar examples in the data abound (Poland to Germany migration; Italy to Switzerland) and perhaps point to the importance of historical flows when trying to predict what is more likely at another time.

7.3.3   UN Flow Data Comparison

Comparison with IMEM data suggests that these new estimates are potentially better than other global estimates currently available, but comparison with alternative flow data confirms this. Using the time series UN flow data from the Kim and Cohen ((2010)) paper and aggregating both the new estimates and these UN flows to 5-year totals, we can compare the flows directly with Abel's 5-year estimates (as opposed to having to disaggregate Abel's data into single years in the European validation exercise). Figure 7.4 shows the initial results of these comparisons for eight 5-year periods between 1970 and 2005. The y-axis on both sets of graphs features the UN flow data, aggregated from the 5 preceding years and displayed on a log scale. Figure 7.4a plots these UN data against the Abel estimates and Figure 7.4b plots the new estimates (again, both on a log-scale). Table 7.4 provides accompanying descriptive statistics for the relationships between the data.

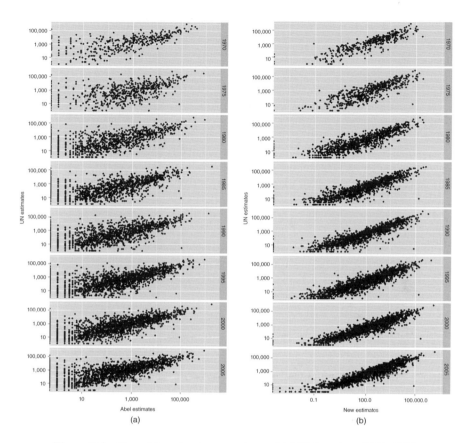

Figure 7.4 Flow data estimate comparison with UN flow data, 1970–2005

Table 7.4 Descriptive statistics for flow data estimate comparison with UN flow data, 1970–2005

5-year period end	n	Total migrants			RMSE		R^2 ($\log_{10}$)	
		UN	New estimates	Abel estimates	New estimates	Abel estimates	New estimates	Abel estimates
1970	399	6,692,562	2,027,441	4,386,639	58,248	47,103	0.77	0.47
1975	586	6,594,886	2,650,419	5,318,863	45,598	43,309	0.68	0.27
1980	1044	6,461,809	3,339,735	6,308,161	24,662	31,769	0.75	0.46
1985	1214	6,522,112	4,359,780	6,763,096	14,513	48,399	0.77	0.47
1990	1364	8,777,383	5,064,034	8,137,808	22,606	41,159	0.79	0.47
1995	1902	10,793,979	7,033,409	12,187,323	18,489	28,914	0.80	0.50
2000	1963	11,584,562	9,270,876	13,712,391	18,384	41,912	0.84	0.57
2005	1908	14,620,087	9,912,514	12,303,871	22,646	26,172	0.84	0.53

Studying Figure 7.4, it is clear for all 5-year periods that the new estimates on Figure 7.4b are generally closer to the UN estimates than Abel's estimates on Figure 7.4a – in all graphs, points are clustered far more closely to the 45° line. In every time period with Abel's estimates, larger flows (towards the right of the graph) are estimated more accurately than smaller flows, whereas with the new estimates, the patterns of over- and underestimation remain relatively stable.

Studying the goodness-of-fit statistics in Table 7.4, a number of points can be made. Firstly, it should be noted that the closer to 1970, the fewer data points are available for comparison – in each case, n relates to the number of origin/destination pairs that are commonly found in each data set – these pairs will be combinations of all possible origin countries and the 17 inflow countries mentioned earlier (inflow data being more reliable than outflow data – although similar patterns are apparent with outflow data as well).

With the new estimates, R^2 values are significantly higher for every 5-year period, although this is also apparent from the graphs. On the whole, the new estimates underestimate the total flows in this section of the system, whereas the Abel estimates get a little closer. However, between 1980 and 2005, the new estimates have a much lower root mean squared error, consolidating the better R^2 values present. Only for the 5-year periods ending in 1975 and 1970 are the RMSE statistics for Abel's data better.

Overall, the impression is that for a whole range of origin destination combinations over quite a lengthy time period, the new estimation methodology described above outperforms the methodology established by Abel. This evidence from empirical UN flow data, taken along with the evidence from the European IMEM estimates, points fairly conclusively to the new methodology, while very simple, outperforming any other flow estimation methodology.

7.4   Discussion

This new methodology is able to generate a long time series of annual migration flows from historical bilateral stock data and estimates of the total migrants moving around the globe in each year. The method is exceedingly simple, but is able to produce better estimates of flows than data that are currently available elsewhere in the public domain.

Because they are better, we should not make the mistake of thinking they are actually very good. With this kind of estimation where we are looking to maximise the plausibility of the estimates across as many origin/destination combinations as possible, then it is inevitable that there will be residuals which are some way from the truth. Rather than dwelling on these individuals, it is perhaps more useful to think about where errors are most likely to occur. Table 7.5 shows the time series of estimated and observed flows between Italy and Australia between 1960 and 2006. The Italians form one of the largest immigrant groups in Australia, but most of this migration has been since World War II (Cresciani, 2003). The observed flows in the data we have record a large numbersof migrants at the beginning of the period, but drop off quite sharply at around 1970. This is a peculiar, but not uncommon in migration histories between countries (e.g. Jamaicans to the United Kingdom) where particular political or economic conditions lead to sudden peaks of migration activity which perhaps do not represent the longer-term picture – this is what Massey et al. (1999) were referring to when commenting that stocks are not necessarily representative of flows. As is apparent in Table 7.5, the result is that we have quite large overestimation of flows the farther away from the real migration peak we get. These distortions are particularly acute as the migrants who moved over from Italy

Table 7.5 Estimates versus recorded flows, Italy to Australia, 1960–2006

Year	New estimate	UN data	Year	New estimate	UN data
1960	3,841	18,360	1984	6,810	562
1961	4,058	16,795	1985	6,835	676
1962	4,284	13,427	1986	6,855	589
1963	4,520	12,903	1987	6,871	530
1964	4,768	10,309	1988	6,881	421
1965	5,030	11,420	1989	6,883	306
1966	5,308	12,888	1990	6,949	353
1967	5,600	15,042	1991	6,950	285
1968	5,906	13,175	1992	6,943	283
1969	6,226	10,224	1993	6,925	286
1970	6,558	7,573	1994	6,898	336
1971	6,605	5,874	1995	6,861	304
1972	6,646	3,426	1996	6,818	272
1973	6,679	2,931	1997	6,766	201
1974	6,704	2,389	1998	6,704	197
1975	6,720	1,365	1999	6,633	168
1976	6,727	1,318	2000	6,554	181
1977	6,725	1,598	2001	6,603	267
1978	6,714	1,282	2002	6,649	139
1979	6,697	1,044	2003	6,692	186
1980	6,672	1,696	2004	6,732	195
1981	6,712	1,381	2005	6,769	187
1982	6,748	553	2006	6,807	204
1983	6,781	492			

to Australia up until 1970 are likely to be still alive and thus boosting the stock. Adjustments could be made for these sorts of trends using information on when these bursts of migration activity occur in parallel with life tables, but we would be rapidly moving away from a quick and generalisable migration estimation model to a far more complex beast.

Further problems with the method as described earlier are the assumed linear relationships and subsequent linear interpolations which have been carried out in order to generate the estimates. For ease of implementation, it has been assumed that a number of migrants active in the world in a given year are linearly related to previous years, although given the superlinear growth in the global population thus far and the projected logistic pattern of the global future population, this may not be wise. The relatively short timescale dealt with here means that population growth looks somewhat linear, but while migration growth will be related in some way to population growth, other factors such as ease of global travel and increasing globalisation mean that a superlinear growth may be more appropriate.

Another issue with this method is that it is obviously deterministic, to the extent that the R^2 values between the stock probabilities and any observed flows will be exactly the same as the R^2 values between any derived migration estimates and any observed flows. This is not ideal. It would be preferable and more realistic to include some random fluctuation in the flows, although the exact method for achieving this would need to be researched. While it would not improve the deterministic nature of the results, incorporating the bilateral stock data into a spatial interaction model with more terms would increase the randomness of the estimates a little and would certainly make the data feel a little less synthetic.

7.5 Conclusions

A new method for estimating a series of global migration flow tables has been outlined here. Through using historical bilateral migrant stock data as flow probabilities, it has been shown that a very simple proportional distribution model can achieve superior migration flow estimates to the only other comparable set of flows currently available: those created by Abel (2013b) following the work of (Abel, 2013a). The new methodology is not perfect – it is shown to underestimate the total flows for the systems where we have comparable data and as stocks reflect historical flows, there are also problems with over- and underestimation between particular country combinations.

This new method should be viewed more as a step in the right direction for migration flow estimation rather than a definitive solution to the problem of knowing which countries people are likely to be migrating between in any given year. As mentioned, the single probability term in the equation means that the estimated flows are directly tied to it, with no variation at all. Including additional predictor terms in the equation such as those used in a number of explanatory migration models (Cohen et al., 2008; Abel, 2010; Fotheringham et al., 2004) would produce more variation and potentially improve the estimates still further. If available, however, historical stocks should definitely be used in the estimation process.

References

Abel, G.J. (2010) Estimation of international migration flow tables in Europe. *Journal of the Royal Statistical Society: Series A (Statistics in Society)*, **173** (4), 797–825.

Abel, G.J. (2013a) Estimating global migration flow tables using place of birth data. *Demographic Research*, **28** (18), 505–546.

Abel, G.J. (2013b) Estimates of global bilateral migration flows by gender between 1960 and 2010. *International Migration Review*. http://www.oeaw.ac.at/vid/download/WP2015_05.pdf.

Cohen, J., Roig, M., Reuman, D. and GoGwilt, C. (2008) International migration beyond gravity: A statistical model for use in population projections. *Proceedings of the National Academy of Sciences*, **105** (40), 15269–15274.

Cresciani, G. (2003) *The Italians in Australia*, Cambridge University Press.

de Beer, J., Raymer, J., van der Erf, R. and van Wissen, L. (2010) Overcoming the problems of inconsistent international migration data: A new method applied to flows in Europe. *European Journal of Population/Revue européenne de Démographie*, **26** (4), 459–481.

Flowerdew, R. and Lovett, A. (1988) Fitting constrained Poisson regression models to interurban migration flows. *Geographical Analysis*, **20** (4), 297–307.

Fotheringham, A.S., Rees, P., Champion, T. et al. (2004) The development of a migration model for England and Wales: Overview and modelling out-migration. *Environment and Planning A*, **36** (9), 1633–1672.

Kim, K. and Cohen, J.E. (2010) Determinants of international migration flows to and from industrialized countries: A panel data approach beyond gravity. *International Migration Review*, **44** (4), 899–932.

Massey, D.S., Arango, J., Hugo, G. et al. (1999) *Worlds in Motion: Understanding International Migration at the End of the Millennium*, Oxford University Press, Oxford.

Özden, C., Parsons, C.R., Schiff, M. and Walmsley, T.L. (2011) Where on earth is everybody? *World Bank Economic Review*, **25** (1), 12–56.

Raymer, J. and Abel, G. (2008a) *Methods to Improve Estimates of Migration Flows - the MIMOSA Model for Estimating International Migration Flows in the European Union*. Geneva: UNECE/Eurostat work session on migration statistics - Working Paper, 8.

Raymer, J. and Abel, G. (2008b) *The MIMOSA Model for Estimating International Migration Flows in the European Union*. In Joint UNECE/Eurostat Work Session on Migration Statistics. Geneva: UNECE/Eurostat.

Raymer, J., Wiśniowski, A., Forster, J.J. et al. (2013) Integrated modeling of European migration. *Journal of the American Statistical Association*, **108** (503), 801–819.

Stillwell, J. (1978) Interzonal migration: Some historical tests of spatial-interaction models. *Environment and Planning A*, **10**, 1187–1200.

UKSA (2009) *Migration Statistics: The Way Ahead*, UK Statistics Authority, London.

UN (2013) *World Population Prospects: 2012 Revision - Metadata, ed*, U. N. P, Division.

Willekens, F. (1999) Modeling approaches to the indirect estimation of migration flows: from entropy to EM. *Mathematical Population Studies*, **7** (3), 239–278.

Wilson, A. (1971) A family of spatial interaction models, and associated developments. *Environment and Planning A*, **3**, 1–32.

Wiśniowski, A., Bijak, J., Christiansen, S. et al. (2013) Utilising expert opinion to improve the measurement of international migration in Europe. *Journal of Official Statistics*, **29**, 583.

Part Four

Security

Part Four

8

Conflict Modelling: Spatial Interaction as Threat

Peter Baudains and Alan Wilson

8.1 Introduction

Outbreaks of conflict, whether stemming from interstate or civil wars, insurgencies or civil unrest, continue to dominate news reports around the globe. Their onset and evolution is traditionally discussed using anecdotal perspectives, rather than by employing explicit models to seek out underlying mechanisms or patterns that might be exploited from a policy perspective. However, there has been a recent dramatic increase in the quantity and quality of explicit models detailing various aspects of conflict. This is partly due to increased data availability, which is crucial for modelling as it enables the development of models that are empirically consistent, and partly due to an increased range of sophisticated modelling techniques. Our understanding of conflict processes can be improved through such models. This may in turn improve the way in which interventions are planned. Some have even suggested that by using modern modelling techniques to investigate problems of crime, war and terrorism, the number of fatalities associated with such events can ultimately be reduced (Helbing et al., 2015).

In the first part of this chapter, we examine some of these models and investigate the extent to which they can be used in a policy context to aid decision-making. In Sections 8.2–8.4, we consider three domains in which models may be utilised: operational decision-making; understanding the underlying causes of conflict; and the identification and forecasting of global conflict hotspots. Within each of these domains, we discuss the state of the art not just from a research perspective (i.e. with a focus on developing the best model) but also from an operational or policy perspective (how the model might be used to help decision-making). These three domains are chosen as examples where the availability of fine resolution data on conflict events can improve the application of models in the policy domain. The studies cited in each of these domains are not intended to represent an exhaustive review of the literature in

Global Dynamics: Approaches from Complexity Science, First Edition. Edited by Alan Wilson.
© 2016 John Wiley & Sons, Ltd. Published 2016 by John Wiley & Sons, Ltd.

each area, but rather are selected to highlight how different modelling approaches might be applied.

In each case, we note that conflict can occur over very different scales. Since small disputes can escalate quickly, the study of global dynamics has to consider conflicts across all of these scales. Complications arise because a disagreement between two individuals – a conflict at the small scale – is likely to have very different underlying causes and implications to a prolonged large-scale interstate war. There is increasing evidence, however, that similar tools can be used to investigate different types of conflict. Johnson et al. (2013), for example, build on a number of previous studies to demonstrate consistent distributions of event severity and timing across a large number of conflicts that occur over a wide range of scales. They suggest that confrontation between different hostile parties can be considered in terms of similar underlying mechanisms, almost regardless of the actual conflict type. Other studies have pointed to particular patterns of conflict in space and time and have used analogies with similar patterning of events in Criminology (e.g. Braithwaite and Johnson, 2015) and Ecology (e.g. Brantingham et al., 2012), amongst others, to draw conclusions about the human behaviour responsible for generating such events.

In the second part of this chapter, we introduce the concept of measuring threat using a spatial interaction approach (Section 8.5). Inspired by the ability for analytical tools to be pertinent over a range of conflict types, we build on the idea that the threat on a particular area can be interpreted as a measure of the likelihood of observing future conflict events in that area. We argue that this framework is capable of representing a broad range of conflict processes and discuss how the model might be applied to each of the three policy domains discussed in this chapter. We conclude by offering some avenues for further research.

8.2   Conflict Intensity: Space–Time Patterning of Events

Quantitative analysis of the variation of conflict intensity in space and time can be used to detect regularities that can be exploited to aid decision-making in an operational setting. In this section, we consider examples of studies that identify such regularities and explain how their insights can be used during decision-making.

During a war or insurgency, certain locations are more likely to experience conflict than others and this conflict might be more likely to occur at certain times. High-frequency events, such as improvised explosive device (IED) attacks, exhibit consistent regularities. A number of studies have considered how event interdependence might be exploited to identify those areas and times more at risk. Townsley et al. (2008), for example, study IED attacks during the Iraq insurgency and show that a further IED attack is more likely to occur within 2 days and 1 km of a prior attack than at any other randomly selected time or location. Such insights have significant implications for security force deployment. Indeed, as the authors describe, their finding 'allows security forces and the police to organise a local response that, while having to mobilise quickly, is not required to remain in situ for periods longer than a few days'.

In Chapter 12 (see also Braithwaite and Johnson, 2012), a similar analytical approach based on tests for space–time interaction originally developed in the study of epidemiology is used to identify patterns of interdependence between insurgent attacks and counterinsurgent response. In particular, the authors identify the types of counterinsurgent actions that were most successful in reducing the likelihood of an insurgent attack. Braithwaite and Johnson find that more discriminatory counterinsurgent activities – such as a geographically targeted raid – were more likely to reduce the risk of further attacks than less discriminatory actions–such as the search and cordon of an area, which were often performed over a comparatively larger area (e.g. city block or equivalent).

In an effort to capture the regularities of these patterns within a predictive framework, Benigni and Furrer (2012) employ a Poisson process model to quantify the spatio-temporal interaction in the data with the explicit aim to aid operational decision-making. They focus on one stretch of road in Baghdad that was well known for being a dangerous route for counterinsurgent forces. Their model generates an evolving risk surface along this stretch of road, composed of space–time Gaussian kernel density estimators calibrated using historical attack data. The authors then demonstrate how this model can be used to plan the deployment of IED surveillance assets and patrol teams. When deploying surveillance assets, the probability that each team will come across an unexploded IED is maximised so that the IED can be identified, defused and removed. When deploying patrol teams, the threat of attack is minimised in an effort to ensure their safety on the route. The approach relies on the spatio-temporal clustering of IED attacks and the authors demonstrate the improvement in the short-term predictability of the model when compared to a model that treats space and time independently.

Zammit-Mangion et al. (2012) also employ spatio-temporal point process models to capture space–time dependencies of events associated with insurgent warfare. Their model is formulated differently from the one of Benigni and Furrer: sophisticated stochastic integro-difference equations are used to model the evolution of a spatial intensity function, which indicates the likelihood that an attack will occur in any given space–time window. The model is calibrated against the Afghan War Diary, a detailed description of events associated with the insurgency in Afghanistan released by Wikileaks. The model enables the presentation of some useful information for leaders in an operational setting: the growth and volatility of event activity in different geographic regions. In addition, the authors demonstrate how the model is able to successfully predict the evolution of the Afghan conflict in a statistical sense, which may be of use for policy (although they do caution that perhaps the reason for this is due to the irregular nature of the Afghan conflict itself and that the model may not be so successful when adversaries are more unified and organised).

In Chapter 10 (see also Baudains, 2015), a different type of point process model is calibrated against conflict data, called a Hawkes process. This model estimates the conflict intensity via a self-excitation mechanism. It is supposed that the occurrence of an event directly increases the likelihood of observing a further event for a certain period of time. The interactions between insurgents and counterinsurgents are explored via this process and the predictive capability of the model tested. Although some of the variation in the event timings cannot be explained by this simple mechanism, the model is able to offer some insight with regard to the prediction of future attacks. Such models could be used to aid decision-making concerning counter-insurgent deployment.

A further example of using the patterns in events during conflict to help decision-making is given by Shakarian et al. (2009), who introduce the Spatial Cultural Abductive Reasoning Engine (SCARE). In the example presented, this model enables the identification of IED weapons cache sites (where materials are collected and the weapon is made and stored until an attempted attack takes place) using only the locations of previous successful IED attacks. A similar approach called Geographic Profiling is used in the field of Criminology to identify likely residential locations of offenders based on the locations of the crimes that they commit (Rossmo, 2000). The model can be used to prioritise the locations of search and cordon operations for counterinsurgent forces.

These models all demonstrate how operational insights can be obtained from sophisticated statistical techniques combined with real-time data. They exploit event interdependency to identify the change in conflict intensity in space and time. This enables forecasting of likely locations of future attacks or cache sites at a resolution that is operationally useful for influencing deployment decisions. Determining the success of such models in this operational setting (e.g., with respect to the improved targeting of adversaries or the reduction in casualties from IED attacks) is a crucial next step for this type of research.

A limitation of the models discussed in this section is their inability to incorporate and test intricate causal mechanisms by which conflict events are thought to arise. They seek out patterns in event data, but rarely test explanations as to why those patterns might be arising. Deploying troops in accordance with some of the models' recommendations might improve the chances of being in the right place at the right time to prevent a potential attack; however, the models are unable to explore theories as to why events are more likely to occur in one area over another. Incorporating such an explanation can help to test our understanding as to why such attacks are being committed and might even help design policy targeted at reducing the grievances of those likely to commit such attacks. In the next section, we consider models that are constructed using a more theoretical perspective than those considered here.

8.3   Understanding Conflict Onset: Simulation-based Models

Models of social systems provide an opportunity to articulate theory within a quantitative framework that can be empirically validated. Simulation-based models in particular are well suited to testing theories concerning how conflict arises and evolves under different circumstances. Agent-based models (ABMs) allow the modeller to specify some mechanism by which individuals choose to engage in violence. It is then possible to test whether the emergent properties of the model, taken over a population of individuals, each acting autonomously and interacting with other agents, match the empirical record. This can then enable the evaluation of whether the proposed mechanisms provide a plausible explanation for the production of conflict events.

Traditionally, ABMs have been used as purely theoretical tools, designed to observe the emergent behaviour that arises from simple interactions. Recently, however, a number of models have been proposed that incorporate empirical data and are calibrated so that their outputs correspond as close as possible to the real world. If the model is considered to be a close fit to the data and a reasonable representation of the real-world process, then it may be possible to use it for policy exploration by considering the impact of proposed policies on the rate at which conflict events are produced. In this section, we discuss examples of such models in the existing literature and consider how each model can contribute to policy-making. In particular, we discuss two examples that have been used to examine the relationship between ethnic segregation and violent conflict.

An ABM is a model in which each component entity, which in many cases is taken to be an individual, is represented as an independent and autonomous agent. The model consists of a set of rules that describe how each agent behaves and, crucially, how each agent interacts with other agents and the environment. ABMs are well suited to exploring how a proposed behaviour at the individual level results in some observed emergent property at the societal level. A number of sophisticated ABMs with empirically driven modelling and validation procedures have explored the role of individual migration and the resulting spatial distributions of ethnic groups as a causal factor in the onset of conflict. Two examples are Weidmann and Salehyan (2013) and Bhavnani et al. (2014).

Weidmann and Salehyan (2013) model the evolution of violence in Baghdad for different periods during the US-led war in Iraq, beginning in 2003. The model is empirically derived in that it is initialised using residential data in Baghdad, which details whether an area is comprised of a majority of Shia or Sunni residents, or whether the region is more mixed. The model proceeds by letting a fixed percentage of insurgents within each area decide whether or not to commit violence against the other ethnic group. If an attack is successful, then it might prompt civilians to migrate to a new area, one with a lower number of attacks. The resulting violence and segregation patterns across the city are then tracked and parameter selection is achieved by minimising the errors associated with these observations between the simulation and the empirical data. The resulting parameter estimates can then tell us something about the violence in Baghdad. Specifically, the authors demonstrate that the patterns of violence and segregation in Baghdad were consistent with a process by which minority ethnic groups are more likely to be attacked when faced with a different majority ethnic group. Migration is also shown to be positively correlated with the level of violence experienced. These insights contribute to social science theories concerning conflict in mixed urban spaces. The models can also be used to address significant policy questions, such as determining whether segregation or migration of minority ethnic groups should be encouraged or, conversely, whether integration should be supported. Weidmann and Salehyan also go on to consider possible policing strategies for handling the resulting levels of ethnic violence, highlighting how the simulation can be used to test proposed strategies.

In Bhavnani et al. (2014), a model of segregation and violence in Jerusalem is presented more directly in the policy domain. The authors explore a number of counterfactual scenarios resulting from different policy decisions that could be made with regard to the way in which different districts within Jerusalem are put under varying levels of segregation and governed under different authorities. The model is distinct from the model of Weidmann and Salehyan in a number of ways, perhaps notably in the way that any civilian is able to commit violence if they find themselves in a particular set of circumstances and in the way that those circumstances can depend on factors such as the social distance and history of violence between two groups. A number of scenarios in which the movement between different regions is reduced are shown to have significant reductions on levels of violence according to the model. The authors suggest that such policies might be considered when social distance between groups is high and violence cannot be reduced by other means. The authors stress the limitations of the model, pointing out that if different policies are adopted it may change the underlying dynamics in a way not captured by the model. Nevertheless, the model demonstrates how empirical ABMs might be used to explore different policy options and contribute to the range of tools used by decision-makers.

8.4 Forecasting Global Conflict Hotspots

Another area in which conflict modelling can contribute to policy decision-making is concerned with global modelling. In the examples discussed so far, the model has typically focused on a small part of a country. This works well when there is an ongoing conflict within that region, but less well when considering policy that is required to address regions with merely a risk of the onset of conflict. There are a number of ongoing efforts to construct models that are capable of forecasting likely future locations of unrest, violence or conflict around the world. The most prominent examples incorporate large data sets within a statistical framework to identify early warning signals of future conflict in a particular country or region.

One example is the Integrated Crisis Early Warning System (ICEWS), supported by the US Department of Defense (O'Brien, 2010). Online news articles are used to automatically generate event data. Each event is coded according to a prescribed set of possible events and information such as the location of the event or the main actors involved in the event are also recorded. The ICEWS project then uses this data, as well as a wealth of other independent data sources, to build models capable of predicting five different events of interest: domestic political crises, rebellions, insurgencies, ethnic/religious violence, and international crises.

The Global Database of Events Language and Tone (GDELT; Leetaru and Schrodt, 2013) is another large data collection effort, whose aim is to construct event data from open media sources and then to make the resulting data freely available. A number of studies are beginning to exploit this information and make predictions concerning conflict around the world. Ward et al. (2013), for example, use online event data similar to GDELT within a mixed-effects logistic regression model to determine the likelihood of the occurrence of civil conflict at the monthly level between 1997 and 2011. Using a wide range of predictive performance metrics, the authors demonstrate high predictive capability, suggesting that the use of online data can indeed be used as an early warning signal for the onset of civil wars (another example is Hegre et al., 2013). A wide range of stakeholders including policymakers and investors could make use of such predictions when considering challenges such as troop deployments and development aid.

8.5 A Spatial Model of Threat

In the policy domains discussed in Sections 8.2–8.4, there are a wide range of geographic scales and motivating policy questions. One consistency, however, is how modelling the likely locations of future conflict events underpins every model. The geographic dependence and spatial interactions need to be accounted for when developing many conflict models and there is often no obvious approach for doing so. In this section, we introduce a spatial measure of threat between two adversaries who are distributed over some area or geographic region. We argue that this measure offers a versatile framework for incorporating spatial dependence in models of conflict. In particular, the measure may be incorporated into many of the models discussed earlier and, provided appropriate data are available, used to model the likelihood of future conflict events within any given region.

We begin by assuming that the conflict takes place in some Euclidean space or manifold, with each location $\mathbf{x}$ defined by a coordinate system so that $\mathbf{x} = (x^{(1)}, x^{(2)}, \cdots, x^{(n)})$. In many cases, for example those where just the geographic nature of the problem is of interest, n will be equal to 2 so that each location represents geographic coordinates. In this derivation, however, we retain generality by considering an arbitrary number of dimensions. These dimensions might later be specified according to a range of factors such as social distance or political similarity (i.e. the conflict may take place over arbitrary landscapes).

Between any two locations $\mathbf{x}$ and $\mathbf{y}$, we define $d(\mathbf{x}, \mathbf{y})$ to be some metric on the space. A metric takes two locations and produces a non-negative abstract measure of 'distance' between those locations. We suppose here that the conflict takes place between two well-defined groups and that one of these, group X, has members located at positions $\mathbf{x}_1, \mathbf{x}_2, \cdots, \mathbf{x}_N$. The adversary to group X, group Y, is assumed to have supporters at positions $\mathbf{y}_1, \mathbf{y}_2, \cdots, \mathbf{y}_M$. In other words, the first adversary X is located at N positions and the second adversary Y is located at M positions in the abstract space in which the conflict takes place.

In addition to a measure of distance between any two adversaries, we also require some measure of aggression or discontent at each location. This might be interpreted as the extent to which the member of each group is willing to engage in conflict against an adversary, with higher values indicating more extreme and hostile group members. We denote a measure of aggression of X at $\mathbf{x}_i$ by u_i and aggression of Y at $\mathbf{y}_j$ as v_j. The specification of this measure may vary depending on the application of the model. In a military setting, it might correspond to the amount of resources X has at $\mathbf{x}_i$ or, for ethnic conflict, it might be a measure of the number of previous attacks that have emanated from $\mathbf{x}_i$ (or, indeed, the ethnic composition at $\mathbf{x}_i$). Broadly, aggression is conceived to model the total level of threat that can emanate from an adversary at a particular location.

We assume that the presence of X at $\mathbf{x}_i$ exerts some threat on Y at $\mathbf{y}_j$ and that this is measured by T_{ij}^{XY}. Threat can be interpreted as the level of intimidation that is due to the presence of that adversary. In the case of ethnic violence, for example, threat may arise from the perceived risk of experiencing violence due to a different ethnic enclave located nearby.

We model T_{ij}^{XY} using an entropy-maximising spatial interaction model. These models have previously been used to study flows of physical objects such as people in the case of modelling migration (Dennett and Wilson, 2013) or money in the case of modelling retail trips (Wilson, 2008). To begin, we assume that the total threat on Y that can come from X at $\mathbf{x}_i$ is constrained by the level of aggression u_i. Thus,

$$\sum_j T_{ij}^{XY} = u_i \tag{8.1}$$

Following Wilson (2008), we assume that there exist constants C^{XY} and B^{XY} such that

$$\sum_{i,j} T_{ij}^{XY} d(\mathbf{x}_i, \mathbf{y}_j) = C^{XY} \tag{8.2}$$

and

$$\sum_{i,j} T_{ij}^{XY} \log \, v_j = B^{XY}. \tag{8.3}$$

Equation (8.2) can be interpreted as the total 'energy' in the threat system that supports the flows and Equation (8.3) is justified by its role in generating more threat between those adversaries that are more aggressive in the final model. Although these constraints do not admit simple interpretation, they have been shown to generate accurate flows in models of retail trip distributions in previous applications.

An estimate of T_{ij}^{XY} is then found by maximising the entropy measure:

$$S = - \sum_{i,j} T_{ij}^{XY} \log T_{ij}^{XY} \tag{8.4}$$

subject to the constraints (8.1)–(8.3). It can be shown that this gives

$$T_{ij}^{XY} = A_i u_i v_j^\alpha \exp\left(-\beta d(\mathbf{x}_i, \mathbf{y}_j)\right) \tag{8.5}$$

where

$$A_i = \sum_{j=1}^{M} v_j^\alpha \exp\left(-\beta d(\mathbf{x}_i, \mathbf{y}_j)\right) \tag{8.6}$$

for parameters α and β.

The model can be written as

$$T_{ij}^{XY} = u_i \frac{\exp\left(\alpha \ln v_j - \beta d(\mathbf{x}_i, \mathbf{y}_j)\right)}{\sum_{k=1}^{M} \exp\left(\alpha \ln v_k - \beta d(\mathbf{x}_i, \mathbf{y}_k)\right)} \tag{8.7}$$

a form that enables clear interpretation. To explain, the aggression u_i is exerted over the adversary Y as a series of flows T_{ij}^{XY}. The amount of u_i that is exerted on Y at $\mathbf{y}_j$ is weighted according to how the term

$$Z_{ij} = \alpha \ln v_j - \beta d(\mathbf{x}_i, \mathbf{y}_j) \tag{8.8}$$

varies over the range of locations of Y. If Z_{ij} is large for a given j (in comparison to the other locations of Y, indexed by k), then a large proportion of the aggression u_i will be directed towards Y at $\mathbf{y}_j$, resulting in a large flow of threat. Z_{ij} can be thought of as the *utility* gained by X at $\mathbf{x}_i$ in directing their aggression towards Y at $\mathbf{y}_j$. Their aggression will generate greater utility if it is directed towards those adversaries that are nearby and towards those adversaries which themselves have high levels of aggression. Furthermore, utility is often considered to represent benefits minus costs. In this setting, $\alpha \ln v_j$ is therefore the associated benefits obtained by directing threat towards $\mathbf{y}_j$ (i.e. the benefit associated with defending the aggression v_j) and the distance metric specifies the corresponding costs (i.e. the cost of exerting threat over some landscape or distance). The parameter α determines how this utility is influenced by the aggression of the adversary and β determines how the resulting threat varies with the distance metric. Large values of β imply that it is difficult to exert threat over any substantial distance and therefore any intimidatory effects will only be felt by those adversaries nearby.

Using this formulation of threat flows, it is possible to calculate the total threat from all locations of X exerted on Y at $\mathbf{y}_j$ by summing over the possible locations of X to obtain

$$\tau_j^Y = \sum_i T_{ij}^{XY} \tag{8.9}$$

$$= \sum_i \frac{u_i v_j^\alpha \exp\left(-\beta d(\mathbf{x}_i, \mathbf{y}_j)\right)}{\sum_k v_k^\alpha \exp\left(-\beta d(\mathbf{x}_i, \mathbf{y}_k)\right)}, \tag{8.10}$$

which holds for $j = 1, 2, 3, \cdots, M$. This measure can be interpreted as a weighted sum of aggression emanating from adversary X, weighted according to the amount of aggression of Y at each location (those locations of Y with a higher level of aggression will attract more threat from X, all other things being equal) and according to the ease with which threat can travel across the spatial region of interest (those locations of Y that are closer to X will experience higher levels of threat, all other things being equal).

A similar expression can be derived for the threat that Y exerts on X at $\mathbf{x}_i$, so that

$$\tau_i^X = \sum_j T_{ji}^{YX} \tag{8.11}$$

$$= \sum_j \frac{v_j u_i^{\gamma} \exp\left(-\delta d(\mathbf{y}_j, \mathbf{x}_i)\right)}{\sum_k u_i^{\gamma} \exp\left(-\delta d(\mathbf{y}_j, \mathbf{x}_k)\right)}, \tag{8.12}$$

for further parameters γ and δ.

The threats τ_j^Y and τ_j^X are location specific and capture spatial dependence via the metric d. In addition, they are general enough to be incorporated into a wide range of conflict models. In what follows, we discuss some of these applications as avenues for future research.

8.6   Discussion: The Use of a Spatial Threat Measure in Models of Conflict

In this section, we propose how the measure of threat introduced in Section 8.5 might be applied to each of the three policy areas discussed in Sections 8.2–8.4. In doing so, we demonstrate the versatility of this measure and present a number of opportunities for further research.

8.6.1   Threat in Models for Operational Decision-Making

Considering the application of threat within an operational setting, we first note that many of the models discussed in Section 8.2 were concerned with identifying the specific locations that might be at risk of experiencing future events such as IED attacks. The threat measures in Equations (8.10) and (8.12) can be used to model this risk for different locations and times. To do so, consideration must be given to defining the spatial configuration of the model, constructing the distance metric d and capturing the aggression measures u_i and v_j of the two adversaries using available data. In many cases, ideal measures for these variables cannot be used due to a lack of information on the locations and activities of adversaries. Nevertheless, proxy measures can be used to give some indication as to how threat might be distributed over space.

Deciding on an appropriate spatial configuration is the first modelling decision to be made. The resolution of the spatial units under consideration should be fine enough to obtain actionable insights, yet should be sufficiently coarse to operationalise the remaining components of the model via other data sources. Available data might be aggregated to certain geographic areas, for example. In terms of operations, the ideal spatial configuration in a model of threat might consist of zones of the right size for the deployment and patrol of surveillance teams.

The distance metric used in the threat measure might be chosen to consist of geographic distance or travel time between two locations. In Section 8.2, evidence was discussed that demonstrates how the occurrence of attacks such as IED explosions in a particular location is likely to lead to other attacks nearby as insurgents attempt to repeat previous successful attacks. Thus, locations nearest to areas with more insurgent attacks will be those in which the threat is highest since these are the areas to which the conflict is most likely to spread. This example demonstrates how threats can dissipate over distance and therefore why geographic distance is often a suitable metric.

Without knowing the locations and productivity of insurgents' weapons caches, it is difficult to directly measure the aggression of each adversary at a particular location, which is required in order to specify the threat measure. However, in Section 8.2, work was described in which the locations of attacks are used to triangulate the likely locations of weapons caches. In this work, it is assumed that the clustering of attacks in a particular area or region is likely to imply that those attacks were initiated by the same group using the same weapons cache. The number of attacks that result from a particular insurgent group is also likely to give some indication of the level of aggression and resources of that group. It follows that one way of measuring the aggression of an insurgent group in a particular area is to use a measure of the rate at which attacks occur in that area.

Targets of insurgent attacks include both counterinsurgent forces and civilians. If the aggression of insurgents in area i, u_i, is measured by the rate at which attacks occur in a particular area, then v_j is required to measure the 'aggression' of those who are in conflict with the insurgents in area j. In the derivation of the threat measure in Section 8.5, it was discussed how, as well as aggression, the measure v_j can be associated with the benefits obtained by (in this case) insurgents from directing threat towards their adversary at j (benefit was taken to be given by $\alpha \ln v_j$). This interpretation of the model is useful in the context of insurgent attacks, as it means that v_j can be operationalised by considering the targets that might be available within area j and the benefits that might be obtained by insurgents should those targets be attacked. Measures of benefit might therefore include the population density in a particular area or the locations of targets such as government buildings or military bases. Braithwaite and Johnson (2015), for example, show that the location of an airport garrison is a significant explanatory factor in a logistic regression model of insurgent attacks in Baghdad.

We emphasise that the measures discussed in this section are designed to offer a proof of concept for the use of spatial interaction models of threat in an operational setting. Further research on designing appropriate proxies for some of the variables would be beneficial. It is feasible, however, that the use of such measures can enable operational insights to be exploited for deployment purposes.

8.6.2   Threat in a Model of Conflict Escalation

As described in Section 8.3, empirical agent-based simulations are becoming popular ways of modelling conflict processes. More traditional models, however, have explored how and why conflict might emerge between adversaries using systems of differential equations. These models are more amenable to analysis using a wide range of mathematical techniques. Consequently, differential equations can sometimes lead to insights that might not have been obtained using a simulation-based model, analysis of which often requires repeated simulations for different parameter values.

A limitation of traditional differential equation-based conflict models is that they are often highly simplified and unable to account for complex behaviour that is often observed in the real world. In this section, we demonstrate how the threat measure developed in Section 8.5 can be used to extend a simple model of conflict escalation, namely, the model of Richardson (1960). Baudains et al. (2015) demonstrate in more detail how this model exhibits complex behaviour and how a mathematical analysis can lead to intricate insights into the spatial dependence of conflict escalation. The use of the threat measure can provide differential equation based models with some of the versatility offered by ABMs, whilst still retaining some degree of analytical tractability.

Richardson's model considers the levels of military spending of two competing nations during the lead-up to war. Supposing that the spending levels of these two nations are given by p and q, respectively, then Richardson's model supposes that these values change as a result of three processes: they increase proportionally to the amount of spending of the opponent; they decrease proportionally to the existing amount of spending; and they increase or decrease according to external events, which are assumed to influence spending similarly over time. The model is given by

$$\dot{p} = \rho_1 q - \sigma_1 p + \epsilon_1 \tag{8.13}$$

$$\dot{q} = \rho_2 p - \sigma_2 q + \epsilon_2, \tag{8.14}$$

where ρ_1 and ρ_2 determine the strength of action–reaction relationship between the two adversaries, σ_1 and σ_2 determine the cost of maintaining existing resources, and ϵ_1 and ϵ_2 represent external grievances.

This model results in two types of long-term behaviour: either the internal restraining dynamics will outweigh the competitive action–reaction dynamics and the system will eventually come to equilibrium and stop changing, or the action–reaction dynamics outweigh the damping in the system and the magnitude of military spending continues increasing indefinitely, resulting in an arms race between the two adversaries.

Baudains et al. (2015) use the threat measure of Equations (8.10) and (8.12) to incorporate spatial interaction into the model of Equations (8.13) and (8.14). Embedding a spatial interaction framework into models that do not explicitly incorporate space increases the range of application and the intricacy with which insights can be obtained. To specify threat, a measure of aggression is required, which is taken to be the dependent variable in the model in Equations (8.13) and (8.14). In Richardson's model, this is often taken to be the level of military spending or available military resources of the two adversaries.

The measure of threat is then embedded within the Richardson model by assuming that the action–reaction term in the model at each location, determining the rate at which each adversary increases their level of expenditure, is in fact proportional to the threat on each adversary at that location. Thus, the model becomes

$$\dot{p}_i = \rho_1 \tau_i^X - \sigma_1 p_i + \epsilon_1 \tag{8.15}$$

$$\dot{q}_j = \rho_2 \tau_j^Y - \sigma_2 q_j + \epsilon_2, \tag{8.16}$$

for $i = 1, 2, 3, \cdots, N$ and $j = 1, 2, 3, \cdots, M$. This model is similar to the model in Equations (8.13) and (8.14). In fact, if we were to take the sum of the different levels of hostility over the distinct locations, then the model is identical. The aggregate system can therefore either tend to be in equilibrium or result in an escalating process. What is different is that the level of expenditure at each location is tracked and depends on the threat from the adversary across all their locations. The contributions to this action–reaction relationship act as a weighted sum over the different locations, and complex dynamics can arise as a result.

Baudains et al. (2015) go on to identify bifurcations that arise due to the non-linearities embedded within Equations (8.15) and (8.16). These bifurcations introduce spatial instabilities into the system as the parameters ρ_1 and ρ_2 increase.

The use of threat in dynamical systems models provides them with greater complexity, making them more capable of incorporating real-world processes by which conflict is thought to arise. In addition, such models are amenable to a wide range of analytical techniques. Differential equations enable the logical exploration of the implications of a proposed mechanism by which adversaries are thought to interact. The threat measure presented in this chapter can be a valuable tool in increasing their range of application.

8.6.3 Threat in Modelling Global Military Expenditure

In Chapter 11 of *Geo-Mathematical Modelling*, an empirical application of the threat measure is given in the context of global arms expenditure. In particular, a threat measure of the form in Equations (8.10) and (8.12) is used in a regression model of global military expenditure and shown to have significant explanatory power. The threat measure used in this study differs slightly to the one developed in Section 8.5. This is because, instead of having just two adversaries who might be located over some geographic space, an arbitrary number of adversaries are incorporated. This means that it is possible to consider how each country changes its military spending based on the threat from all other nations who might be considered adversaries.

Alliances are incorporated by defining the metric d to be a measure of both geographic distance and political similarity between every two countries. If two states have broadly similar foreign policy portfolios (as determined by the alliances they choose to make), they are considered to be close allies and the value of d between them will be large (since they are far apart in conflict space and unlikely to react to each other's military spending). Two countries with dissimilar alliance portfolios, who are also near to each other, will be the closest countries in the conflict space since they will be most likely to react to each other's military spending and to ultimately engage in conflict with each other.

This model outlines how the mathematical formulation of threat can be employed in a global setting. Further research might consider whether threat measures can also be constructed to investigate the risk of conflict hotspots, in addition to military spending. In particular, some of the data sources described in Section 8.4 might be usefully employed in developing a global measure of threat.

8.6.4 Summary

In this chapter, we have reviewed a number of ways in which models can provide insights into human conflict. We separated our discussion to consider three different policy domains. The first identified a number of studies that use event interdependency to provide insights that can aid operational decision-making during wars or insurgencies. The second considered the application of simulation-based models to ethnic violence in contested urban spaces. Such models provide insights into difficult social questions and enable the flexible exploration of different policy options. The third considered how different sources of data are being incorporated into global models constructed with the direct aim of forecasting outbreaks of conflict around the world.

We then introduced a mathematical model of threat, which we argue can offer benefits to the way in which spatial interactions are handled in a wide range of conflict models. We offered three ways in which this model of threat can contribute to the mathematical modelling of conflict: proposing how the measure might indicate how the threat of insurgent attacks varies in space; showing how traditional models of conflict processes can be extended by incorporating threat acting over space; and by discussing a global arms race model in which the metric d takes a novel form that does not depend on geography.

In the chapters that follow and in the chapters of the companion volume, *Geo-Mathematical Modelling*, further examples of modelling conflict such as rioting, rebellions and piracy are given. We hope that some of the discussion of this chapter can inspire future research in conflict modelling.

References

Baudains, P.J. (2015) Spatio-temporal modelling of civil violence: four frameworks for obtaining policy-relevant insights. Doctoral Dissertation, University College London, London.

Baudains, P., Fry, H.M., Davies, T.P., Wilson, A.G., and Bishop, S.R. (2015) A dynamic spatial model of conflict escalation. *European Journal of Applied Mathematics*, doi: 10.1017/S0956792515000558.

Benigni, M. and Furrer, R. (2012) Spatio-temporal improvised explosive device monitoring: improving detection to minimise attacks. *Journal of Applied Statistics*, **39** (11), 2493–2508.

Bhavnani, R., Donnay, K., Miodownik, D., Mor, M. and Helbing, D. (2014). Group Segregation and Urban Violence. *American Journal of Political Science*, **58**(1), 226–245.

Braithwaite, A. and Johnson, S.D. (2012) Space-time modeling of insurgency and counterinsurgency in Iraq. *Journal of Quantitative Criminology*, **28**, 31–48.

Braithwaite, A. and Johnson, S.D. (2015) The battle for Baghdad: testing hypotheses about insurgency from risk heterogeneity, repeat victimization, and denial policing approaches. *Terrorism and Political Violence*, **27**, 112–132.

Brantingham, P.J., Tita, G.E., Short, M.B., and Reid, S.E. (2012) The ecology of gang territorial boundaries. *Criminology*, **50** (3), 851–885.

Dennett, A. and Wilson, A.G. (2013) A multi-level spatial interaction modelling framework for estimating inter-regional migration in Europe. *Environment and Planning A*, **45**, 1491–1507.

Hegre, H., Karlsen, J., Nygård, H.M., Strand, H., and Urdal, H. (2013) Predicting armed conflict, 2010-2050. *International Studies Quarterly*, **57**, 250–270.

Helbing, D., Brockmann, D., Chadefaux, T., Donnay, K., Blanke, U., Woolley-Meza, O., Moussaid, M., Johansson, A., Krause, J., Schutte, S., and Perc, M. (2015) Saving human lives: what complexity science and information systems can contribute. *Journal of Statistical Physics*, **158** (3), 735–781.

Johnson, N.F., Medina, P., Zhao, G., Messinger, D.S., Horgan, J., Gill, P., Bohorquez, J.C., Mattson, W., Gangi, D., Qi, H., Manrique, R., Velasquez, N., Morgenstern, A., Restrepo, E., Johnson, N., Spagat, M., and Zarama, R. (2013) Simple mathematical law benchmarks human confrontations. *Scientific Reports*, **3**, 3463–.

Leetaru, K. and Schrodt, P.A. (2013) GDELT: global data on events, location and tone, 1979-2012. Presented at: International Studies Association, San Francisco, April 2013. Available online at: http://data.gdeltproject.org/documentation/ISA.2013.GDELT.pdf (accessed 30 December 2015).

O'Brien, S.P. (2010) Crisis early warning and decision support: contemporary approaches and thoughts on future research. *International Studies Review*, **12**, 87–104.

Richardson, L.F. (1960) *Arms and Insecurity*, The Boxwood Press, Pittsburgh, PA.

Rossmo, K.D. (2000) *Geographic Profiling*, CRC Press LLC, Boca Raton, FL.

Shakarian, P., Subrahmanian, V., and Sapino, M.L. (2009) SCARE: a case study with Baghdad. Proceedings of the Third International Conference on Computational Cultural Dynamics, AAAI.

Townsley, M., Johnson, S.D., and Ratcliffe, J.H. (2008) Space time dynamics of insurgent activity in Iraq. *Security Journal*, **21**, 139–146.

Ward, M.D., Metternich, N.W., Dorff, C.L., Gallop, M., Hollenbach, F.M., Schultz, A., and Weschle, S. (2013) Learning form the past and stepping into the future: toward a new generation of conflict prediction. *International Studies Review*, **15**, 473–490.

Weidmann, N.B. and Salehyan, I. (2013) Violence and ethnic segregation: a computational model applied to Baghdad. *International Studies Association*, **57** (1), 52–64.

Wilson, A.G. (2008) Boltzmann, Lotka and Volterra and spatial structural evolution: an integrated methodology for some dynamical systems. *Journal of the Royal Society, Interface / the Royal Society*, **5** (25), 865–871.

Zammit-Mangion, A., Dewar, M., Kadirkamanathan, V., and Sanguinetti, G. (2012) Point process modelling of the Afghan War Diary. *Proc. Natl. Acad. Sci. U.S.A.*, **109** (31), 12414–12419.

9

Riots

Peter Baudains

9.1 Introduction

Riots involve groups of people at a given location engaging in or threatening acts of violence for a common purpose. Outbreaks of rioting can have significant implications for global dynamics. Locally, they can disrupt economic activity and drastically alter the social cohesion of an area. At the regional level, news reporting of rioting in one location has been shown to provide the stimulus for rioting in other locations (Myers, 2000; see also Chapter 13 of this volume, which considers contagion of ethnic conflict across national borders via media channels). A cascading effect can then elevate regional instability and unrest into the domain of global dynamics (Braha, 2012).

As well as being relevant to the study of global dynamics, riots provide an archetypal example of a complex social system. The behaviour of individuals during rioting is influenced by interaction with those nearby. Ongoing riots act as situational precipitators, serving to prompt, pressure, permit or provoke individuals to engage in the disorder (Wortley, 2008). Individual behaviours, influenced by such interaction, combine to generate emergent patterns in the intensity, timings and locations of rioting. Such emergent behaviour is a defining feature of a complex system (Newman, 2011) and was particularly evident during the 2011 riots in London, during which outbreaks of rioting occurred in various locations across the capital. Due to the natural way in which riots can be interpreted as complex systems, riots can be used as a test-bed for the application of new analytical techniques.

In this chapter, we give an overview of different analytical frameworks that can be applied to rioting. We restrict our focus to four recent and novel studies,[1] not to imply that these represent an exhaustive or definitive list of such work (there are many other studies that could have been included), but in an effort to highlight a wide range of analytical frameworks, all of which have implications for policy and operational decision-making. Furthermore, all four studies

[1] These are Baudains et al. (2013a). Baudains et al. (2013c). Baudains et al. (2013b). Davies et al. (2013).

discussed in this chapter have been applied to the same case study: the August 2011 riots in Greater London. This enables a comparison to be made between the frameworks presented and, moreover, to consider how the policy implications vary with the analytical framework used. When choosing a particular framework with which to investigate and model a complex system, a trade-off often exists between choosing an approach that has the potential to offer great insight and choosing an approach that minimises uncertainty with the insights that are obtained. The discussion that follows each study highlights this trade-off and its importance in the modelling of complex systems.

We begin in Section 9.2 by outlining the case study of interest: the 2011 riots in Greater London. We explain why this case is of interest from a modelling perspective and then present some statistical summaries that help understand what happened. In many policy applications, summary statistics can quickly and reliably inform a decision-maker about the system of interest. However, in many cases, the insights obtained do not tell the full picture and do not allow the decision-maker to extrapolate into the future particularly in cases where the underlying conditions conducive to rioting quickly change.

We then introduce a number of analytical techniques which seek to go beyond simple statistical summaries. These through the use sophisticated modelling approaches and aim to generate improved insights. In Section 9.3, we explore with more intricacy the prominence of particular space–time patterns of rioting, leading to a greater understanding of the role of contagion. In Section 9.4, a statistical model of rioter target choice is summarised, which identifies the regions in London that may have been more susceptible to rioting due to certain features or characteristics of those areas. In Section 9.5, a generative model of riot scenarios is outlined that is capable of replicating some of the general patterns in the data, thereby providing a policy tool for planning police deployment strategies. We conclude by highlighting the insights that each model framework affords and by proposing new avenues for future research.

9.2   The 2011 Riots in London

Between the 6th and 10th of August 2011, riots occurred at numerous locations across the United Kingdom. Violence initially broke out after a peaceful protest by family, friends and members of the community of Mark Duggan, who was shot and killed by police officers in Tottenham, North London, on 4 August. On 6 August, riots broke out in neighbouring communities. For five nights, the riots continued, initially throughout the capital and subsequently across the country. After the initial disturbances, the unrest on subsequent nights grew in intensity, before a large number of police were deployed across the capital and in other cities, leading to a restoration of order. In London, it is estimated that there was in excess of £200 million of damage to public and private property, over 200 injuries to police and two deaths (Riots Communities and Victims Panel, 2011). Over 4,000 arrests were made in London alone (Metropolitan Police Service, 2012), many of which were identified via CCTV footage in the days following the disorder.

There are many policy questions directly relevant to the 2011 UK riots. For example, studies have sought to identify the underlying sociological causes of the rioting (Solomos, 2011; Kawalerowicz and Biggs, 2015) and have examined whether the criminal justice response was appropriate (Bell et al., 2014). The policy question considered here is concerned with the spatial and temporal dependency of the riots. Considering first the spatial aspect of the riots, it is notable that the riots predominantly took place in the highly populated areas of London,

Birmingham and Manchester. However, even within these cities, and particularly in London, riots occurred in some areas but not in others. Several geographically distinct areas, such as Hackney, Brixton and Croydon, experienced large-scale violence, looting and arson; whereas some of the areas in between – including Central London, Shepherd's Bush and Leyton – experienced comparatively few events. We seek to understand why rioting happened in some places but not others and investigate whether studying these aspects of the riots might lead to operational insights such as strategies for police deployment.

The temporal dependence of the riots, when considered alongside their spatial dependence, also exhibits striking patterns. The first locations to experience rioting were around the Tottenham area in North London. Over the next days, riots occurred in South London, before also occurring in other UK cities. This gave the impression that the riots were spreading geographically and many commented how the onset of rioting in one location was imitated by others in different locations, implying some form of event dependency (Gross, 2011; BBC, 2011). In addition to a copycat effect, there may well have been different complementary causes for the riots such as deprivation or social exclusion (Kawalerowicz and Biggs, 2015). Nevertheless, this apparent event dependency implies that traditional modelling techniques assuming event independence are likely to be inappropriate. The interactions between events form an avenue for academic enquiry that is well suited to a complex systems perspective.

The Metropolitan Police Service provided data on arrests associated with the riots ($N = 3,914$), many of which were detected in the days following the disorder via CCTV footage. Figure 9.1 shows the locations of those offences for which a location was recorded within the boundary of Greater London ($N = 2,868$). Figure 9.2 shows the timing, to the nearest hour, of those offences which also included a recorded time ($N = 2,593$). This figure also shows how the number of Metropolitan police officers responding to the disorder on the streets of London increased as the riots progressed (Metropolitan Police Service, 2012). The availability

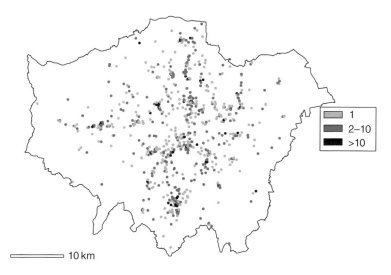

1
2–10
>10

10 km

Figure 9.1   The geography of the 2011 riots in Greater London. The shading of each point indicates the number of offences that occurred at each location.

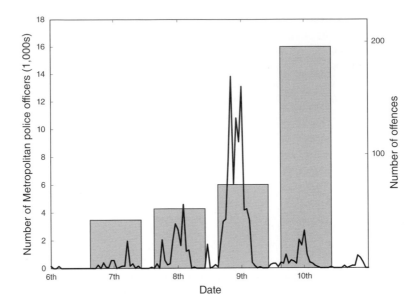

Figure 9.2   The timing of the 2011 riots in Greater London with reported police numbers on London streets for each day of unrest

of this data at such a fine scale of resolution has provided a unique opportunity to study rioting as a complex system. This has been achieved by modelling individual behaviours while also accounting for interaction between those individuals and by studying the patterns that emerge from such interactions.

Statistical descriptions of this data can immediately provide policy-relevant insights. A detailed statistical overview of the riots across the United Kingdom is given by the Home Office (2011). Such analysis can be used to get a general sense of what happened and may be used to have an informed debate at a policy level. Statistical descriptions can also be used in an operational context. In what follows, we present a number of statistical summaries of the data provided by the Metropolitan Police Service in order to demonstrate the insights that might be obtained. We restrict our focus to summaries of the data that describe and help explain the spatial and temporal distribution of the riots (it being our principal policy question of interest).

9.2.1   Space–Time Interaction

Figures 9.1 and 9.2 show the spatial and temporal heterogeneity in the data, respectively. Both show distinctive levels of clustering: events appear to occur closer to each other in space and time than might be expected if events were occurring independent of time or location. In Figure 9.1, there are a number of locations containing more than 10 offences, but there are also large areas of Greater London with no recorded offences. In Figure 9.2, a daily oscillation to the times at which riots occur can be observed, with the majority of crimes occurring in the evening each day.

The spatial and temporal clustering of these events can be quantified through the use of statistics designed to capture autocorrelation: the extent to which events are more or less clustered when compared to a random Poisson process. Baudains et al. (2013a) show how this can be done in the case of spatial clustering. In particular, this study demonstrates that offences were far more likely to occur either in the same area or nearby to areas containing offences than any other location selected uniformly at random.

When event occurrence varies with both space and time, it can be important to determine whether there is also space–time interaction. Tests for space–time interaction are distinct from tests that identify the presence of purely spatial or temporal dependency: they focus on the likelihood of a further event occurring in close proximity in both space and time to a prior event. From an operational perspective, this type of statistical summary of the data can be more useful than examining the spatial or temporal distributions of offences independently. During outbreaks of rioting in a city, for instance, police leaders face decisions concerning the allocation of limited resources of police officers in real time. Insights into the space–time interaction of riot events can help to answer questions such as whether police resources should remain at sites recently rioted or whether these resources would be better deployed elsewhere in the city, for example, at perceived attractive targets that have not yet experienced rioting.

One method for determining the level of space–time interaction in event data is via the Knox (1964) test. This compares the distances in both space and time between pairs of events by allocating each pair to a spatio-temporal window of pre-specified resolution. The resulting categorisation of pairs of events over spatio-temporal windows of different resolutions can be compared against a Monte Carlo simulation in which event times are randomly permuted over the locations in the empirical data. It can then be determined whether events are more or less likely to occur near to each other in both space and time when compared to a scenario in which the spatial and temporal distributions of the data are independent.

To operationalise a Knox test for the data on the London riots, we overlay a spatio-temporal grid over the geographic and temporal ranges of interest. Each event is then mapped into the space–time window within which the event occurred. The number of pairs of events that occur within a particular space–time window of each other is calculated and compared to a null model in which space and time are independent. We take the space–time window to consist of a grid cell and its neighbours over a time period of 1 hour.

Figure 9.3 plots the number of pairs of events, denoted by S_K, occurring within this space–time window for different spatial resolutions δs for both the actual data (shown by the black dots) and for 499 Monte Carlo simulations in which space and time are independent (shown by the white dots). Pairs of events are much more likely to occur near to each other in both space and time than when compared to a scenario in which there is no space–time interaction for all grid sizes tested.[2] In fact, the effect is extremely strong: around four times more pairs of events occurred nearby to each other in the data compared to a scenario with no space–time interaction.

Statistical tests like this confirm the level of space–time interaction in the data. For timescales of around 1 hour, areas up to 650 m away from ongoing rioting were likely to contain a much higher likelihood of further rioting. Extending the analysis to different space–time windows may further improve the insights obtained, for example, by understanding the times

---

[2] Different temporal grid sizes were also tested and showed similar results.

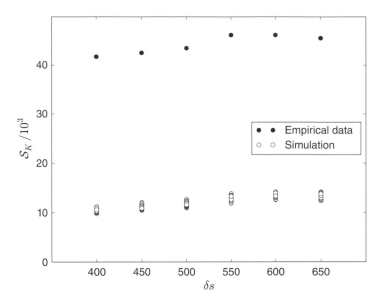

Figure 9.3 Results of the Knox test for space-time interaction. Six different spatial grid sizes are used and all demonstrate significant levels of space-time clustering.

and distances at which an increased risk of rioting diminishes back to marginal levels (i.e. the levels given by the overall riot intensity).

9.2.2   Journey to Crime

In addition to the information on the offences that occurred, the data on the London riots contained an identifier for each offender residential location. This makes it possible for the presentation of a statistical summary of the data that has been considered extensively in Criminology literature: the journey to crime curve (Townsley and Sidebottom, 2010). Figure 9.4 plots the distribution of distances travelled by 2, 550 rioters from their residential location to the location of their offence. The resulting curve is consistent with the curves for different types of crime (Rossmo, 2000). In particular, the majority of offences occurred within a relatively short distance of the offender's residential address.

Figure 9.4 shows how over 50% of rioters travelled within just 3 km of their residential address to engage in rioting. This information can have immediate policy implications. During the rioting, one of the policy options that might have been considered was to close down transport links such as the London underground network in order to prevent rioters from travelling to vulnerable targets. Figure 9.4 implies that such policies would have been unlikely to prevent a large number of rioters from reaching their target, since most rioters committed their offence a short distance from where they lived.

In addition, the finding that many rioters were rioting within a short distance of their home has implications for understanding the underlying causes of the riots. The inability of rioters to identify a sense of ownership of their local area, as implied by the damage caused to nearby facilities and infrastructure within their neighbourhood, suggests a degree of social exclusion

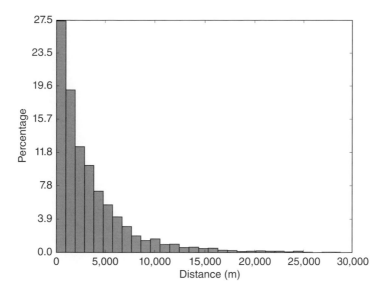

Figure 9.4   The distribution of distance travelled between rioter residence and offence location

among those who rioted. Subsequent policies may seek to improve the sense of engagement within communities that might be more susceptible to rioting.

9.2.3   Characteristics of Rioters

Given that the majority of rioters travelled short distances from their residential location to engage in the rioting, an important factor in generating the resulting space–time profiles of the riots is where the rioters were coming from. This is also important from a policy perspective. If similar characteristics can be identified in the areas where rioters were more likely to come from, then policies might be directed at addressing shared grievances in these areas. Similarly, policy insights may also be obtained by considering the characteristics of rioters themselves. For instance, it may be that a particular subgroup within the population was more likely to be rioters, a finding which could guide future policies.

A relationship appears to exist between the deprivation of an area and the number of rioters that each area produced (Kawalerowicz and Biggs, 2015; Davies et al., 2013; Home Office, 2011). The areas in which rioters were residing were among those in the United Kingdom that were more deprived. However, the causal mechanism behind this relationship is unclear. This is because there were rioters that came from areas with low levels of deprivation, and, in many of the most deprived areas, there were no rioters.

The age–crime curve (see Figure 9.5) shows the age distribution of offenders during the riots. It implies that arguments for the causes of the riots must not focus solely on just one subsection of the population: there were offenders represented across the age spectrum. The majority of offenders are under the age of 25, suggesting that young people may have been particularly susceptible to any contagion or copycat behaviour. It is important, however, to recognise that the shape of this curve is consistent with age–crime curves for offending outside

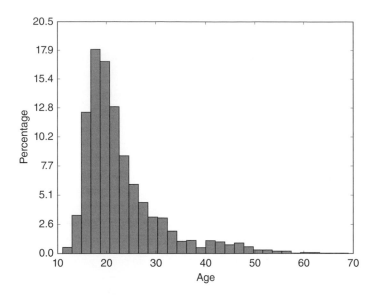

Figure 9.5   The distribution of rioter age

the unusual conditions of rioting. In addition, any explanation of the riots by which young offenders are more likely to engage in co-offending should be treated with caution since curves for solo offending have been shown to exhibit similar shapes (Stolzenberg and D'Alessio, 2008).

We have presented three statistical summaries of the riots, all of which lead to policy insights. Although a great deal of insight can be obtained, sometimes the picture that emerges becomes complicated, as a range of possible explanations might account for the regularities observed in the data.

In the sections that follow, we give an overview of three modelling frameworks that have been applied to the 2011 riots in London. These frameworks all seek to go beyond the statistical summaries of the data presented in this section and, in turn, seek to generate more useful policy and operational insights. It is also interesting to compare how these three frameworks differ from each other, both with respect to the types of policy insight that might be obtained and with respect to how uncertain those insights might be.

9.3   Data-Driven Modelling of Riot Diffusion

A key objective for law enforcement officers during rioting is to minimise the extent of the area at risk. Indeed, a common public order policing strategy is one of containment. The Metropolitan Police Service have stated that containment tactics were initially used to counter the riots in London (Metropolitan Police Service, 2012). It was claimed that this was due to a combination of the severe and unprecedented scale of the riots together with a lack of resources (in terms of the number of police officers) available to react to the disorder. The objective of this strategy is to retain the rioters within a small but well-defined geographic area. Although this inevitably increases the risk of violence and property damage within that area, it minimises the opportunity for the riot to spread to neighbouring areas, thereby minimising the number of victims and properties that may be at risk.

In Section 9.2.1, the interdependence between offences was analysed by considering events that occurred nearby to one another in space and time. This analysis had no link to the underlying geography. In this section, we demonstrate how a similar exploratory and data-driven analytical framework can be used to analyse the riots in a way that retains the geographic scope of the rioting in the analysis and, therefore, contributes insights towards this operational objective. This procedure is described in more detail in Baudains et al. (2013c).

To begin, we define a spatio-temporal grid over the area and time period of study. This is done to establish a link between the events and the underlying geography. Each event is mapped into the space–time grid cell within which it took place. Then, the dependent variable of interest is given by a binary tuple $(X, Y)_{(j,k)}$ for each space–time unit, indexed by the tuple (j, k). The index j denotes the spatial grid unit of interest and the index k denotes the temporal window. For each (j, k), $X \in \{0, 1\}$ indicates whether at least one offence occurred in the focal space–time window of interest and $Y \in \{0, 1\}$ indicates whether at least one offence occurred within any of the focal area's neighbouring units.

Since the variables of interest are binary, they do not distinguish between the number of events occurring in each space–time window: the occurrence of a single event is recorded as being equivalent to the occurrence of many events. Although this comes with some limitations (e.g. it means that intensity of riot outbursts is not considered), it allows the primary subject of analysis to be the geographic scope of the rioting.

Considering how $(X, Y)_{(j,k)}$ changes for a spatial unit j over time quantifies the evolution of local patterns of rioting in space. In Baudains et al. (2013c), the prevalence of four local patterns of riot events in space and time are investigated. Each of these patterns is associated with a different value of the variable $(X, Y)_{(j,k)}$ for two time periods. The first type of pattern is termed containment. This is defined by the transition $(1, 0)_{(j,k)} \rightarrow (1, 0)_{(j,k+1)}$ and occurs when areas already affected by disorder in one time period are also affected in the next, but when the disorder does not spread to neighbouring areas. Second, relocation is defined by $(1, 0)_{(j,k)} \rightarrow (0, 1)_{(j,k+1)}$ and occurs when the disorder moves from one locality to another, without persisting in the original location. Third, processes of escalation are given by the transition $(1, 0)_{(j,k)} \rightarrow (1, 1)_{(j,k+1)}$ and occur when rioting continues for a prolonged period in a certain area and also spreads to contiguous areas. Finally, flashpoints are given by $(0, 0)_{(j,k)} \rightarrow (1, 1)_{(j,k+1)}$ and are outbursts of co-occurring offences located in areas that are geographically distinct from areas that had recently experienced offences. In other words, they occur when areas and their neighbouring areas suddenly experience widespread disorder. In Figure 9.6, examples of these four patterns are illustrated.

These four patterns are chosen since they are likely to correspond to mechanisms of interest during rioting. For instance, scenarios in which groups of individuals mobilise themselves and target a particular time and place at which to riot would increase the prevalence of flashpoints. Situational precipitators of rioting, whereby individuals are more likely to engage in the disorder due to the precipitating influence of ongoing rioting nearby, would correspond to an increased prevalence of containment, relocation and escalation. Differences in the number of these three patterns can be explained by considering how much the rioting is tied to the underlying geography. If environmental factors, such as the presence of a particular retail store, tied the rioting to specific areas, and if neighbouring areas did not offer suitable alternatives at which to offend, then containment would be observed. A prevalence of relocation and

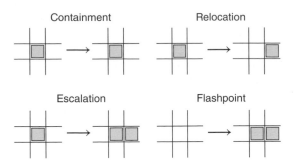

Figure 9.6   Four patterns of geographic diffusion

escalation, however, would imply that the location of rioting is dynamic and is not tied to the underlying geography. In addition, relocation and escalation processes would imply that the police are unable to contain the rioting and had limited ability to control its geographic scope. If the police were employing containment tactics, we would expect to see a large number of containment patterns.

In Baudains et al. (2013c), a null model is described that enables a comparison between the count of each type of diffusion pattern in the empirical data and the counts that would have been obtained if the geographic locations of the rioting were independent from the times of the offences. A comparison between the distribution of counts obtained from a null model and the empirical count of each diffusion pattern (as indicated by the dashed line) for a particular grid size[3] (400 m and 4 hours) is shown in Figure 9.7. They show that instances of containment were more prevalent in the empirical data than in the null model. Relocation occurred less than expected and there were more of both escalation and flashpoints than might have been anticipated.

The results offer a number of perspectives for policing from an operational perspective. The prevalence of flashpoints during the riots, in which widespread disorder occurs at a given location quite spontaneously, is a phenomenon that is difficult to predict. It is therefore unlikely that

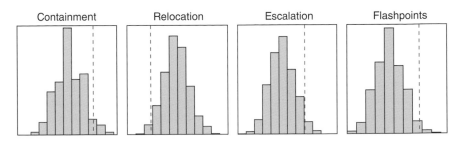

Figure 9.7   Example of null model output compared against the empirical data (dashed line) for the four diffusion patterns in Figure 9.6

---

[3] Results for different grid sizes are presented in Baudains et al. (2013c) in order to address the modifiable areal unit problem. They are consistent with the results shown here.

police officers could have been present at such locations antecedently. The occurrence of flash-points is inherently difficult to police. On the other hand, the results suggest that instances of containment and escalation were more prevalent than would have been expected assuming that the events were independent. These patterns provide more opportunity for policing. A general finding is that the rioting appeared to be fairly static and rooted in the underlying geography for timescales over which police may be deployed. The adoption of reactive strategies by police officers – by which officers are quickly deployed to locations where rioting is ongoing – as opposed to proactive strategies – by which officers are deployed to locations which are not experiencing disorder but at which disorder may be anticipated – is perhaps a good strategy to adopt. The prevalence of flashpoints, however, suggests that police allocation should also be dynamic, and that there is a balance that needs to be struck.

These results demonstrate how data-driven analytical approaches, in which model assumptions are derived from empirical data, can lead to operational and policy insights. Another reason for employing data-driven approaches to modelling, particularly when first faced with an empirical data set, is that it can often suggest assumptions that might be used to construct more descriptive or complex models. For example, two important considerations in this study are the impact of interdependency between events and the relationship between the locations of the riots and the underlying geography. In Section 9.4, these are incorporated into a model of rioter target choice.

With respect to informing policy, data-driven models have the advantage that they rely almost exclusively on empirical data, rather than restrictive assumptions which might have little bearing on the real world. This is likely to make them less uncertain and more trusted by stakeholders. While data-driven approaches are well suited to establishing stylised facts of phenomena that have occurred in the past, they are less good at prediction. Prediction with data-driven approaches that are informed by observations relies on the sample data containing sufficient information to enable extrapolation. In contrast, mechanistic or generative approaches, in which a proposed mechanism generates model outputs that are thought to be responsible for the empirical data, are likely to be further removed from the real world, but more likely to be able to account for qualitative changes in the underlying data-generating process.

9.4   Statistical Modelling of Target Choice

Recent literature on criminality has been concerned with how the choice of target for an offence reflects the motivations of the offender and the opportunities that are present when the offence takes place. A popular way to investigate such factors is by employing spatial discrete choice models (e.g. Bernasco and Block, 2009). These are suitable for situations in which an actor is faced with a choice in which each option has characteristics that are quantitatively distinguishable. Estimation of discrete choice models using empirical data can highlight the relative importance of the characteristics of an area in influencing the choice of target.

In Baudains et al. (2013b), a discrete choice model is derived and estimated and used to gain insights into the behaviours of rioters in August 2011. The model estimates the effect that a range of variables has on the target choice of offenders. The variables are defined at the Lower Super Output Area level, a geographic area designed to contain around 1,500 residents.

Table 9.1   Odds ratios of selected parameter values

Variable	7 August			8 August			9 August		
	Lower	Mean	Upper	Lower	Mean	Upper	Lower	Mean	Upper
Contagion	1.12	1.14	1.16	1.06	1.07	1.07	1.03	1.04	1.05
O–D distance ($\geq$ 18)	0.61	0.65	0.69	0.63	0.64	0.66	0.60	0.63	0.66
O–D distance ($<$ 18)	0.55	0.61	0.67	0.56	0.58	0.60	0.44	0.49	0.54
Retail centre	1.24	1.31	1.38	1.26	1.28	1.32	1.19	1.24	1.29
Deprivation	1.37	1.57	1.80	1.55	1.63	1.72	1.16	1.27	1.40

In Table 9.1, the odds ratios for four of the variables tested are shown over 3 days of rioting.[4] The table also shows the lower and upper bounds of a 95% confidence interval associated with each odds ratio. The estimates indicate the change in the odds of a target being selected if the corresponding variable were to increase by one unit while all other variables remain fixed.

The results in Table 9.1 shed further light on some of the statistical summaries given in Section 9.2. The variable measuring contagion of the riots is given by the number of offences that occur in a target in the 24 hours prior to the time of each offence and is designed to capture the increased attractiveness of a target due to ongoing or recent rioting at that target. The parameter estimates are consistently positive and significant, providing evidence that recent rioting acts as a situational precipitator, in which rioters are encouraged to engage in the disorder more so than they otherwise would due to the presence of that rioting. To illustrate, on 7 August, the odds of an area being targeted by an offender increased by a factor of 1.14 for every additional (detected) offence that occurred in that area in the previous 24 hours.

The distance between a rioter's residential address and the location of their offence is significantly negatively associated with the choices made by rioters for each of the days presented (since the odds ratio is less that one, representing diminishing attractiveness with increasing distance). Areas further away from a rioter's residence were less likely to be selected, confirming and quantifying the relationship shown in Figure 9.4. The effect of distance on the target choice of rioters is more pronounced for juvenile offenders (those under the age of 18) than for adult offenders. Put simply, adults tended to travel further to commit their crimes, all other things being equal.

The effect of retail centres was positively associated and significant with the likelihood of an area being selected for all days considered. For every additional 250 m^2 of retail floorspace in an area, the odds that it was selected as a location in which to riot increased by a value of between 1.24 and 1.31, all other things being equal. This finding suggests that retail centres were attractive to rioters, supporting the argument that rioters were driven largely by the opportunity to take part in looting.

Areas with a higher level of deprivation, as measured by the Index of Multiple Deprivation for 2010 (McLennan et al., 2011), were more likely to be selected on each day of the riots. The odds of an area selected increased by a factor between 1.27 and 1.63 for each unit increase in this measure. In addition to the relationship between the level of deprivation in the areas

---

[4] Other variables are also included in the calibration. These are: the presence of an underground station, the number of schools, whether the target is the same side of the Thames as the residence of the offender, distance to the city centre, population density, churn rate, and a measure of ethnic diversity.

where rioters were coming from (as described in Section 9.2), a relationship appears to also exist between deprivation and the locations of offences, even after controlling for the distance travelled.

There are two distinct ways of using a parametric statistical model in a policy or operational context. The first is to consider the qualitative findings of such a model, such as the identified association between target choice and deprivation, to broadly inform policies. The second is to use them as a predictive tool. Ward et al. (2010) argue that statistical modelling of social phenomenon should change focus from identifying statistically significant findings, which may be used in a qualitative way to inform policies, to improving the ability for such models to predict what will happen. In Chapter 10 of *Geo-Mathematical Modelling*, it is demonstrated how the model of target choice presented here can be translated into a predictive tool and incorporated into a dynamic algorithm for police allocation problems. In what follows, we present a different approach to developing quantitative policy models of rioting: one based on proposed mechanisms for the way in which rioters and police behave and interact.

9.5   A Generative Model of the Riots

Davies et al. (2013) propose a generative simulation-based model of the riots in London (see also Chapter 9 of *Geo-Mathematical Modelling*). This model is distinct from those considered so far. The model explicitly specifies causal mechanisms for the behaviour of rioters and police in a deterministic way. The task is then to evaluate whether the mechanisms proposed are capable of adequately reproducing some of the observed features of the riots. If it is, then the proposed mechanisms may represent a plausible explanation of the phenomenon and the model can be relied upon to reproduce scenarios that can be used for policy exploration. The model is also dynamic, capable of generating entire riot scenarios for a particular set of parameter values. As we will explain, this increases the range of potential insights that can be obtained from a policy perspective.

The simulation begins with a geographic partition of Greater London. Rioters can originate from any of these locations with the number of rioters that can originate from an area constrained by the population of that area. Retail centres within London act as possible targets or destinations for motivated offenders. Each rioter origin–destination pair has with it an associated attractiveness function. This depends on the size of the retail centre in the target, the distance between the origin and the destination, and the amount of ongoing rioting (and the number of associated police officers) at the destination. Each of these factors have associated with them a parameter that can be calibrated in a similar way to the model described in Section 9.4.

The dynamics of the model have three processes that are updated at each time step. The first process determines how likely residents in different areas are to become motivated to join the riots. This is a function of the attractiveness of each possible target given the origin and also the deprivation of the origin area. Residents in more deprived areas are modelled to be more likely to join the riots, in accordance with the positive relationship described in Section 9.2. The expected number of rioters emanating from each origin is then calculated using the population of each origin and the likelihood that a given resident at that origin is to become motivated.

The second process in the model then assigns each of these motivated residents a target site to which they travel and begin to riot. This part of the model again relies on the attractiveness of origin–destination pairs and is based on a well-known retail model (see, e.g. Wilson 2008). After allocating motivated rioters to their targets, the model then allocates police officers to retail sites, but with a slight lag to represent the time taken by the police to learn about the movements and actions of rioters. Police are allocated to large retail sites where there is ongoing rioting in a similar way to rioters.

Once both rioters and police have arrived at their locations, the police are able to arrest a number of rioters. This number depends on the ratio of rioters to police. The arrest mechanism represents the final component of the model.

These three processes – riot participation, spatial assignment and the interaction between rioters and police – are informed by data and are derived by analogy with other systems and models. The model described here is able to generate riot scenarios for a given set of parameter values. Davies et al. (2013) chose parameter values in order to minimise the difference in the observed and simulated resulting riot patterns. For each simulation, an estimate of the total riot severity is made. The total riot severity is found to exponentially decrease with increasing police numbers. A rapid reduction to very low riot severity is identified with around 10,000 police officers. Although this number depends on a wide range of factors including the particular specification of the model, it is interesting to compare this finding with the actual police numbers present in Greater London during the riots, as shown in Figure 9.2. Between 9 August and 10 August, police numbers increased from 6,000 to 16,000. The model of Davies et al. suggests that just 10,000 officers would have been enough to largely suppress the disorder.

This example demonstrates the power of the insights that can be obtained by generative models. As a consequence, such models can be very useful in a policy-making context. In addition to establishing an estimate of the number of police officers required, the model is well suited to testing the effectiveness of police deployment strategies by allocating a number of police officers to different locations across London. Riots that occur near to where the police are located are then quelled more quickly than they otherwise would be and those riots are consequently unable to attract a large number of motivated rioters, thereby reducing the total riot severity. Such a model offers great potential as a tool for police leaders from a training or planning perspective. The limitation of this modelling approach is the large number of assumptions required in its construction. A large amount of testing of model outputs is required before a high level of confidence can be given to the plausibility of riot scenarios. The training or planning obtained from such a model should recognise that this is just one explanation of a riot process and that there is no guarantee that future riots will evolve in a similar way. Such challenges remain opportunities for future research.

9.6   Discussion

We have presented a range of different analytical frameworks, each investigating the 2011 riots in London. In this section, we emphasise the differences between these frameworks, particularly with regard to the policy-relevant insights that might be obtained.

Data-driven frameworks, which include both simple statistical summaries of a data set and more complex approaches such as that presented in Section 9.3, can lead to robust findings but have two limitations: a frequent lack of data availability at the desired resolution and an over-reliance on cases that have happened in the past (and on which there is data). While the first of these can start to be addressed with new technology for data collection, the second means that some insights may not be generalised to different scenarios. Such data-driven frameworks do, however, relate closely to the case of interest and, without a reliance on a large number of assumptions, have the advantage that they are relatively easy to understand.

Statistical models require a number of assumptions, which are sometimes implicit in the modelling procedure and more difficult to communicate to those who might use their outputs in decision-making. In contrast to data-driven approaches, however, the objective of statistical models is to capture some mechanism in the underlying data-generating process. This is achieved by using sample data to determine whether variables associated with this mechanism covary with the empirical data in the expected direction. This, in turn, invites insights into the proposed mechanisms corresponding to those variables. In many cases, these insights can contribute to policy in a qualitative way. Statistical models also enable probabilistic predictions of specific events. The success of a model in terms of its prediction can then be measured, leading to a continual process of model improvement.

Proposing specific and quantifiable mechanisms for the way in which rioters and police behave led to a model outlined in Section 9.5 that the police can directly use for the testing of deployment strategies. The limitation of such models is that they rely less on empirical data to inform them and are therefore more at risk of having little relation to the actual underlying mechanisms. This limitation should be borne in mind during any policy-making process utilising similar models.

The presentation of a wide range of modelling frameworks, all investigating the distinctive spatial and temporal patterns in the riots, has demonstrated the benefits of applying a plurality of models to a particular case study. A plurality of modelling frameworks can help to increase trust in model conclusions and ultimately improve the accuracy of inferences about the real world. With advances in technology, real-time tracking of events such as rioting is becoming a possibility. Combining some of the above modelling frameworks with real-time data on rioting and law enforcement activity could help make sense of what is happening and aid police deployment. Future research might consider the technology required to optimally allocate police officers (or, indeed, rioters or protesters) in real time.

References

Baudains, P., Braithwaite, A., and Johnson, S.D. (2013a) Spatial patterns in the 2011 London riots. *Policing*, **7** (1), 21–31.

Baudains, P., Braithwaite, A., and Johnson, S.D. (2013b) Target choice during extreme events: a discrete spatial choice model of the 2011 London riots. *Criminology*, **51** (2), 251–285.

Baudains, P., Johnson, S.D., and Braithwaite, A.M. (2013c) Geographic patterns of diffusion in the 2011 London riots. *Applied Geography*, **45**, 211–219.

BBC (2011) London Riots: Looting and Violence Continues, Available online: London riots: http://www.bbc.co.uk/news/uk-england-london-14439970 (Accessed 12 March 2015).

Bell, B., Jaitman, L., and Machin, S. (2014) Crime deterrence: evidence from the London 2011 riots. *The Economic Journal*, **124**, 480–506.

Bernasco, W. and Block, R. (2009) Where offenders choose to attack: a discrete choice model of robberies in Chicago. *Criminology*, **47** (1), 93–130.

Braha, D. (2012) Global civil unrest: contagion, self-organization, and prediction. *PLoS ONE*, **7** (10), e48596.

Davies, T.P., Fry, H.M., Wilson, A.G., and Bishop, S.R. (2013) A mathematical model of the London riots and their policing. *Scientific Reports*, **3**, 3103.

Gross, M. (2011) Why do people riot? *Current Biology*, **21** (18), 673–676.

Home Office (2011) An Overview Of Recorded Crimes and Arrests Resulting from Disorder Events in August 2011, Available at: https://www.gov.uk/government/publications/an-overview-of-recorded-crimes-and-arrests-resulting-from-disorder-events-in-august-2011 (Accessed 6 May 2015).

Kawalerowicz, J. and Biggs, M. (2015) Anarchy in the UK: economic deprivation, social disorganization, and political grievances in the London riot of 2011. *Social Forces*, Available at: http://sf.oxfordjournals.org/content/early/2015/03/05/sf.sov052.full.

Knox, E.G. (1964) The detection of space-time interactions. *Journal of the Royal Statistical Society, Series C (Applied Statistics)*, **13** (1), 25–30.

McLennan, D., Barnes, H., Noble, M., Davies, J., Garratt, E., and Dibben, C. (2011) *The English Indices of Deprivation 2010*, UK Department of Communities and Local Government, London.

Metropolitan Police Service (2012) 4 Days in August: Strategic Review into the Disorder of August 2011, Available at: http://www.met.police.uk/foi/pdfs/priorities_and_how_we_are_doing/corporate/4_days_in_august.pdf (Accessed 7 May 2015).

Myers, D. (2000) The diffusion of collective violence: infectiousness, susceptibility, and mass media networks. *American Journal of Sociology*, **106** (1), 173–208.

Newman, M. (2011) Complex systems: a survey. *American Journal of Physics*, **79**, 800–810.

Riots Communities and Victims Panel (2011) 5 Days in August: An Interim Report on the 2011 English Riots, Available at: http://riotspanel.independent.gov.uk/wp-content/uploads/2012/04/Interim-report-5-Days-in-August.pdf (Accessed 7 May 2015).

Rossmo, K.D. (2000) *Geographic Profiling*, CRC Press LLC, Boca Raton, FL.

Solomos, J. (2011) Race, rumours and riots: past, present and future. *Sociological Research Online*, **16** (4), 20.

Stolzenberg, L. and D'Alessio, S.J. (2008) Co-offending and the age-crime curve. *Journal of Research in Crime and Delinquency*, **45** (1), 65–86.

Townsley, M. and Sidebottom, A. (2010) All offenders are equal, but some are more equal than others: variation in journeys to crime between offenders. *Criminology*, **48**, 897–917.

Ward, M.D., Greenhill, B.D., and Bakke, K.M. (2010) The perils of policy by p-value: predicting civil conflicts. *Journal of Peace Research*, **47** (4), 363–375.

Wilson, A.G. (2008) Boltzmann, Lotka and Volterra and spatial structural evolution: an integrated methodology for some dynamical systems. *Journal of the Royal Society, Interface / the Royal Society*, **5** (25), 865–871.

Wortley, R. (2008). Situational precipitators of crime, in *Environmental Criminology and Crime Analysis* (eds R. Wortley and L. Mazerolle), Willan Pub., Portland, OR, pp. 48–69.

10

Rebellions

Peter Baudains, Jyoti Belur, Alex Braithwaite, Elio Marchione and
Shane D. Johnson

10.1   Introduction

Rebellions frequently arise around the globe and have received considerable attention from
the academic community. Data-driven studies have traditionally analysed such civil conflicts
at the country-year level (e.g. Gurr, 1970; Fearon and Laitin, 2003) and rarely explicitly model
the interaction between rebels and the state (although there are notable exceptions, such as
Toft and Zhukov, 2012, 2015). In this chapter, we contribute to an emerging literature on the
temporal dynamics of rebellions by using a point process framework to model the evolution
of the Naxal rebellion within the Indian state of Andhra Pradesh at a daily level of resolution.
We model the occurrence of Naxal attacks and police counterinsurgent actions as a coupled
point process, which enables us to consider the level of interaction between the two sides of the
conflict.

The Naxal movement, whose name is taken from the small village of Naxalbari in West
Bengal, where a peasant revolt took place in 1967, is a left-wing extremist group who have
engaged in numerous attacks against civilians and the state in recent decades. Grievances of
the Naxal movement initially stemmed from economic inequality and rural agricultural workers' inaccessibility to land ownership (Ahuja and Ganguly, 2007). After being quashed by the
Indian government in the 1970s through the use of police and paramilitary forces (Basu, 2011),
several factions of the Naxal movement were formed, many of which had militant groups who
engaged in insurgency against the state. In the early 2000s, various Naxal groups merged to
form both militant (the People's Liberation Guerrilla Army) and political groups (the Communist Party of India (Maoist)). Insurgent violence continues to present day, but, in recent years,
tends to be restricted within localised regions in Eastern and North Eastern India.

The states of Andhra Pradesh and Telangana, the latter of which was formed in 2014 when
Andhra Pradesh bifurcated, experienced high levels of violence during the 2000s. Police peri-
odically adopted various counterinsurgency measures in an attempt to quell the rebellion,

Global Dynamics: Approaches from Complexity Science, First Edition. Edited by Alan Wilson.
© 2016 John Wiley & Sons, Ltd. Published 2016 by John Wiley & Sons, Ltd.

including the formation of an aggressive paramilitary group called the Greyhounds. On numerous occasions, the police were drawn into armed conflict with the rebels, resulting in both Naxal and police loss of life. Police counterinsurgent measures in Andhra Pradesh have been claimed to be effective in reducing levels of violence, despite limited quantitative studies (Sahni, 2007).

In this chapter, we analyse the temporal dynamics associated with the Naxal rebellion, using data on insurgent actions and counterinsurgent response. We employ a multivariate Hawkes process model and investigate how the rebellion changed by comparing parameter estimates and overall model fit for two time periods. Our findings identify differences in the dynamics of the rebellion as it evolved. The time periods considered fall on either side of a reported truce on the part of the police, which was called in response to the formation of the Communist Party of India (Maoist) in the hope that a diplomatic solution to the insurgency could be found. However, no solution was forthcoming and insurgent attacks continued, which led, eventually, to a resumed counterinsurgent campaign. We discuss how the temporal dynamics changed between the two time periods, considering any potential effect of this truce.

We begin by outlining the data used in this chapter and explain why it is of interest to consider distinct time periods before presenting the multivariate Hawkes process model. Similar models have recently been used to model a wide range of different temporal dynamics, including, in many cases, the occurrence of events related to crime and security. We outline how the parameters in the model can be obtained from the data using a maximum-likelihood procedure and explain how they might be interpreted. We then present the results for the two time periods considered. These consist of the maximum-likelihood parameter estimates associated with each of the two time periods, bootstrapped distributions for each of those estimates and also a residual analysis to examine overall model fit. We conclude by discussing the implications of these findings with respect to our understanding of the Naxal movement and to rebellions more generally.

10.2   Data

Police recorded crime data detailing hostile events associated with the Naxal rebellion for the 10-year period 2000–2010 in the state of Andhra Pradesh were provided by the Andhra Pradesh police. The data consisted of official police records of Naxal-related violence or threat recorded in the 1,642 police stations within the state. Counterinsurgent activities of police were not detailed explicitly in the data; however, fieldwork described in Belur (2010) and subsequent fieldwork conducted in left wing extremist affected areas suggests that aggressive police action took place and sometimes involved the killing of unwanted persons (in this case Naxals) during shootouts. Such events were recorded as an "exchange of fire" in the data. As a result, it is assumed that events described in the dataset as an "exchange of fire" between Naxal and police, and which resulted in at least one Naxal fatality, were largely caused by strategic counterinsurgency activities. On this basis, it is possible to partition the data set into events initiated by Naxals and events initiated by police.

The temporal distribution of the 4,820 incidents in the data set is shown in Figure 10.1, where the shading of each bar indicates the number of events that occur each week (darker shading corresponds to more events). This figure also distinguishes between the 4,234 Naxal events (in red) and the 586 police events (in blue). In what follows, we analyse the temporal dependence between these two sets of events. In particular, we examine the extent to which

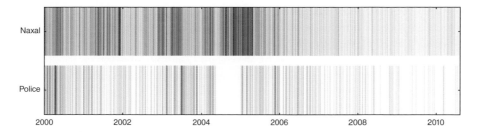

Figure 10.1   Temporal distribution of the event data

the temporal signatures of these events exhibit self- or mutually exciting dynamics with a multivariate Hawkes process.

As shown in Figure 10.1, midway through 2004, no police actions were identified for a period of around 6 months. This period coincides with a reported cease fire adopted by the Indian police. The number of Naxal attacks during this period does not have a similar reduction, and Figure 10.1 suggests that they may have even increased.

To examine the effect that this truce had on the dynamics of the insurgency, and in an attempt to set up a natural experiment, we split the data to be examined into two distinct time periods. The first, from the beginning of the data set on 1 January 2000 to 31 December 2003, contains the period of time before the truce and during which the conflict was at its most intense. Although the truce did not begin until May 2004, we truncate the first time period to 31 December 2003 to discount any dynamics in the build up to this truce. The second time period is taken to start once police actions resume after the truce on 1 January 2005. To ensure that the two time periods are of the same length, the second interval is taken to end on 31 December 2008. In what follows, we explain the model to be applied to these two time periods before calibrating it on them separately. We compare the results obtained and discuss any differences that arise.

10.3   Hawkes model

A number of studies have investigated the timings of events associated with issues in crime and security using a point process modelling framework. A wide range of structural variables can be incorporated into a point process model; however, recent examples have typically only included information on the timing of historic events. The reason for this is that many types of events, such as urban crime (e.g. Pease, 1998; Johnson et al., 2007) and insurgency (Townsley et al., 2008), have been shown to cluster in both space and time. In other words, the timing of one event often signals an elevation or "excitation" in the risk of others in the near future. For example, Holden (1986) uses the so-called Hawkes process (Hawkes, 1971), which accounts, for excitation in the rate at which events occur, to determine whether a contagion effect can be observed in the frequency of aircraft hijackings in the United States between 1968 and 1972. More recently, Hawkes processes have been employed to model the timings of events associated with gang rivalries (Egesdal et al., 2010) and civilian deaths during the Iraq war (Lewis et al., 2011). Extensions of this same model have also been used to consider the timings of terrorist attacks in Southeast Asia (Porter and White, 2012; White et al., 2013). Short et al. (2014)

propose a multivariate point process model to account for possible interaction effects arising from the behaviours of different gangs. Spatio-temporal models of point processes, in which the locations as well as the timings of future events are modelled, have been used to model burglary (Mohler et al., 2011) and insurgent warfare (Zammit-Mangion et al., 2012). Such studies demonstrate that, when calibrated, point process models can generate more accurate predictions than contending alternatives.

We define a point process as a collection of random events $\{(t_i, c_i)\}_{i=1,2,3,\cdots,N}$ ordered so that $t_i \leq t_{i+1}$, where t_i denotes the time at which event i occurred and c_i is a component index. A point process is simple if this inequality is strict for all values of i. In the present study, $c_i = 1$ denotes an event initiated by Naxals and $c_i = 2$ denotes an event initiated by police. If the timings of events for each component is modelled as a separate process, as will be the case in the models that follow, then the point process is multivariate. More formally, a multivariate point process is defined as a series of counts

$$Z_j : \mathcal{T} \to \{\mathbb{Z} | z \geq 0\}, \tag{10.1}$$

on some temporal domain $\mathcal{T} = [0, T)$ (for some maximum time $T \in \mathbb{R}$) defined by

$$Z_j(t) = \sum_{\substack{t_i < t \\ c_i = j}} \mathbf{1}_{t_i}([0, t)), \tag{10.2}$$

where $\mathbf{1}_{t_i}([0, t))$ is an indicator function, which is equal to one if $t_i \in [0, t)$ and equal to zero otherwise. The subscript j is used to denote the type of event (with $j = 1$ denoting Naxals and $j = 2$ denoting police).

The history of the system until some time t, $\mathcal{H}(t)$, is defined to be the set of events that have occurred before time t, so that

$$\mathcal{H}(t) = \{(t_i, c_i) | t_i < t\}. \tag{10.3}$$

The conditional intensity function associated with the count Z_j is defined as

$$\lambda_j(t | \mathcal{H}(t)) = \lim_{\delta t \to 0} \frac{\mathbb{E}(Z(t + \delta t) - Z(t) | \mathcal{H}(t))}{\delta t}. \tag{10.4}$$

For a given j, if $Z_j(t)$ is simple and finite for all $t \in \mathcal{T}$, then the associated conditional intensity function λ_j is unique (Daley and Vere-Jones, 2003). It follows that in order to define a particular simple and finite point process given by Z_j, it is sufficient to specify the function λ_j. Many models of point processes specify a functional form for the conditional intensity function, rather than for the count, and this is also the approach that is taken here.

We employ a two-dimensional Hawkes process with conditional intensity functions for the Naxal events, λ_N, and police events, λ_P, given by

$$\lambda_N(t | \mathcal{H}) = \mu_N + \sum_{\substack{t_i < t \\ m_i = 1}} \alpha_{NN} e^{-\omega_N(t - t_i)} + \sum_{\substack{t_i < t \\ m_i = 2}} \alpha_{PN} e^{-\omega_N(t - t_i)} \tag{10.5}$$

$$\lambda_P(t | \mathcal{H}) = \mu_P + \sum_{\substack{t_i < t \\ m_i = 1}} \alpha_{NP} e^{-\omega_P(t - t_i)} + \sum_{\substack{t_i < t \\ m_i = 2}} \alpha_{PP} e^{-\omega_P(t - t_i)}, \tag{10.6}$$

for parameters μ_N, α_{NN}, α_{PN} and ω_N, which describe the way in which the rate of Naxal attacks depends on the history of the system, and for parameters μ_P, α_{PP}, α_{NP} and ω_P, which describe the way in which the rate of police events depends on the history of the system.

The Hawkes models in Equations (10.5) and (10.6) suppose that events arise as a result of two possible processes: a background rate, which is modelled as a Poisson process with constant intensity, and excitation, which is modelled as a sum of terms that exponentially decay with the time from each event.

Background events are supposed to occur naturally throughout the duration of study with constant intensity. In the case of the Naxals, the rate of background events might indicate some baseline level of underlying grievances over the time period studied, where these grievances are equally likely to result in attacks at any point in time. In the case of the police, background events might occur due to a consistent policy directed to ensure a certain number of counterinsurgent operations take place in a given time frame. The parameters μ_N and μ_P define the background rate of the Naxal and police processes, respectively.

Excitation arises when a number of events occur more closely together than is considered to be plausibly represented by a constant Poisson process. As a result, these terms account for any temporal clustering in the process. In Equations (10.5) and (10.6), the occurrence of either type of event can contribute to an increase in the intensity functions. The magnitude of this increase is determined by the parameters α_{NN}, when Naxal events cause an increase in the Naxal intensity function; α_{PN}, when a police event causes an increase in the Naxal intensity function; α_{NP}, when a Naxal event causes an increase in the police intensity function; and α_{PP}, when police events increase the police intensity function. The decay terms ω_N and ω_P describe the rate at which this increased excitation decays, with higher values representing a more rapid decline in excitation.

Temporal clustering of events—which the excitation terms in Equations (10.5) and (10.6) are designed to capture—can arise as a result of a number of mechanisms. It may be that the occurrence of an event directly influences the chance of further events occurring. In the case of the Naxals, this can arise if insurgents are inspired to carry out further attacks once they have observed or taken part in a successful attack (Johnson and Braithwaite, 2009). Similarly, responding to prior police events, insurgents might retaliate as a direct result of police action. Indeed, tit-for-tat behaviour between opponents during insurgencies has been identified in a number of previous studies (Linke et al., 2012; O'Loughlin and Witmer, 2012; Braithwaite and Johnson, 2012). Clustering may also arise as a result of confounding factors that make certain time periods more likely to experience events, creating risk heterogeneity in time. For example, improved communication or a change in leadership within the rebellion group might coincide with an increase in the rate of attacks. This type of clustering can produce similar patterns in the event data to self-excitation and it can be difficult to disentangle the causal effect (Mohler, 2013).

In a similar way to insurgent attacks, clustering of police actions can be caused, on the one hand, by prior events (e.g. police may retaliate to Naxal attacks or be more incentivised to undertake further action after successful police actions) or, on the other, by confounding factors that generate risk heterogeneity (e.g. if the police are adhering to a recently introduced policy supporting the increased rate of counterinsurgent activity).

The Hawkes process does not distinguish between explanations of temporally varying factors and direct excitation; instead, only distinguishing between time-stable factors that might influence the rate of attacks—as described by the background rate—and the change in risk intensity due to clustering that is modelled as excitation. Nevertheless, the model is useful for a number of reasons. First, it can be used to identify temporal regularities in the data, which may then be studied further to consider the causal mechanisms that might be responsible. Second, the model can quantify the rate with which events are occurring in a given period, which can be used for assessing the severity of the disorder or assessing whether responses to any violence were appropriate. Third, the model is well suited to prediction: clustering of recent events can indicate whether events are more or less likely to occur in the future.

Parameter estimates can be obtained via a maximum-likelihood procedure. The log-likelihood function for the multivariate process in Equations (10.5) and (10.6) is given by

$$\log \mathcal{L}(\theta_N, \theta_P) = \sum_{\substack{t_i < t \\ m_i = 1}} \log \lambda_N(t_i; \theta_N) + \sum_{\substack{t_i < t \\ m_i = 2}} \log \lambda_P(t_i; \theta_P) - \int_0^T \lambda_N(s; \theta_N)ds - \int_0^T \lambda_P(s; \theta_P)ds,$$

(10.7)

where T denotes the end point of the time period of interest (Embrechts et al., 2011). The parameters $\theta_N = (\mu_N, \alpha_{NN}, \alpha_{PN}, \omega_N)$ and $\theta_P = (\mu_P, \alpha_{NP}, \alpha_{PP}, \omega_P)$ that maximise the value of the function in Equation (10.7) are the parameters that most closely fit the models in Equations (10.5) and (10.6) to the data.

The uncertainty of the parameters θ_N and θ_P can be estimated via a parametric bootstrap procedure (Wang et al., 2010). This is performed by simulating a large number of histories of a system defined by the models in Equations (10.5) and (10.6), specified with the parameters as obtained from the maximum likelihood procedure. For each simulated history, the parameters are then re-estimated. If these values are consistently close to the parameters estimated from the data, then there is confidence that those parameters specify a model that is close to what is observed. If there is a large range in re-estimated parameter values from these simulated processes, then the observed process might have been plausibly reproduced with quite different parameters, therefore indicating greater uncertainty.

While the bootstrap procedure enables the assessment of uncertainty in the model at the parameter level, a residual analysis can be used to determine how well the model fits the data overall. A residual process is obtained by probabilistically selecting events from the data set that occur when the modelled intensity is at its lowest. This process therefore selects events that were poorly predicted by the model (i.e. such events represent a departure between the model and the data that it attempts to recreate). If the model is a good fit to the data, then the residual events would be expected to form a Poisson process over the study period. Considering the extent to which the residual process deviates from a Poisson process, therefore, indicates the goodness of fit of the model.

In the next section, results are presented for two time periods during the Naxal rebellion. These results consist of point estimates for the models proposed in Equations (10.5) and (10.6); the distribution of these estimates as obtained from the bootstrap procedure; and the comparison between the residual processes obtained for each time period, compared to a constant Poisson process. The results provide a number of insights into the nature of the rebellion and these are then discussed.

Table 10.1   Point estimates for parameters in Equations (10.5) and (10.6). The calibration is performed separately for two different time periods

Time interval	μ_N	α_{NN}	α_{PN}	ω_N	μ_P	α_{PP}	α_{NP}	ω_P
2000–2003	0.5992	0.6054	0.0000	0.1592	0.0424	0.5684	0.0390	0.0633
2005–2008	0.0851	0.7461	0.8790	0.1394	0.0442	0.2528	0.0718	0.0599

10.4   Results

The vectors θ_N and θ_P that maximise [1] the log-likelihood over the two time periods and which therefore most closely fit the model in Equations (10.5) and (10.6) are shown in Table 10.1. When calculating these estimates, a randomisation procedure was performed on concurrent events, whereby a random time interval of less than 1 day is added to each event that occurs on the same day. This is performed to ensure that the data form a simple point process (with no two events occurring at the exact same point in time), thereby ensuring the uniqueness of the conditional intensity function. Although the procedure to remove concurrent events was random, meaning that slightly different event histories could be used to calibrate the parameters over repeated optimisations, there was very little variance in the optimal parameters obtained across 20 such procedures. The results reported are the means from these 20 optimisation procedures.

The point estimates in Table 10.1 show some striking differences between the two time periods, but there are also a number of similarities. The decay rates, given by ω_N and ω_P, are found to be quite similar for the two time periods. Any increased intensity due to the occurrence of events is estimated to decay over a period of around $0.15^{-1} = 6.67$ days in the case of Naxal events and around $0.06^{-1} = 16.67$ days in the case of police events. The background rate for police events, μ_P, is also similar over the two time periods considered. Around 0.04 police background events are expected to occur each day on average.

Background Naxal events occur with a rate of around 0.6 between 2000 and 2003 and with a rate of around 0.09 between 2005 and 2008, signifying quite a departure from the dynamics over the first time period. Perhaps to account for this severe reduction, excitation of Naxal events from both Naxal and police events is found to be much greater during the second time period considered. During the first time period, no excitation effect is found from police events on the Naxal intensity function, but quite a large excitation is observed in the second time interval.

The excitation of police events also varies over the two time intervals considered, although not to such a large extent. The size of self-excitation of police events decreases in the second time interval from the first, but retaliation from Naxal events increases.

The results in Table 10.1 are just point estimates and so, in order to determine the level of uncertainty associated with them, we plot the density of the distribution of bootstrap estimates for each parameter for 2000–2003 and for 2005–2008 in Figure 10.2. These densities are constructed by re-estimating the parameters from a simulation of the model in Equations (10.5) and (10.6), specified with the parameters in Table 10.1, a process which is repeated 100 times.

---

[1] The Nelder–Mead optimisation procedure is used (Nelder and Mead, 1965).

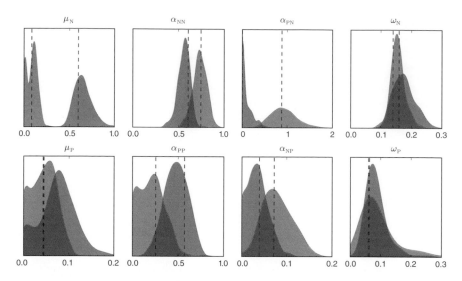

Figure 10.2   The distribution of each parameter, as obtained from the bootstrap procedure, for the period 2000–2003 (in blue) and 2005–2008 (in red). The shaded distributions are density plots obtained from 100 re-estimations of the parameters from simulations of the process in Equations (10.5) and (10.6). The dashed lines represent the point estimates in Table 10.1 and correspond to the parameters with which these simulations are specified

The point estimates for each time interval are plotted by a dashed line in each case. By examining these figures, it is possible to determine the degree of uncertainty associated with each estimate. If the distribution of bootstrap estimates spans a wide region on the x-axis of each figure, then there is a large uncertainty associated with the corresponding point estimate. If, on the other hand, the density plot only covers quite a small region and has a well-defined and sharp maximum, then there is low uncertainty associated with the estimate.

The distributions of the parameter values obtained from the bootstrap procedure are, on the whole, consistent with the point estimates. A number of the distributions appear to be truncated at zero, as can be detected by the smaller local peak on the left of the global maximum in a number of the distributions. This suggests that, in these cases, the parameters are likely to be indistinguishable from zero or from very small positive values. In addition, there appears to be greater uncertainty associated with the parameters of λ_P than the parameters of λ_N for both time periods, perhaps owing to the relatively smaller number of police events in comparison to Naxal events.

A number of the conclusions made above with regard to the differences between the two time intervals are supported by the distributions in Figure 10.2. For μ_N, the lack of any intersection in the two distributions supports the finding that the background rate of Naxal attacks decreased in the second time period. The distributions for α_{NN} and α_{PN} are consistent with the finding that excitation of the Naxal intensity function increased in the second time period, although the distributions do intersect over a small interval. The remaining parameters have distributions that intersect over quite a wide range of their possible values, although the modal values in the density plots are consistent with the change identified in the point estimates (with the exception of μ_P; although very little change in the point estimates was identified in this case).

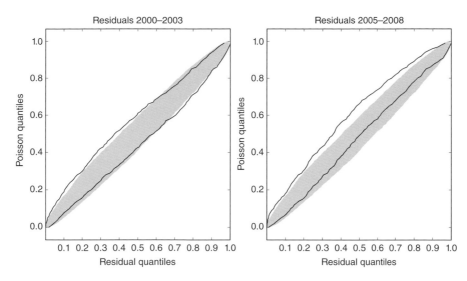

Figure 10.3   QQ-plots comparing the residual process to a Poisson process for the two time periods. The black lines represent the 95% confidence interval of the quantiles of the residual process in comparison to a Poisson process. These are obtained from 1,000 samples of the residual process. The grey shaded region represents an equivalent Poisson process

In order to determine the overall model fit, comparison of the residual process against a Poisson process is made in two QQ-plots in Figure 10.3. This figure is constructed by randomly simulating 1,000 residual processes of length 107 for the model of both Naxal and police events in Equations (10.5) and (10.6), using the parameters in Table 10.1 for the two time periods under consideration. Length 107 is chosen since this would be the expected number of Poisson distributed events (of both types) as calculated by the smaller of the background rates for the two periods (the second time period being the smallest). For each quantile value on the x-axis, the proportion of events in each residual process remaining is calculated, and the 95% confidence interval of these values is plotted against what would be expected assuming a Poisson process. Thus, the black lines in Figure 10.3 represent the 95% confidence of the residual process compared against a Poisson process. The grey shaded region plots what would be observed if the process was exactly a Poisson process.

The model fit is considered to be good if the residual process resembles a Poisson process, and thus if the black lines in Figure 10.3 lie within the grey shaded region. The process associated with the first time interval lies more closely within the grey shaded region than the second time interval and so we conclude that the model fits the first time period better overall.

10.5   Discussion

In this chapter, we have examined the temporal dynamics associated with the Naxal rebellion and the police response to it. We fitted a multivariate Hawkes process to events associated with both Naxal insurgent attacks and police counterinsurgent actions to two periods on either side of a truce undertaken by the police. For both periods, evidence was found for self-excitation of both Naxal and police events: Naxal events increased the likelihood of further Naxal events and police events increased the likelihood of further police events. In the second time period, the strength of self-excitation was increased for Naxal events and decreased for police events. The second time interval appeared to contain higher levels of mutual excitation: police actions increased the likelihood of Naxal attacks and vice versa and more so in the second period than in the first. In addition, the background rate of Naxal attacks decreased significantly in the second period.

The sensitivity of these findings has been explored via a bootstrap procedure to obtain an indication of parameter uncertainty, as well as by employing a residual process to ascertain overall model fit. When interpreting these findings, it should be borne in mind that the model fits the data better in the first time interval, when more of the events can be attributed to the background process, than in the second time interval, in which excitation appears to be more dominant.

Despite limitations associated with the sensitivity of the model, our findings suggest that the second time interval had distinct dynamics from the first. There were fewer events (as can be seen from Figure 10.1), but they tended to be more susceptible to clustering via excitation processes. There are a number of mechanisms that might account for this pattern. It may be that the grievances of the Naxals were, on the whole, much reduced in the second time interval, but that sporadic outbursts of events arose in response to a particular counterinsurgent action or other policy aimed at the insurgents. The increased clustering of events during the second time period is consistent with a smaller but more organised insurgent campaign, which seeks to maximise impact by committing a number of attacks in a short period of time (Johnson and Braithwaite, 2009). These regularities might also be explained by an effective counterinsurgent campaign, which reduced the likelihood of events for the majority of the second time interval, but which suffered periodic setbacks in the form of a number of closely clustered attacks.

Despite uncertainty with respect to the causal mechanisms, the pattern identified in this exploratory approach can help stimulate further, more detailed analysis. Further work might investigate the differences identified and construct models that directly incorporate causal mechanisms and structural covariates. Baudains (2015) disaggregates the model presented here to incorporate geography and examine the resulting spatial excitation. He also utilises a predictive analysis to determine the extent to which regularities in the data might be used as a predictive tool. This demonstrates how point process frameworks might be used in a policy setting (see also Zammit-Mangion et al., 2012; Mohler et al., 2011).

In conclusion, we have shown that regularities exist within the social dynamics of rebellions that can be exploited in a statistical framework to obtain insights. Importantly, using the history of the system as the only data in the model has generated these insights, highlighting the utility of the point process framework. Our findings emphasise the increased prominence of excitation as the conflict progressed. In addition to self-excitation, we conclude that mutual excitation plays an important role, indicating the presence of retaliation or tit-for-tat behaviour, particularly during the latter stages of the insurgency. The analysis of other rebellions using a similar framework might further improve our understanding of these types of social dynamics.

References

Ahuja, P. and Ganguly, R. (2007) The fire within: naxalite insurgency violence in India. *Small Wars and Insurgencies*, **18** (2), 249–274.

Basu, I. (2011) Security and development - are they two sides of the same coin? Investigating India's two-pronged policy towards left wing extremism. *Contemporary South Asia*, **19** (4), 373–393.

Baudains, P.J. (2015) Spatio-temporal modelling of civil violence: four frameworks for obtaining policy-relevant insights. Doctoral Dissertation, University College London, London.

Belur, J. (2010) *Permission to Shoot? Police Use of Deadly Force in Democracies*, Springer-Verlag, New York.

Braithwaite, A. and Johnson, S.D. (2012) Space-time modeling of insurgency and counter-insurgency in Iraq. *Journal of Quantitative Criminology*, **28** (1), 31–48.

Daley, D. and Vere-Jones, D. (2003) *An Introduction to the Theory of Point Processes, Volume I: Elementary Theory and Methods in Applied Sciences*, No. 1, 2nd edn, Springer-Verlag, New York, pp. 114–122.

Egesdal, M., Fathauer, C., Louie, K., and Neuman, J. (2010) Statistical and stochastic modeling of gang rivalries in Los Angeles. *SIAM Undergraduate Research Online*, **3**, 72–94.

Embrechts, P., Liniger, T., and Lin, L. (2011) Multivariate Hawkes processes: an application to financial data. *Journal of Applied Probability*, **48**, 367–378.

Fearon, J.D. and Laitin, D.D. (2003) Ethnicity, insurgency, and civil war. *American Political Science Review*, **97** (1), 75.

Gurr, T. (1970) *Why Men Rebel*, Princeton University Press, Princeton, NJ.

Hawkes, A.G. (1971) Spectra of some self-exciting and mutually exciting point processes. *Biometrika*, **58** (1), 83–90.

Holden, R.T. (1986) The contagiousness of aircraft hijacking. *American Journal of Sociology*, **91** (4), 874–904.

Johnson, S.D., Bernasco, W., Bowers, K.J., Elffers, H., Ratcliffe, J., Rengert, G., and Townsley, M. (2007) Space-time patterns of risk: a cross national assessment of residential burglary victimization. *Journal of Quantitative Criminology*, **23**, 201–219.

Johnson, S.D. and Braithwaite, A. (2009) Spatio-temporal distribution of insurgency in Iraq, in *Countering Terrorism through SCP, Crime Prevention Studies*, vol. **25** (eds J.D. Freilich and G.R. Newman), Criminal Justice Press, New York, pp. 9–32.

Lewis, E., Mohler, G., Brantingham, P.J., and Bertozzi, A.L. (2011) Self-exciting point process models of civilian deaths in Iraq. *Security Journal*, **25** (3), 244–264.

Linke, A.M., Witmer, F.D., and O'Loughlin, J. (2012) Space-time granger analysis of the war in Iraq: a study of coalition and insurgent action-reaction. *International Interactions*, **38** (4), 402–425.

Mohler, G. (2013) Modeling and estimation of multi-source clustering in crime and security data. *The Annals of Applied Statistics*, **7** (3), 1525–1539.

Mohler, G.O., Short, M.B., Brantingham, P.J., Schoenberg, F.P., and Tita, G.E. (2011) Self-exciting point process modeling of crime. *Journal of the American Statistical Association*, **106** (493), 100–108.

Nelder, J.A. and Mead, R. (1965) A simplex method for function minimization. *The Computer Journal*, **7**, 308–313.

O'Loughlin, J. and Witmer, F.D. (2012). The diffusion of violence in the North Caucasus of Russia, 1999–2010. *Environment and Planning A* **44**: 2379–2396.

Pease, K. (1998). Repeat victimization: taking stock. *The Home Office: Police Research Group: Crime Detection and Prevention Series Paper 90.*

Porter, M.D. and White, G. (2012) Self-exciting hurdle models for terrorist activity. *The Annals of Applied Statistics*, **6** (1), 106–124.

Sahni, A. (2007) Andhra Pradesh: the state advances, the Maoists retreat. *South Asia Intelligence Review: Weekly Assessments and Briefings*, **6** (10), 1.

Short, M.B., Mohler, G.O., Brantingham, P.J., and Tita, G.E. (2014) Gang rivalry dynamics via coupled point process networks. *Discrete and Continuous Dynamical Systems - Series B*, **19** (5), 1459–1477.

Toft, M.D. and Zhukov, Y.M. (2012) Denial and punishment in the North Caucasus: evaluating the effectiveness of coercive counter-insurgency. *Journal of Peace Research*, **49** (6), 785–800.

Toft, M.D. and Zhukov, Y.M. (2015) Islamists and Nationalists: rebel motivation and counterinsurgency in Russia's North Caucasus. *American Political Science Review*, **109** (2), 222–238.

Townsley, M., Johnson, S.D., and Ratcliffe, J.H. (2008) Space time dynamics of insurgent activity in Iraq. *Security Journal*, **21** (3), 139–146.

Wang, Q., Schoenberg, F.P., and Jackson, D.D. (2010) Standard errors of parameter estimates in the ETAS Model. *Bulletin of the Seismological Society of America*, **100** (5A), 1989–2001.

White, G., Porter, M.D., and Mazerolle, L. (2013) Terrorism risk, resilience and volatility: a comparison of terrorism patterns in three Southeast Asian countries. *Journal of Quantitative Criminology*, **29** (2), 295–320.

Zammit-Mangion, A., Dewar, M., Kadirkamanathan, V., and Sanguinetti, G. (2012) Point process modelling of the Afghan War Diary. *Proceedings of the National Academy of Sciences of the United States of America*, **109** (31), 12414–12419.

11

Spatial Interaction as Threat: Modelling Maritime Piracy

Elio Marchione and Alan Wilson

11.1 The Model

Spatial interaction modelling is usually associated with trips or other kinds of flows. In this Chapter, we aim to show how it can be used to represent threat. We illustrate this with a model of maritime piracy: the pirates threaten shipping, naval vessels can offer some defence. We are aware, however, that this idea can be applied much more widely.

Our study area is shown in Figure 11.1. This shows the broader geographical context and then, specifically, we focus on Somalia and the Gulf of Aden. This work complements Marchione et al. (2014).

The specific area is shown in Figure 11.2 – divided into a kilometre-square grid. In this figure, we show the likely density of vessels passing through the Gulf to and from the Suez Canal. We show below how this is constructed as part of the model.

To formulate the model, we define the following variables: X_i, measure of propensity to attack from i; $\hat{W}_j^k$, volume field of type k ship in cell j; $\hat{N}_j$, volume field of naval units in cell j; L_{ij}^k, level of threat from origin i to cell j for ship k; $T_j^k(t)$ total threat at cell j of type k ship; c_{ij}, distance from origin i to cell j.

The model can then be constructed as follows. In Equation (11.1), $L_{ij}^k(t)$ is the level of threat delivered from a pirate base at zone i to vessels in cell j – measured as X_i. $\hat{W}_j^k(t)$ is a measure of the volume of shipping in cell j and $\hat{N}_j(t)$ is a measure – in commensurable units (see below, through the parameter α in Equation (11.3)) – of the defence that can be offered by naval vessels.

$$L_{ij}^k(t) = X_i \cdot (\hat{W}_j^k(t) - \hat{N}_j(t)) \cdot e^{-\beta c_{ij}} \tag{11.1}$$

Global Dynamics: Approaches from Complexity Science, First Edition. Edited by Alan Wilson.
© 2016 John Wiley & Sons, Ltd. Published 2016 by John Wiley & Sons, Ltd.

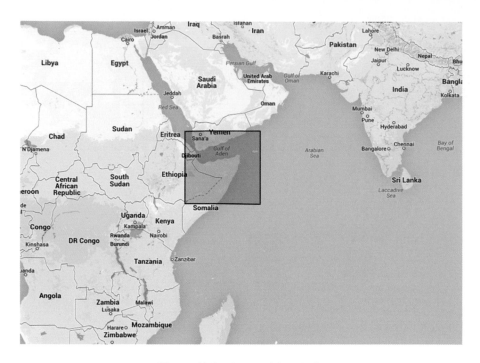

Figure 11.1   Area under analysis

A spatial deterrence function with parameter β is added. $\hat{W}_j^k(t)$ and $\hat{N}_j(t)$ are calculated from Equations (11.2) and (11.3), respectively.

$$\hat{W}_j^k(t) = \sum_z W_z^k(t) \cdot e^{-\mu c_{jz}} \tag{11.2}$$

$$\hat{N}_j(t) = \alpha \sum_z N_z(t) \cdot e^{-\nu c_{jz}} \tag{11.3}$$

These also include spatial deterrence functions with parameters μ and ν. β determines the range of the pirates from bases, μ the extent to which ships are likely to be found at some distance from what is regarded as the optimal route, and ν the effectiveness of naval deterrence in a cell some distance away from the actual vessel.

The total threat at cell j can then be calculated by summing over all pirate sources, i, and this is done in Equation (11.4).

$$T_j^k(t) = \sum_i L_{ij}^k(t) \tag{11.4}$$

The next step is to test the model for an idealised but plausible situation.

11.2   The Test Case

We construct a test case by assigning shipping to an optimum route and then using Equation (11.2) to allocate ships away from this route (as happens in practice). The set of lines that

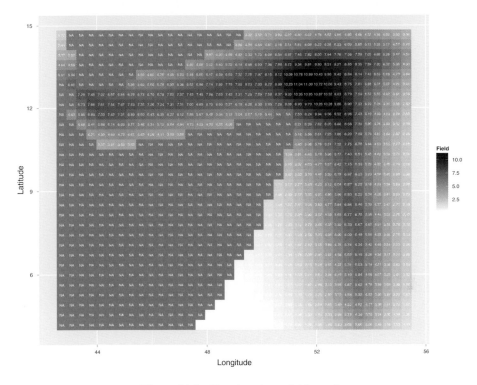

Figure 11.2 Vessels volume field $\mu = 8$

determine the optimum route and the means of generating the 'volume of shipping' field are shown in Appendix. These volumes by cell can be interpreted as forming a 'volume of shipping' field – indeed as a field of probabilities. This field has already been shown in Figure 11.2. We have now been able to explain how it was constructed. In the case of naval vessels, we assign three to particular points for this test and the areas of influence are described by Equation (11.3). These generate a 'naval defence' field via Equation (11.3) (see Figure 11.3).

These terms can be fed into Equation (11.1) to generate the level of threat from a particular pirate source. For the test case, we assume three port bases in Somali, the three mother ships, which act as bases at the sea. The summation in Equation (11.4) then generates a threat field–, see Figure 11.4. We use the following parameters to produce the fields:

$\beta = 10$
$v = 6$
$\mu = 8$

11.3 Uses of the Model

It is interesting to compare the threat field with the observed attacks in 2010 in Figure 11.5.

There is a reasonable degree of fit – by eye, since this is not a formal process. However, it does indicate that it should be possible to calibrate a model of this kind against data. More

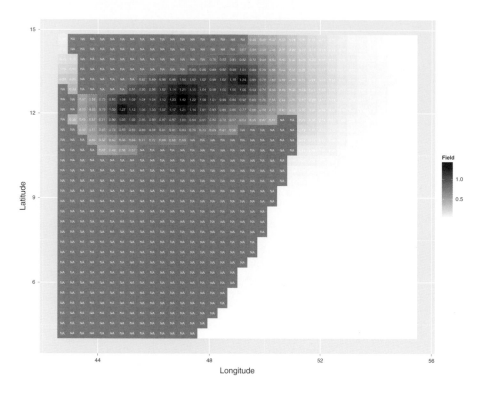

Figure 11.3 Navy field $v = 6$

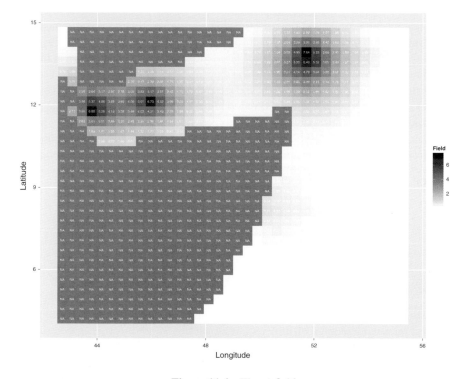

Figure 11.4 Threat field

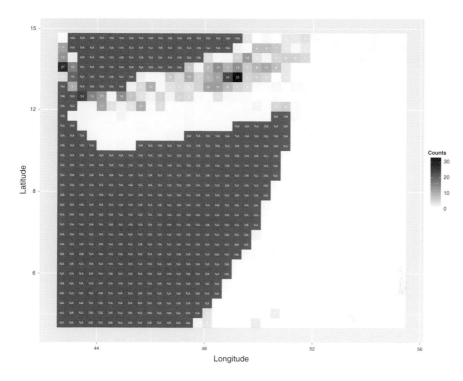

Figure 11.5   Observed attacks in 2010

importantly, it should then be possible to optimise naval strategies on a real-time basis –, given appropriate intelligence on the likely location of pirate sources. Such optimisation might be along the lines: let the naval units available be 3, then the $\min \left(\sum_{ik} T_j^k (t) \right)$ by varying $N_z(t)$ where $\sum_z N_z(t) \in \{3\} \subset \mathbb{N}$ and keeping $W_z^k(t)$ and pirates' port and mother ships' locations constant would identify the best naval units location to deter the threat coming from pirates.

This may be a particularly interesting problem because it would have to be run in real time as possible pirate positions shift or new intelligence is collected.

Reference

Marchione, E., Johnson, S., and Wilson, A. (2014) Modelling maritime piracy: a spatial approach. *Journal of Artificial Societies and Social Simulation*, **17** (2), 9.

Appendix

A.1   Volume Field of Type k Ship

Figure 11.2 shows the volume field of type k ship. It is obtained assuming that W_z^k are all those cells that touched the polyline Lin1 + Line2 or Line1 + Line3 (see Figure A.6):

- Line1
    1. point: Longitude = 42.75476, Latitude = 13.57411
    2. point: Longitude = 44.04895, Latitude = 11.76767
    3. point: Longitude = 47.93152, Latitude = 12.49025
    4. point: Longitude = 50.51990, Latitude = 13.21282
- Line2
    1. point: Longitude = 50.51990, Latitude = 13.21282
    2. point: Longitude = 55.26526, Latitude = 13.93539
- Line3
    1. point: Longitude = 50.51990, Latitude = 13.212820
    2. point: Longitude = 53.10828, Latitude = 11.767672
    3. point: Longitude = 53.53967, Latitude = 4.180644

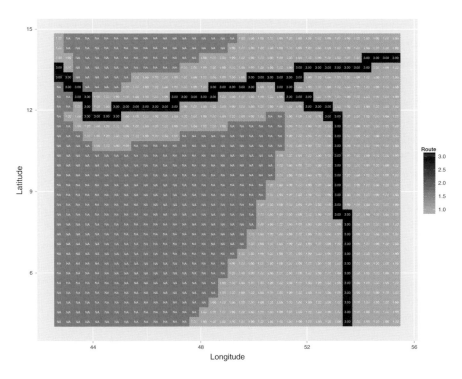

Figure A.6   Vessels ideal route

A.2 Volume Field of Naval Units

Three naval force units at Longitude/Latitude

1. 44.87579 / 12.12896
2. 47.03278 / 12.49025
3. 49.18976 / 13.21282

A.3 Pirates Ports and Mother Ships

Three pirates ports at Longitude / Latitude

1. 47.857500 / 4.654444 (Harardhere)
2. 49.815066 / 7.980800 (Eyl)
3. 48.524975 / 5.351592 (Hobyo)

and three mother ships at Longitude / Latitude

1. 43.725586 / 11.931852
2. 45.791016 / 12.039321
3. 51.789551 / 13.859414

Figure A.7 shows their locations and Figure A.8 shows their distances.

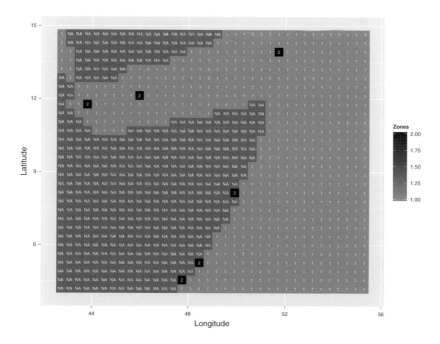

Figure A.7   Pirates ports and mother ships

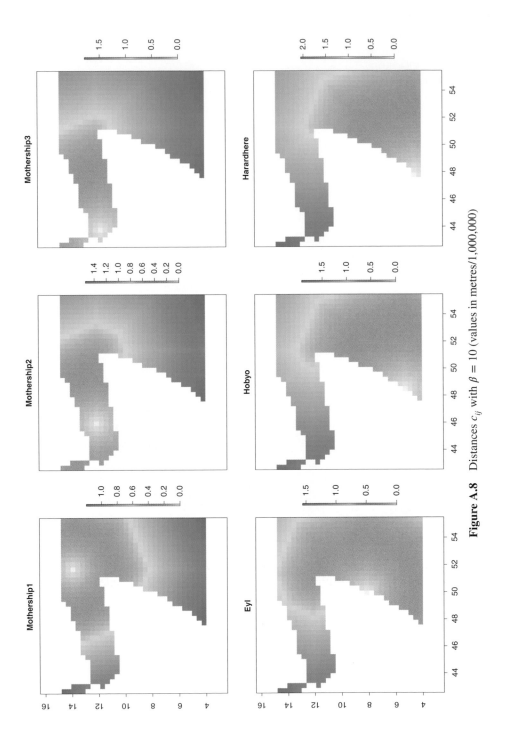

Figure A.8   Distances c_{ij} with $\beta = 10$ (values in metres/1,000,000)

12

Space–Time Modelling of Insurgency and Counterinsurgency in Iraq

Alex Braithwaite and Shane Johnson

12.1 Introduction

The United States and its Coalition partners concluded combat operations in Iraq in August 2010. Rather surprisingly, little empirical evidence exists as to the factors that contributed to localized patterns of Improvised Explosive Device (IED) attacks. To compensate somewhat for this dearth in evidence, this chapter investigates an increasingly relevant yet understudied phenomenon in counterinsurgency (COIN) policy-making: the dilemma governments face in determining how best to balance the use of various coercive actions when attempting to minimize the threat posed by campaigns of violence carried out by non-state actors. This study is carried out by assessing the co-evolving space *and* time (hereafter, space–time) distributions of insurgency and counterinsurgency in Iraq in 2005. To do so, we employ a novel analytic technique that helps us to assess the sequential relationship between these two event types.

Forty years have passed since the emergence of a modern wave of transnational terrorism and insurgency (Hoffman, 2006; Wilkinson, 2001) with the rise of hijackings of Israeli airliners by Palestinian organizations. Israel's preferred approach to countering the threat of Palestinian violence in the four decades since has been to retaliate with military strikes against terrorist cells, munitions stores and leadership hideouts. Israel has commonly compared their approach to the 'tit-for-tat' strategy (cf., Axelrod, 1984), while critics have argued that it is, in fact, more accurately, disproportionate in scope and deadliness. Empirical evidence suggests that episodes of heightened retaliation by Israel are associated with short-term decreases but long-term increases in subsequent violence (Maoz, 2007).

Global Dynamics: Approaches from Complexity Science, First Edition. Edited by Alan Wilson.
© 2016 John Wiley & Sons, Ltd. Published 2016 by John Wiley & Sons, Ltd.

The Israeli–Palestinian case neatly parallels the expectations of the body of formal, game-theoretic models that identify a proactive response dilemma faced by governments countering the threat of terrorism and insurgency (see, e.g. Arce and Sandler, 2005; Bueno de Mesquita, 2005a, 2005b, 2007; Bueno de Mesquita and Dickson, 2007; Rosendorff and Sandler, 2004; Siquiera and Sandler, 2007). These studies derive propositions suggesting that when responding to being attacked by non-state actors, governments are expected by their constituents to respond proactively in order to demonstrate that they are making an active effort to counter the threat. At the same time, however, it has been shown that if the government's response is too harsh, it is likely to increase antigovernment grievances, aid recruitment to violent non-state organizations and, ultimately, increase the number of spectacular attacks against the state (e.g. Rosendorff and Sandler, 2004). Take the ongoing COIN operations in Afghanistan and Iraq, for instance. A number of commentators (e.g. Ryu, 2005) and scholars (e.g. Pape, 2003, 2005) have posited that continued use of coercive COIN tactics (such as the cordoning off, searching and raiding of city blocks and buildings) help to explain the apparent increase in grievances among local populations and the observed increase in recruitment to insurgent and terrorist organizations in both countries. This process is highlighted, for instance, in the case of Northern Ireland (LaFree et al., 2009a).

However, with very few exceptions (e.g. LaFree et al., 2009a), the literature lacks empirical testing of the propositions garnered from game-theoretic models. Moreover, there is a distinct absence of robust and reliable recommendations regarding the costs and benefits of maintaining significant coercive presence on the ground in such theatres of combat. Accordingly, military and political analysts attach great value to information about empirical trends in ongoing campaigns of terrorism and insurgency. The aim of the present study is to inform decision-making regarding the optimal allocation of coercive and conciliatory measures in COIN operations, and to examine whether coercive COIN policies motivate new attacks in the vicinity of their deployment. The longer-term goal of the research is to advance efforts to predict the location of hot spots and to explain, more specifically, what timing and location of COIN operations are most effective in countering the emergence of new attacks and hot spots.

The chapter proceeds as follows. First, we offer a discussion of the dynamics of the US/Coalition counterinsurgency in Iraq. This discussion draws upon some common claims about the interaction between insurgent violence and COIN operations and is used to derive some testable hypotheses. We then offer up a discussion of the rich source of data to be utilized in the remainder of the study before moving on to specify a number of analytical techniques for modelling the spatial, the temporal and, most crucially, the space–time distributions of insurgency and counterinsurgency, along with the co-evolution of these two types of events. Finally, we offer a discussion of the findings, touching upon possible implications for military operations and future avenues for research on this topic.

12.2   Counterinsurgency in Iraq

A total of 4,738 Coalition troops (4,420 US troops) died during the 7 years of combat operations in Iraq. Reports suggest that by the end of 2007, approximately 63% of casualties resulted from IED attacks.[1] IED attacks have presumably risen to the fore in the Iraqi insurgency and

---

[1] http://www.washingtonpost.com/wp-dyn/content/graphic/2007/09/28/GR2007092802161.html Accessed September 12, 2010.

have become the preferred choice of insurgents because of their ease of manufacture and use, as well as their considerable affordability. IEDs typically require little more than modest amounts of materials that are common to agriculture and the production of paint, such as potassium chlorate and ammonium nitrate aluminium mix. Easy access to these materials also provides an opportunity for insurgents to get around the otherwise significant restrictions on and monitoring of sales of commercial explosives.

As with earlier work (Johnson and Braithwaite, 2009), we consider terrorist and insurgent violence in Iraq to be sufficiently similar in form as to not look to differentiate between them – choosing to focus, instead, upon the population of IED attacks. The reason for this is that where differences do exist, they tend to be in terms of the intended target of violence. Traditionally, the intended target of terrorism is the public, while insurgent violence is employed against military and government targets. Given that many violent tactics employed by non-state actors across Iraq commonly have a joint impact upon public, military and government targets, any effort to draw a clear distinction between these two strategic options would likely prove futile. Thus, while we will refer to insurgency and counterinsurgency, we are actually focused more broadly upon all violent attacks carried out by non-state actors (and catalogued by US/Coalition forces) and the operations the US/Coalition forces have employed in response to this violence.

Building upon a tradition of criminology scholarship (e.g. Johnson and Bowers, 2004; Johnson et al., 2007) recent work (Townsley et al., 2008; Johnson and Braithwaite, 2009) has demonstrated that insurgent attacks in Iraq in 2004 and 2005 clustered in space–time, spreading much like an infectious disease (and crime). These studies were motivated by basic principles regarding human mobility and theories of insurgent strategy (see Johnson and Braithwaite, 2009), but all converge on the prediction that insurgent attacks cluster in space–time. In essence, such conclusions parallel those found for campaigns of terrorism and insurgency in Spain (LaFree et al., 2009b), for general global patterns of transnational terrorism (Braithwaite and Li, 2007) and for militarized interstate disputes (Braithwaite, 2010). In the current study, for the purpose of comparison with the novel analyses discussed later, we present findings that demonstrate that IED explosions cluster simultaneously in space *and* time (these results were previously reported in Johnson and Braithwaite, 2009) and illustrate the precise patterns, but this is not our primary focus.

One weakness of previous studies is that while they examine insurgent activity, they do so without reference to other factors that might influence insurgent decision-making, such as counterinsurgent action. Consequently, it is difficult to ascertain whether the patterns observed – that IED attacks cluster in space-time – are the result of insurgents adopting particular strategies, or whether they emerge as a result of tit-for-tat interactions whereby an IED attack provokes a swift military response in the local area, which subsequently provokes further insurgent activity, and so on.

Accordingly, the present study looks to extend extant work by examining a key correlate of these observed patterns – the individual activities of the COIN operations of the Coalition forces. Specifically, we hypothesize that similar to attacks, COIN operations will occur closer in time and space than would be expected on a chance basis. Such an expectation is clearly in line with doctrinal approaches to COIN practice. For instance, an unclassified field manual on IED defeat operations clearly states that it is a working assumption of US/Coalition COIN practices that insurgents will continue to target locations at which they have previously been successful and that this behaviour will continue unabated until such a time as COIN actions force a change. Thus, logically, doctrine dictates that COIN efforts be aimed at disrupting insurgent activities at the locale of prior attacks (Dept. of the Army, 2007).

H1: *Given their strategic deployment in response to or pre-empting IEDs (which are known to cluster in space–time), COIN operations also cluster in space–time.*

Naturally, of course, we might also anticipate that insurgent attacks cluster in areas proximate to and at times after various COIN operations are carried out. Such an expectation would certainly align with the tit-for-tat hypothesis. To explain when and under what conditions COIN operations are likely to prove effective in countering the threats they are designed to tackle (i.e. when and where they are more or less likely to successfully reduce subsequent levels of insurgent attacks in the vicinity), in what follows we combine the concept of the Proactive Response Dilemma derived from game-theoretic work on terrorism and insurgency with the central tenets of Situational Crime Prevention (SCP) theory. Crucially, both approaches – from distinct disciplinary backgrounds – are grounded in rational choice.

Many common characterizations of insurgent campaigns point towards a strong element of strategic interaction between insurgents and the governments that they challenge (Pape, 2005; Kydd and Walter, 2006). It is important to note, for instance, that insurgent decisions to engage in violent activities are not the result of a singular desire for destruction but, rather, are intended to alter the decision-making and capacity of their opponent (Kydd and Walter, 2006; Bueno de Mesquita, 2007). IEDs are favoured, in particular, because they reduce the mobility of the opponent's troops and, therefore, reduce their ability to engage with the local population, thereby undermining the more holistic ambitions of COIN operations.

Evidence of an evolution in campaigns of insurgency and of a strategic interaction between state and non-state actors would align with certain expectations from the terrorism literature. A debate pervades this literature in which it is argued, on the one hand, that military force as part of a COIN campaign works. The opposing view identifies a distinct tension related to such applications of force in which they are seen as merely provoking additional subsequent attacks. These studies identify a dilemma faced by governments that need, on the one hand, to engage in some proactive kinetic COIN practices in order to provide a basic level of national security, yet are conscious of the fact, on the other hand, that uses of military force can have the effect of mobilizing popular support for the non-state actors they are attempting to eradicate (Bueno de Mesquita and Dickson, 2007; Rosendorff and Sandler, 2004; Faria and Arce, 2005).

Rosendorff and Sandler (2004) address the costs and benefits of proactive policy responses to terrorist activities. They develop a two-player proactive response game in which the government's level of proactivity and the terrorist's subsequent choice of target (normal or spectacular) are endogenized. If the government responds too harshly, it runs the risk of inadvertently empowering the terrorists by motivating a wider aggrieved population – ultimately leading to the potential for spectacular events. Bueno de Mesquita (2005b) demonstrates, however, that while aggressive responses can have the effect of increasing support for terrorists, counterterrorist policies occasionally do not radicalize popular support. This varying response depends upon the amount of damage caused by the government's actions and, perhaps most crucially, the perceptions and reactions of the 'aggrieved population'. Kalyvas (2006) reiterates this logic by noting, specifically, that the state's reliance upon the collective targeting of combatants was responsible for increased levels of insurgent violence in 45 historical cases of insurgencies.

In the specific context of the Iraqi insurgency, it is possible to identify this very dilemma in the strategic choices of the US-led Coalition. In the early period of the insurgency (from the intervention in 2003 to the surge in 2007), Coalition forces faced violence based upon both attrition and intimidation strategies (Kydd and Walter, 2006). These strategies combined sustained general violence designed to wear down the willingness of the foreign occupying forces, with targeted assassinations and attacks, designed to intimidate coalition and local security forces and encourage them to retreat from specific, high-value targets. The Coalition-led forces had a range of possible options that they could have employed in response. These included making concessions on inessential issues, narrowly targeting their retaliation against the leadership of the insurgency, hardening key targets, and engaging in public education to undermine the prevalent fear among the populations local to the violence (see Kydd and Walter, 2006, pp. 64–67). In other words, the Coalition forces had a range of proactive and reactive, coercive and conciliatory options available.

Kydd and Walter (2006, pp. 67) quote US Secretary of State, Condoleezza Rice, in October 2005, saying that the appropriate response to the insurgent use of intimidation in Iraq, ' … is to clear, hold, and build: clear areas from insurgent control, hold them securely, and build durable national Iraqi institutions'. In particular, this COIN strategy of 'Clear and Hold' manifested itself in the employment of the following key tactics: cordon searches of areas and buildings, raids on buildings, the clearing of weapons caches and the search for and clearing (where possible) of deployed IEDs.

In extending previous work, the current study not only examines the underlying distribution of COIN operations individually but also examines the co-evolution of COIN operations alongside IED attacks. This interest is derived from a simple question as to whether or not COIN operations are designed and executed to simply respond to observed insurgent attacks or also with a view towards reducing subsequent levels of insurgent attacks – as is clearly argued in the US military doctrine (Dept of the Army, 2007). In the aggregate, we anticipate that in addition to clustering individually, the two types of events – IEDs and COIN – will cluster sequentially.

II2: *By virtue of their design to deal with IEDs, COIN operations will cluster in a manner that follows that of IEDs.*

Given the logic of the literature on the proactive response dilemma, however, we anticipate that certain disaggregated COIN operation choices may reduce or exacerbate subsequent levels of insurgent attacks in the vicinity of their operation. In particular, three key characteristics of a COIN operation might, according to this logic, help determine whether COIN operations are followed by higher or lower levels of insurgency. First, operations can be distinguished along a spectrum from discriminate (targeted against combatants) to indiscriminate (more likely to affect non-combatants). Our expectation, drawing upon the terrorism and insurgency literature discussed earlier, is that less discriminating tactics are more likely to provoke a backlash in the form of active or passive support for the insurgency (see, cf. Rosendorff and Sandler, 2004; Kydd and Walter, 2006; Bueno de Mesquita, 2007). Second, operations can be identified by whether or not they directly or indirectly reduce the capacity of the insurgent forces to conduct additional attacks. More specifically, some COIN activities directly remove resources in the form of weapons and IED components from the stockpiles of insurgent groups, functionally reducing the force that they can bring to bear against the Coalition forces. Third, COIN operations can be compared in terms of whether or not they require the deployment of additional troops at a particular location for any extended period. This third aspect matters, we claim, because additional troops on the ground means additional Coalition assets that can be targeted by subsequent violence. Given that each of these three characteristics of COIN activities can vary, we anticipate that certain configurations of these characteristics should correlate with higher or lower subsequent levels of insurgency in the vicinity of COIN operations as follows:

H3: *COIN operations have the effect of reducing levels of IEDs at proximate locations and times when those operations have the effect of directly reducing insurgent capacity and when those operations are more discriminatory.*

H4: *COIN operations have the effect of provoking increased levels of IEDs at proximate locations and times when those operations involve the placement of observable troops on the ground and when those operations are less discriminatory.*

12.3 Counterinsurgency Data

Herein, our priority is to build upon extant research by more directly testing the empirical relationship between insurgency and COIN operations. To do so, we utilize five data series, each of which is drawn from Significant Activity (SIGACTS) reports covering the period January 1–June 30, 2005. Johnson and Braithwaite (2009) offer a detailed discussion of the SIGACTS reports and their level of classification. All of the data that we draw upon herein are taken from this same data source, although with one exception all of the analyses presented here are novel. It is worth noting at this time that each entry in this database is the product of on-the-ground reporting by members of the Coalition forces. Accordingly, one could be concerned that any observed patterns of clustering reflect the distribution of locations at which Coalition forces happen to be deployed and from which they are able to observe and report events, rather than from some systematic and purposive distribution of capabilities by insurgents. We think that this concern is unfounded, however, given that members of the armed forces engaged in reporting through SIGACTS reports are not limited in the scope of the insurgent events that they are permitted to report. Given the significant emphasis placed upon countering the threat of IEDs, in particular, there remains a high likelihood that individuals will report even those events to which they are not local.

In previous work (Townsley et al., 2008; Johnson and Braithwaite, 2009), space–time patterns in both the use of IEDs and non-IED insurgent attacks across Iraq have been explored. Herein, we examine the coincidence of IED attacks and COIN operations across Iraq. In particular, as noted in Table 12.1, we examine four COIN operation data series alongside the data on IED attacks. All data were recorded accurate to a spatial resolution of 100 m.

There are a total of 3,775 observations of 'IED Explosion'. IEDs take a variety of forms, including being Vehicle Borne (VBIED), Remote Controlled (RCIED), Personnel Borne (PBIED) and taking the more familiar form of a roadside bomb, triggered by moving vehicles. Their common characteristic is that they are 'booby traps'. Contrary, in many aspects, to popular portrayals over recent years, such tactics have been employed in conflict zones ever since the first explosives were invented. At this time, the use of IEDs is second-to-none, the most common tactic of choice among insurgents in Iraq (as well as Afghanistan and elsewhere). As has been previously shown (Johnson and Braithwaite, 2009), these events cluster in space *and* time more than would be anticipated if their timing and location were independent.

In addition to the IED data, Table 12.1 indicates that we are employing four additional series that represent the most common forms of COIN operation employed by the US/Coalition forces in Iraq in 2005. We also analyze these data in the aggregate to reflect overall trends in COIN activity. The four disaggregated series can be differentiated in terms of whether

Table 12.1 Counts and characteristics and summary hypotheses regarding insurgency and counterinsurgency events, January–June 2005

Event type	Count	Troops required	Target discrimination	Insurgent capacity reduction	Hypothesis re: past IEDs	Hypothesis re: future IEDs
Insurgency						
IED explosion	3,775					
Counterinsurgency						
IED found/cleared	3,333	Normal	High	High	+	−
Cordon/search	1,637	Elevated	Low	Low	+	+/−
Cache found/cleared	1,614	Normal	High	High	+	−
Raid	1,416	Elevated	Medium	Medium	+	−

they (i) discriminate between combatants and non-combatants, (ii) directly reduce the capacity of insurgents to conduct additional attacks and (iii) require observable troops on the ground. Given their general strategic deployment – in response to and in preemption of IED attacks – we anticipate that these events will also cluster in space and time more than would be expected according to chance.

The 'IED found/cleared' series acknowledges a successful outcome from COIN operations, with an IED successfully identified and cleared before it explodes. There are a total of 3,333 such events in our data series. IEDs may be discovered in a variety of ways. However, perhaps the most frequent scenario is that they will be discovered near to and shortly after other devices explode due to the military searching such localities for further IEDs (see, e.g. Dept. of the Army, 2007). Consequently, we anticipate these events in particular to closely follow the distribution of IED attacks.

The 'cordon/search' data series summarizes US/Coalition directed operations designed to search for insurgents and their weapons caches. These 1,637 coded observations represent a relatively discriminatory tactic that, nonetheless, involves the location of troops on the ground in potentially vulnerable positions, as most cordon/search operations are designed to directly access suspected insurgent strongholds and bases of operation, as well as to enable the defence against and defeat of additional IED attacks against previously targeted areas (Dept. of the Army, 2007). Accordingly, we anticipate that these events will be shown to follow IED attacks closely as they are designed to respond to them but will also be associated with either (i) fewer subsequent IED attacks in the vicinity shortly afterwards insofar as they reduce insurgent capacity or (ii) more attacks because they place more troops on the ground for an extended period, providing the potential targets for subsequent attacks.

The 'Cache found/cleared' data captures those occasions on which US/Coalition forces have successfully identified, located, secured and disposed of an insurgent weapons cache. Given that such caches are often considered bases from which the deployment of new attacks – including IED attacks – are launched, it is reasonable to suggest that each weapons cache found and cleared represents a dent in the capacity of the insurgents to carry out additional subsequent attacks. There are a total of 1,614 such events in this data series. Given the unique capacity-reducing qualities of these particular operations, we expect these to show the greatest dampening effect upon the subsequent observation of IED attacks.

   The 'raid' data series also captures US/Coalition directed operations designed to search for insurgents and weapons caches. In this instance, however, the 1,416 coded observations capture events that are likely more discriminatory than those carried out as cordon/search operations. In this instance, the operation is targeted, for instance, against a specific building rather than – as may be the case in cordon/search – a city block or equivalent. For the same reason as was listed for cordon/search operations, we anticipate an increase in these operations following IED attacks in a particular locale. We also suggest that raids will reduce subsequent levels of IED attacks. Given that raids are targeted more discriminately, we also expect that they would provoke fewer attacks in the vicinity relative to the patterns observed for the slightly less geographically focused cordon/search operations.

12.4   Diagnoses of Space, Time and Space–Time Distributions

12.4.1   Introduction

Given the hypotheses derived earlier, it is clear that our primary focus for analysis is the characterization of the space–time distribution of these series and their co-dependence. Analysis of this joint distribution will, specifically, afford an opportunity to neatly specify the local-level interaction of IED and COIN events, which would be lost in the analysis of spatial hotpots without reference to temporal patterns, or vice versa. Nonetheless, for context it is informative to first examine how the concentration of these event types varies in space, and how they vary in time at the national level over the period for which data are available. A range of approaches of varying levels of sophistication could be employed to do this, but as our primary focus is on space–time patterns, we use simple approaches here so as not to distract the reader from what follows.

12.4.2   Spatial Distribution

Figure 12.1 shows thematic maps for each of the five event types and for the aggregation of all COIN operations, with the exception of IEDs that are found. The reason for excluding the latter in the aggregated counts is because, as previously discussed, this activity most likely reflects the most reactive form of COIN activity. Each map was generated using a grid of regular sized cells (5 km × 5 km) and simply counting the number of events within each cell. These counts were then logged and, to generate a standardized index, divided by the (logged) maximum count across all cells. The greyscale used to shade each cell was calibrated using this index. Thus, as there were more COIN events overall ($N=4,665$) than anything else, the cells for that map are generally darker than those for the other event types. Due to the sensitivity of the data, we exclude any reference to the precise locations of events and do not display any landmarks (e.g. roads) on the maps. Such detail is also excluded because it is not necessary for the approach to hypothesis testing adopted here.

   Figure 12.1 demonstrates that each series of events appears to cluster in space, and the different series of events follow quite similar patterns. Some of this patterning likely reflects the demographic, communications and infrastructure skeleton of the state of Iraq, with heavy intensity of events in Baghdad and, for instance, along the Tigris river to the North of Baghdad

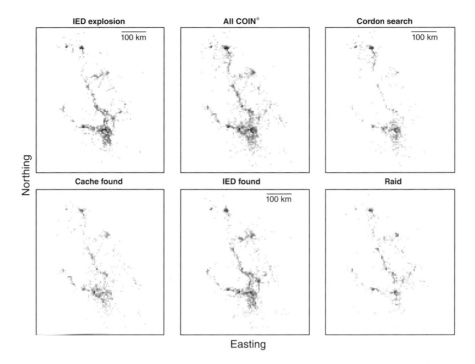

Figure 12.1 Spatial distribution of IED and COIN events (projected relative to an arbitrary origin point). *The 'All COIN' set excludes the IED found/cleared events (see text for details)

through Tikrit and Mosul. Informally, these patterns fit with what we would anticipate from similar studies elsewhere (including, Braithwaite and Li, 2007; Townsley et al., 2008; Johnson and Braithwaite, 2009), which highlight the non-random distribution of terrorist and insurgent violence across space.

12.4.3   Temporal Distribution

In characterising the temporal distribution of each series, our approach is to simply graph them along time axes. Figure 12.2 does just this. Each of the series displays some variation but, for example, the series for IEDs exhibits first-order serial correlation (auto-correlation function (ACF) at lag 1, $ACF(1) = 0.43$, $p < 0.05$), indicating that the pattern is not random. For the COIN events, none of the ACF values are statistically significant, although a trend is apparent for cordon searches.

12.4.4   Space–Time Distribution

In the remainder of this study, we employ an approach developed in previous studies to examine patterns of space–time clustering for the different event types (Johnson et al., 2007; Townsley et al., 2008; Johnson and Braithwaite, 2009) and a novel extension to this. In the case

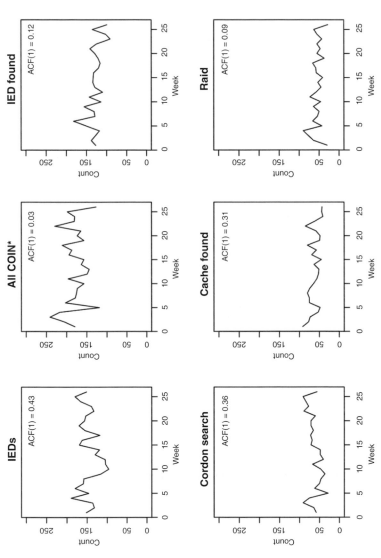

Figure 12.2 Temporal distribution of IED and COIN events (auto-correlation function (ACF) values for lag 1 indicate the strength of serial correlation in the series). *The 'All COIN' set excludes the IED found events (see text for details)

of the latter, we simultaneously examine the relationship between the space–time distribution of insurgency attacks *and* COIN operations. This novel technique is especially valuable, because it enables us to directly analyse the space–time dynamics of the two event types at the micro-level, and uncover any patterns that would be lost using data aggregated to the national level.

To examine whether events do cluster in space and time, we use the approach originally developed by Knox (1964) to detect disease contagion, which has been subsequently developed for the study of crime patterns (e.g. Johnson et al., 2007; Johnson et al., 2009). The null hypothesis is that the timing and location of events are independent. In the case of space–time clustering, more events will be observed to occur close to each other in both time *and* space than would be expected if their timing and location were independent. To examine this, the first step is to generate the distribution of interest for the observed events. In the univariate case, each event is compared to every other and the distance and time between them recorded. A contingency (or Knox) table is then populated to summarize the $n*(n-1)/2$ comparisons (where n is the number of events). The dimensions of the table, and the bandwidths used, are at the discretion of the researcher but should be selected so as to allow a sensitive test of the hypothesis under investigation. For comparison with prior work on IEDs (e.g. Johnson and Braithwaite, 2009), in the current study we use spatial intervals of 500 m and temporal intervals of 1 week. However, sensitivity analyses suggest that – within reason – alternative bandwidths generate qualitatively identical results.

To compute the expected distribution and the probability of obtaining the observed results if the null hypothesis were correct, we use a permutation test. In this case, the dates on which events occur are 'shuffled' across incidents using a uniform random number generator. Thus, for each permutation of the data, the same locations occur with the same frequency as they do for the observed data, which takes account of the fact that such events cannot happen anywhere and in fact occur at some places more than others. Likewise, the same dates occur with the same frequency as they do in the observed data, reflecting, for example, any general seasonal variation and general temporal trends associated with IED attacks. What does, however, vary (randomly) across the permutations is the association between when *and* where events occur (recall that the null hypothesis is that the timing and location of events are independent). For each permutation, a new contingency table is populated and the results are compared with those for the observed distribution. A full permutation will be almost impossible and so a simple Monte Carlo (MC) simulation is used to (re)sample (in this case 99 samples) from all possible permutations.

Where the frequencies of the cells which enumerate the number of event pairs that occur close to each other in space and time are higher for the observed than permuted data, space–time clustering is said to exist. To estimate the size of this effect, the observed frequency for each cell may be divided by the mean derived across all permutations (see Johnson et al., 2007). Values above 1 indicate that there were more observed pairs of events within a particular time interval of each other than would be expected if the timing and location of events were independent (the null hypothesis). The statistical significance of any observed effect can also be derived using the formula specified by North et al. (2002):

$$P = \frac{r+1}{n+1} \qquad (12.1)$$

In Equation (12.1), n is the number of realizations of the MC simulation (in this case 99) and r is the number of permutations for which the value of the test statistic is at least as large as the observed value. In the case where none of the values generated by the MC simulation exceed the observed count for the test statistic, the pseudo-probability of observing that value is estimated as (1/100=) 0.01 or less.

While useful, this does not allow us to determine if events of one type are more likely to occur near to, and follow those of another, or vice versa. To do this, we develop an approach articulated in Johnson et al. (2009) and generate two Knox contingency tables. In the first, we enumerate the number of event pairs for which an IED attack follows a COIN operation. In the other, we do the reverse. The distributions expected on a chance basis (assuming the timing and location of events are independent) are generated and the results are interpreted, in exactly the same way as for the univariate case. However, in the bivariate case, for each permutation, there will be $N \times M$ comparisons – where N and M are the counts for the two data sets. Such analysis allows us to determine if events of one type are more likely to occur within particular space–time intervals of another than would be expected if the two types of event were independent. For example, we can estimate whether it is the case that more IED attacks and raids occur within (say) 500 m and 7 days of each other than would be expected if the two types of event were independent.

12.4.5   Univariate Knox Analysis

Figure 12.3 shows the univariate Knox analyses for each of our six data series – IED explosions, an aggregate of the COIN events, and each of the four individual COIN operations of interest. Each plot is essentially a contingency table, with the columns representing different temporal intervals and the rows identifying different spatial ones. Rather than listing the Knox ratios, to facilitate the easy identification of patterns in the data, each cell is proportionately shaded to indicate the size of the Knox ratios. As the scale bar to the right-hand side shows,

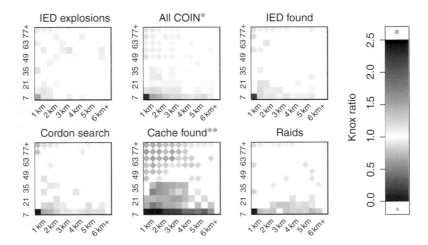

Figure 12.3   Univariate Knox analyses of six event types. *The 'All COIN' set excludes the IED found events. **For caches found, the Knox ratios for event pairs that occurred up to 1500 m and 1 week of each other exceed 2.5 (the actual values are 4.49, 2.99 and 2.73 – in that order), but we truncate the scale bar at 2.5 to make observed patterns clearer

both those Knox ratios that indicate higher than expected cell frequencies (those that are above 1) and those that indicate lower than expected frequencies (those in the range between 0 and 0.999) are represented in the table. In each case, cells that are shaded darkest are those for which the Knox ratio departs the most from 1 (recall that a Knox ratio of 1 indicates that the observed and expected values are equal). Knox ratios that are above 1 are distinguished from those that are below 1 using different symbols; a square indicates that a Knox ratio was greater than 1; a diamond indicates that it was less than 1. Furthermore, cells are only shaded where the observed Knox ratio is statistically significant (i.e. where $p \leq 0.01$).

As discussed in the introduction and elsewhere (Johnson and Braithwaite, 2009), we find that following an IED explosion, subsequent IED explosions are more likely to occur within a short period of time *and* distance of the original explosion than would be expected if the timing and location of events were independent. We can see, for instance, that the Knox ratios for pairs of explosions occur within 1.5 km and 2 weeks of each other are positive and relatively large. The plot also suggests that the effect decays in space and time. For example, the Knox ratios decrease along the bottom row as one looks from left to right. Furthermore, we find that fewer pairs of IED explosions occur near to each other in space but far apart in time (11 weeks or more) than we would expect, assuming that events were independent.

In the case of COIN operations, while there was no obvious pattern in the simple time series analyses discussed earlier, for the analyses presented here – which consider clustering in both space *and* time – the general trend is similar to that for IED explosions. In particular, the highest Knox ratios tend to be found in the bottom left corner of each plot. However, there are variations across the different types of activity. In the case of raids, we find that the clustering observed for events that occur near to each other in time appears to extend over a larger spatial range. And, when pairs of raids occur 2–3 weeks apart, it appears that they are more likely to occur slightly further away than if they occurred within 1 week of each other. Again, relative to chance expectation, fewer pairs of events occur near to each other in space but far apart in time.

For caches found, the pattern observed is really quite stark. We can clearly see that relative to what would be expected if events were independent, substantially more caches are found within 5.5 km and 1 week of each other. This elevation is most pronounced at proximate locations but is still high for those up to 5.5 km away. This effect also appears to be sustained for a period of up to 3–5 weeks, particularly within a radius of about 3.5 km of discovery locations. After 5 weeks has passed, however, the likelihood of a cache being discovered near to a previous find appears to be substantially lower than would be expected, assuming the timing and location of events were unrelated. This may suggest that the 'Clear and Hold' strategy outlined by Condoleezza Rice has some significant effect, as those areas in which caches have been removed appear to experience lower than chance levels of further finds in subsequent periods, likely because of a consequent dearth of available IED components. The alternative is that the military do not return to locations where caches have previously been discovered for sometime.

Finally, compared to chance expectation, we also find that significantly more pairs of 'cordon/search' operations occur near to each other in time and space. As with raids, at locations closest (up to 500 m) to previous troop deployments the elevation is short-lived, but at locations a little farther away (1–3.5 km) it endures for 2–3 weeks.

In combination, these findings suggest that there is significant space–time clustering for each event type. It is perhaps noteworthy that this appears to extend (but decay) over relatively large spatial ranges, but generally only in the short term. This would appear to support the conjecture that COIN operations result in local areas being successfully cleared and held, with an observable reduction in the perceived need for follow-up events of the same type at close proximity for much more than a week or two. However, this hypothesis can be more directly evaluated by assessing space–time correlations between IED attacks and COIN operations.

12.4.6 Bivariate Knox Analysis

This modified Knox procedure is used to generate 10 contingency tables, each of which is plotted in Figure 12.4. Once again, we provide a scale bar on the right-hand side that details the greyscale used to illustrate statistically significant Knox ratios (at the $p \leq 0.01$ level). As earlier, darker squares are used to highlight higher than expected ratios and darker diamonds are used to highlight lower than expected ratios. In this instance, the plots indicate whether, for example, IED explosions appear to occur close to and shortly after COIN operations (the left-hand column) or (say) the reverse is true (the right-hand column). In other words, these plots enable us to examine the space–time distribution of IED attacks in relation to COIN operations, and vice versa.

Recall that the logic underlying the Coalition strategy of 'Clear and Hold' leads us to anticipate (i) that COIN operations will typically be shown to follow IED explosions at targeted locations and nearby more consistently than is shown to be true of the reverse; (ii) that more discriminatory and capacity reducing COIN operations (including raids, cache found and IED found) are expected to have the effect of reducing subsequent levels of IED explosions at proximate locations (at least temporarily) and (iii) that less discriminate and troop-intensive operations (cordon/searches) are likely to be followed by increasing levels of IED explosions nearby in the short-to-medium term.

In the first instance, there is immediately less systematic evidence across the 10 plots of the typical space–time decay than was shown to be the case with the univariate analyses. We do feel comfortable suggesting, however, that some patterns exist. This is most notable for the comparison between IED explosions and IEDs found where we see greater evidence of COIN operations clustering in space and time following IED explosions. In addition to being in line with expectation, this result is useful insofar as it illustrates the utility of the test in detecting patterns and what these might look like.

It would appear that there is some evidence that cordon/search operations also follow IEDs more consistently than vice versa (see the Knox ratios in the bottom left of the plots), but that the timing and location of these COIN operations are influenced by factors other than that of IED explosions. In the case of cordon/searches, which reflect a more strategic action carried out on the basis of hard intelligence, this is perhaps hardly surprising.

In terms of the second general expectation, we find only marginal support for the claim that discriminatory and capacity-reducing operations swiftly reduce subsequent levels of IED explosions in the vicinity of those operations. Indeed, for IEDs found and cordon/search operations, we find that a reduction (relative to chance expectation) in IEDs at locations proximate to these operations is observed only after 11 weeks has elapsed – suggesting that in the longer term, Clear and Hold strategies may have the intended impact. Conversely, we can see that relative to what would be expected if events were independent, there is an increase in subsequent levels of IED explosions around identified cache locations after 11 weeks of their discovery. This is particularly the case at locations that are nearest to those locations where

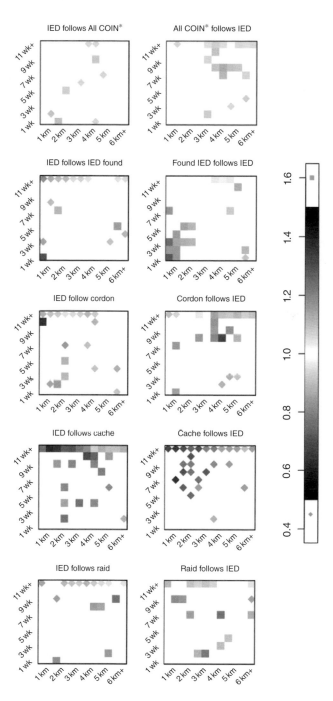

Figure 12.4 Bivariate Knox analyses of six event types. *The 'All COIN' set excludes the IED found events

caches were previously discovered. Notably, prior to the 11th week, any increases appear only to be observed outside of a 1.5–2 km radius of the discovery locations.

With respect to the third general expectation, relative to chance, we find an increase in IED attacks within 1.5–2 km of raids, but the general pattern is hardly unequivocal. In fact, contrary to our hypotheses, we interpret the results as suggesting that at least in the short-to-medium term, raids have little effect on the space–time distribution of IED attacks (either positive or negative). The timing and location of raids also appear to be imperfectly related to that of IED explosions.

12.5 Concluding Comments

The present study uncovers yet further evidence to demonstrate that insurgency events (specifically, attacks involving the deployment of IEDs), cluster in space, in time and in space–time. More uniquely, the present study has also shown that COIN operations have distinct space–time signatures, and that for some types of COIN event there is a relationship between when and where they take place and when and where IED explosions occur. Moreover, the results suggest that it does not appear to be the case that IED attacks are more likely to be observed in the vicinity of recent COIN activity. If anything, the reverse appears to be a little more likely. These findings are inconsistent with the idea that COIN activity provokes insurgent action, at least at those locations near to such operations. Importantly, these results also suggest that the finding that IED attacks cluster in space *and* time is unlikely to be explained in terms of insurgents reacting to COIN action. Instead, it more likely appears to reflect rational decision-making on the part of insurgent actors, perhaps along the lines outlined in the introduction to this chapter.

How might these findings inform military operations on the ground? They offer moderate support for the claim that the Coalition's 'Clear and Hold' strategy has borne fruit. In other words, in addition to the fact that IED explosions do not occur with an elevated frequency shortly after and around those locations that COIN operations take place, there is some evidence to suggest that there may be positive longer-term effects of COIN operations at and near to targeted locations. To elaborate, at the locations that these operations (with the exception of the discovery of caches) take place there appear to be fewer occurrences of IED explosions at proximate locations in the long term than would be expected if the timing and location of events were independent. The exception to this rule – caches found – warrants further investigation. It logically follows that the removal of caches would fundamentally alter the resources available to insurgents to engage in additional IED attacks in the local area. We find that locating and removing caches do not have a systematic impact (positive or negative) on IED deployment within the vicinity in the short term, suggesting that, perhaps, insurgents have alternative caches nearby from which to draw resources or that they are able to redeploy resources from more distant caches.

In the preceding discussion we have referenced the actors involved as if they were essentially two groups who coordinate their own actions. However, the insurgent groups involved are unlikely to represent a top-down organized group (Sageman, 2004). Thus, it is conceivable that the patterns observed here are also somewhat contingent upon the distribution and variety of groups carrying out IED attacks against the Coalition and the local communities (for a related discussion, see Bohorquez et al., 2009). Moreover, perhaps it is the case that the pockets of raised Knox ratios at more distant locations and times reflect the spatial and strategic positions taken by competing insurgent groups. On this particular dynamic, further research could usefully be employed to inform the ongoing debate within terrorism and insurgency literature regarding the strategy of 'outbidding'. Outbidding occurs, some claim, where two or more groups are competing for a position of authority within a community and where the public remains uncertain as to which of the groups is most likely to be able to deliver the goods that the public demands (Kydd and Walter, 2006). The ongoing competition between Fatah and Hamas for the support of the Palestinian populations is perhaps the best exemplar of this strategy at play. Bloom (2004) notes that in this case, increased support among the Palestinian public for the use of suicide bombings has resulted in greater levels of competition between groups for militant recruits. Given the apparent competition both between and within Shia and Sunni communities in Iraq for a role in the governance of Iraq, it appears likely that future research might similarly identify evidence of active outbidding.

A number of other considerations perhaps warrant attention in future research in order to ensure that these speculative conclusions are thoroughly tested against alternative accounts (ideally using data for a different period of time or region) and are, accordingly, more valuable to those designing COIN policy and operations. It is important to note that COIN types are not mutually exclusive (as currently portrayed). Indeed, it is the case that a range of COIN operations are commonly employed in combination and/or in sequence, rather than as substitutes for one another. Similarly, it would perhaps be valuable to assess patterns of different kinds of IEDs. As discussed earlier, there are in excess of six or eight varieties of IEDs commonly deployed in Iraq. Differentiation between these types is not possible using the data from SIGACTS reports, but may be possible through the use of data drawn from the Global Terrorism Database (START, 2011).

A further issue regarding the types of patterns explored is that while it is possible that they are homogeneous across space, it is also possible that they are not. To elaborate, in the current study (and studies reported elsewhere, e.g. Johnson et al., 2007) we have analysed data for a large region and reported upon patterns observed across the entire space. In methodological terms, this is not problematic for the analysis of space–time patterns (it would be for the analysis of spatial patterns alone) as the Knox test was developed for analyses conducted at this kind of spatial scale (see Knox, 1964). The issue to consider is that while, for example, space–time clustering may be a feature of events in some of the region, it may not be (or be less so) in others. Examining variation in observed patterns across the region was beyond the scope of the current study, but doing so would be a useful next step.

The analyses employed above focus upon the dynamic distribution and co-evolution of events in space and time. An alternative strategy would focus, instead, upon a fixed grid-cell approach. This would involve the disaggregation of the region into (say) 5 km × 5 km grid cells and the specification of a series of event count models for each event type within these units of observation. This second approach would be designed to facilitate further testing of the sequencing of different types of events, thereby examining cause and effect relationships – so far as this is possible using methods of correlation. A similar approach is employed with increasing frequency in the study of civil violence (see, e.g. Buhaug and Rod, 2006). Advantages of this approach are that it would enable us to explore the relationship between the types of patterns observed and different populations-at-risk, to examine associations with targets of perceived utility to insurgents, such as iconic buildings or bases, and to examine the extent to which patterns vary across space.

The study of spatial and temporal patterns of insurgent activity is gaining increasing interest in the academic literature. In the current study, we examine both insurgent and counterinsurgent activity to explore the extent to which there are regularities in the coincidence of these two types of action and introduce new methods for studying these interactions. It is important to reiterate that the approach taken represents a departure from studying spatial patterns or temporal trends in isolation. What the approach taken uncovers are the space–time dynamics of events at the micro-level – patterns that cannot be identified using techniques such as spatial hotspot analysis or time series methods. In future work, we aim to continue this trend by exploring, for example, the association between the risk of IED attacks and the recent locations of such events and also the influence of more time-stable factors such as the accessibility or centrality of places, the presence of vulnerable targets and so on.

References

Arce, D. and Sandler, T. (2005) Counterterrorism: A game-theoretic analysis. *Journal of Conflict Resolution*, **49**, 183–200.

Axelrod, R. (1984) *The Evolution of Cooperation*, Basic Books, New York.

Blank, L., Enomoto, C., Gegax, D. et al. (2008) A dynamic model of insurgency: The case of the war in Iraq. *Peace Economics, Peace Science, and Public Policy*, **14**, 1–26.

Bloom, M. (2004) Palestinian suicide bombing: Public support, market share, and outbidding. *Political Science Quarterly*, **119**, 61–88.

Bohorquez, J.C., Gourley, S., Dixon, A. et al. (2009) Common ecology quantifies human insurgency. *Nature*, **462**, 911–914.

Braithwaite, A. (2010) *Conflict Hotspots: Emergence, Causes, and Consequences*, Ashgate Press, Aldershot.

Braithwaite, A. and Li, Q. (2007) Transnational terrorism hot spots: Identification and impact evaluation. *Conflict Management and Peace Science*, **24**, 281–296.

Bueno de Mesquita, E. (2005a) Conciliation, counterterrorism, and patterns of terrorist violence. *International Organization*, **59**, 145–176.

Bueno de Mesquita, E. (2005b) The quality of terror. *American Journal of Political Science*, **49**, 515–530.

Bueno de Mesquita, E. (2007) Politics and the suboptimal provision of counterterror. *International Organization*, **61**, 9–36.

Bueno de Mesquita, E. and Dickson, E. (2007) The propaganda of the deed: Terrorism, counterterrorism, and mobilization. *American Journal of Political Science*, **5** (2), 364–381.

Buhaug, H. and Rod, J.K. (2006) Local determinants of African civil wars, 1970–2001. *Political Geography*, **25**, 315–335.

Department of the Army, Headquarters. (2007) *Combined Arms Improvised Explosive Device Defeat Operations (FM 3-90.119/MCIP 3-17.01)*. Accessed online <www.us.army.mil> (accessed 07 January 2016).

Faria, J.R. and Arce, D. (2005) Terror support and recruitment. *Defence and Peace Economics*, **16**, 263–273.

Hoffman, B. (2006) *Inside Terrorism*, Columbia university Press, New York.

Johnson, S. and Bowers, K. (2004) The burglary as a clue to the future: The beginnings of prospective hot-spotting. *European Journal of Criminology*, **1**, 235–255.

Johnson, S., Berdasco, W., Bowers, K. et al. (2007) Space-time patterns of risk: A cross-national assessment of residential burglary victimization. *Journal of Quantitative Criminology*, **23**, 201–219.

Johnson, S. and Braithwaite, A. (2009) Spatio-temporal distribution of insurgency in Iraq, in *Countering Terrorism through SCP. Crime Prevention Studies*, vol. 25 (eds J.D. Freilich and G.R. Newman), Criminal Justice Press, pp. 9–32.

Johnson, S., Summers, L. and Pease, K. (2009) Offender as forager? A direct test of the boost account of victimization. *Journal of Quantitative Criminology*, **25**, 181–200.

Johnson, S. (2010) A brief history of the analysis of crime concentration. *European Journal of Applied Mathematics*, **21**, 349–370.

Kalyvas, S. (2006) *The Logic of Violence in Civil War*, Cambridge University Press, New York.

Knox, G. (1964) Epidemiology of childhood leukaemia in Northumberland and Durham. *British Journal of Preventive Social Medicine*, **18**, 17–24.

Kydd, A. and Walter, B. (2006) The strategies of terrorism. *International Security*, **31**, 49–79.

LaFree, G., Dugan, L. and Korte, R. (2009a) The impact of British counter terrorist strategies on political violence in Northern Ireland: Comparing deterrence and backlash models. *Criminology*, **47**, 17–46.

LaFree, G., Dugan, L., Xie, M. and Singh, P. (2009b) *Spatial and Temporal Patterns of Terrorist Attacks by ETA*, University of Maryland, Typescript.

Maoz, Z. (2007) Evaluating Israel's strategy of low-intensity warfare, 1949–2006. *Security Studies*, **16**, 319–349.

North, B.V., Curtis, D. and Sham, P.C. (2002) A note on the calculation of empirical p values from Monte Carlo procedures. *American Journal of Human Genetics*, **71**, 439–441.

Pape, R.A. (2003) The strategic logic of suicide terrorism. *American Political Science Review*, **97**, 343–361.

Pape, R.A. (2005) *Dying to Win*, Random House, New York.

Rosendorff, B.P. and Sandler, T. (2004) Too much of good thing? The proactive response dilemma. *Journal of Conflict Resolution*, **48**, 657–671.

Ryu, A. (2005) *Roadside Bombs Cause Increasing Concern in Iraq*. Voice of America. 28th September.

Sageman, M. (2004) *Understanding Terror Networks*, University of Pennsylvania Press, Philadelphia, PA.

Siquiera, K. and Sandler, T. (2007) Terrorist backlash, terrorism mitigation, and policy delegation. *Journal of Public Economics*, **91**, 1800–1815.

National Consortium for the Study of Terrorism and Responses to Terrorism (START). (2011) *Global Terrorism Database [Data file]*. Can be retrieved from http://www.start.umd.edu/gtd (accessed 07 January 2016).

Townsley, M.T., Johnson, S.D. and Ratcliffe, J.R. (2008) Space-time dynamics of insurgent activity in Iraq. *Security Journal*, **21**, 129–146.

Wilkinson, P. (2001) *Terrorism versus Democracy: The Liberal State Response*, Frank Cass, London.

13

International Information Flows, Government Response and the Contagion of Ethnic Conflict

Janina Beiser

13.1 Introduction

In today's world, the interdependence between states is ever-increasing. Contemporary states are linked by complex trade networks, by development aid and by mutual memberships in intergovernmental organisations. This increasing international interdependence is also reflected by increasing global information flows (Mowlana, 1997). The consequences of increasing information from even distant regions of the world with respect to security-related questions have been investigated little. This chapter explores global dynamics through a study of the impact of mass media based information flows on two forms of political violence that threaten the security and well-being of people around the world: armed intrastate conflict and government repression.

Since the end of the Second World War, armed conflict within the borders of states has become an increasingly prevalent phenomenon (Beiser, 2016). Those conflicts are often referred to as intrastate or civil conflicts. A number of data projects are dedicated to measuring and classifying conflicts. The UCDP/PRIO Armed Conflict Data is one of the most commonly used sources of data (Uppsala Conflict Data Program, 2014) and defines conflict as '(...) a contested incompatibility that concerns government and/or territory where the use of armed force between two parties, of which at least one is the government of a state, results in at least 25 battle-related deaths' (Harbom, 2010, 1). These conflicts are classified as intrastate conflict if the government's adversary is a non-government actor and the conflict is fought on a state's territory and as interstate conflict if adversaries are state governments (Harbom, 2010). As Figure 13.1 from Beiser (2016) shows, the number of ongoing armed interstate conflicts has

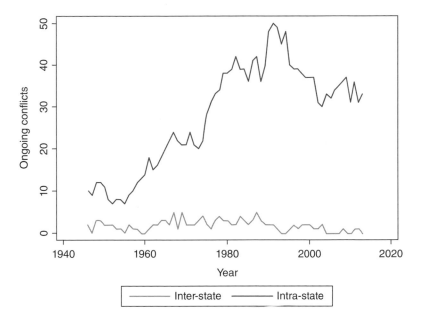

Figure 13.1   The yearly number of ongoing inter- and intrastate conflicts in the world

barely changed since the end of the Second World War (as Figure 13.1 below). However, the incidence of ongoing armed intrastate conflict has increased notably since then.[1]

Similarly, despite the '(...) right to life, liberty and security of person' (UN General Assembly, 1948) in the Universal Declaration of Human Rights and the work of human rights groups, government repression is a common phenomenon (Beiser, 2016). 'By most accounts, repression involves the actual or threatened use of physical sanctions against an individual or organization, within the territorial jurisdiction of the state, for the purpose of imposing a cost on the target as well as deterring specific activities and/or beliefs perceived to be challenging to government personnel, practices or institutions' (Davenport, 2007, drawing on Goldstein, 1978, xxvii). Figure 13.2 from Beiser (2016) illustrates this with the yearly mean of the physical integrity rights index from the CIRI Human Rights data (Cingranelli et al., 2014b) (as Figure 13.2 below). This measure is '(...) an additive index constructed from (...) Torture, Extrajudicial Killing, Political Imprisonment, and Disappearance indicators. It ranges from 0 (no government respect for these four rights) to 8 (full government respect for these four rights)' (Cingranelli et al., 2014a, 3).

Vibrant academic fields investigating the causes of both armed intrastate conflict and government repression have developed. Researchers have found that structural factors such as the regime type of states and interactions between opposition and government (for an overview see Davenport, 2007) help explain government repression (Beiser, 2016). In the case of armed civil conflict, a country's wealth (e.g. Collier and Hoeffler, 2004) and inequalities between groups (e.g. Cederman et al., 2010) have been found to explain conflict (Beiser, 2016). More recently, academics have become increasingly interested in the role of global interdependence

---

[1] Beiser (2016) draws on data from the UCDP/PRIO Armed Conflict Dataset V4-2014 (Gleditsch et al., 2002; Themnér and Wallensteen, 2014).

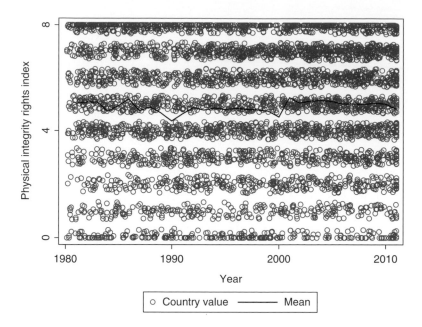

Figure 13.2   Government repression in the world

and dynamics for explaining both government repression and armed intrastate conflict (on this point in the context of conflict, see Beiser, 2016). However, researchers have focused mostly on the effect of events in geographically proximate states (for an overview, see Beiser, 2016). It has been found that close-by armed civil conflicts increase the likelihood of both government repression (Danneman and Ritter, 2014) and armed civil conflict in a state (e.g. Buhaug and Gleditsch, 2008; Forsberg, 2008). In contrast, the effect of global information flows on the likelihood of civil conflict as well as repression has only recently been considered (Weidmann, 2015; Beiser, 2012, 2013). This chapter provides an overview over recent findings unpacking the relationship between foreign information transmitted via mass media and these forms of political violence.

This chapter proceeds as follows. In the following section, empirical data and conceptual considerations introduced in Beiser (2012) that allow measuring information flows via mass media is discussed. In a third section, how global information flows can affect the likelihood of armed civil conflict and how far theoretical expectations have been found to be supported by empirical evidence is discussed, partly drawing on Beiser (2012). Drawing on theory and findings from Beiser (2013), the fifth section does the same in the context of government repression before the final section concludes.

13.2   Global Information Flows

This section discusses how global information flows via mass media can be measured using empirical data in order to be able to analyse their effect on subsequent events. This measure of information was first introduced in Beiser (2012).

The flow of information from a country j to another country i is difficult to observe directly (Beiser, 2016). Following previous literature, Beiser (2012) identifies four factors that could have an effect on the likelihood of information from one country reaching another via mass media. Beiser (2012) uses a number of data sources to measure each of these factors empirically and combines them in an overall index in an attempt to measure information flows via mass media overall.

Firstly, Beiser (2012) notes that a population's access to mass media devices such as radios, print media, TVs as well as Internet access have an effect on the likelihood of information flows about foreign events via the mass media (also see e.g. Hill and Rothchild, 1986). In Beiser (2012), media availability is measured empirically as a country's average rate of TV and radio per capita. Data for this measure is provided by Banks (2011). Beiser (2012) log-transforms this rate as there are some particularly high values.

By focusing on information transmitted via radios and TVs in particular, this measure of information flows is biased towards long-standing technologies and news disseminated via conventional mass media. Print media are ignored here as their effect is moderated by literacy rates that complicate the relationship. In addition, the availability of print media is likely correlated with the availability of devices considered. Moreover, and importantly, citizens' access to the Internet is excluded here. Data on global Internet access exists but it is difficult to place the Internet in a common index with older technologies such as radios and TVs because it is not clear in how far the new technology replaces the use of older ones and in how far it constitutes a complement. Thus, the measure here can be understood as a reflection of the likelihood of information flows via conventional mass media. There is little reason for concern that findings using the measure on radios and TVs discussed here might actually reflect the effect of the Internet as the media availability data used – with very few exceptions – only covers years up until 1999.

Secondly, Beiser (2012) notes that the availability of media sources may not be the only factor affecting the likelihood of information flows between countries. If governments have control over the content of news, they can for instance prevent information about foreign events from entering a country (Hill and Rothchild, 1986). Therefore, Beiser (2012) expects that media censorship can prevent international information flows about unwanted foreign events[2]. Beiser (2012) uses Freedom House's Freedom of the Press (Freedom House, 2015) historical data to distinguish countries with free and unfree media.

Thirdly, Beiser (2012) draws on the literature suggesting that news from economically stronger countries may be more likely to reach other countries. Those include, for example, Kuran (1989) who suggests that economically stronger states have more influence on international information flows and Elkins and Simmons (1998) who suggest that prominent countries have a greater effect on other states via learning mechanisms. Beiser (2012) measures economic prominence by a country's real GDP per capita (2000 prices) from Gleditsch (2002).

Fourthly, Beiser (2012) expects that events in geographically closer countries are more likely to be reported in mass media as has been argued in previous literature such as Hill and Rothchild (1986) drawing on others such as (Katz and Wedell, 1977), and in Mowlana (1997). Geographic proximity is measured by the inverse distance between two states' capitals from Cshapes (Weidmann et al., 2010).

As each of these four factors may exert an effect on interstate information flows via mass media, Beiser (2012) combines all factors in a summary measure on information flows from a

---

[2] This is suggested by findings in Hill and Rothchild (1986).

state j to a state i. Beiser (2012) expects the absence of censorship to be a potentially necessary factor for unbiased foreign information to reach a country. The other factors, on the other hand, may make information transmission between countries more likely without being necessary (Beiser, 2016). The four factors are combined in a measure of media based information flows between two countries. For this, Beiser (2012) sums percentile scores on the variables on logged media availability in country i, GDP of country j and inverse capital distance between i and j for each directed dyad i, j. Subsequently, the measure is multiplied by the measure on media freedom that takes value 0 or 1 (Beiser, 2012). More specifically, the information index measuring information flows from country j to country i is built as follows:

$$\text{info}_{i,j} = (p(\text{Radio/TV}_i) + p(\text{GDP}_j) + p(1/\text{Distance}_{i,j})) * \text{Media freedom}_i \qquad (13.1)$$

where p stands for the percentile measure of each of the factors introduced earlier.

This index measures the likelihood of information from a country j entering country i via mass media, even if this is unwanted by the government. The index takes value 0 in countries with no media freedom and values up to 300 in countries with media freedom. While the distance between states' capitals as well as states' economic output is likely to be fairly stable over time, the availability of media sources such as radio and TV as well as a country's media freedom can change over time.

Figure 13.3 shows factor 1, countries' per capita rate of TV and radio over time. An overall trend is difficult to establish as the number of cases available in each year differs. Between 1950 and 1960, data is mostly available for wealthy countries such as the United States. After 1999, data is only available for Zambia and South Africa. Thus, only the mean level of radio and TV per capita between 1960 and 1999 can be considered representative. Here, we see an

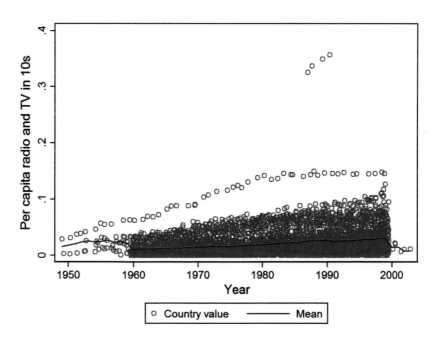

Figure 13.3   The per capita average of radio and TV in the world

increasing trend in media availability. But even in this period, new countries enter the data which can be expected to have a lower rate of media source availability. As a result, the trend would likely appear much steeper if data was available for all countries across all years. In addition, as the four outliers illustrate, there are some cases in the data where the reliability of information is unclear. Unfortunately, to the best of the author's knowledge, no alternative data source is available on the availability of these devices.

Figure 13.4 shows information on factor 2, media freedom. We see the average rate of media freedom in the international system over time. Apart from a drop in 1989, the proportion of countries with free media in the international system seems to be fairly stable. This drop is likely caused by a change in the coding scheme of the data but again, to the best of the author's knowledge, no alternative source of data on media freedom is available (Beiser, 2012).

Information factors 3 and 4 on a country j's GDP and the distance between countries are omitted here as this information can be expected to be fairly stable over time. Figure 13.5 shows the values of the overall information index constructed from the data on media source availability, media freedom, economic prominence of the foreign country and capital distance. Each data point reflects the value on the information index for a directed dyad and can be interpreted as a measure of the expected likelihood of a country i gaining information about events in a particular other country j via mass media. Years after 1999 are omitted as there are very few cases because of limitations in the media data discussed earlier.

As Figure 13.5 illustrates, the information index reflects the trend in the data on media freedom. This is not surprising as media freedom is expected to be necessary for other factors such as the availability of media devices to exert an effect on the likelihood of information flows from other countries. In addition, the likelihood of information flows does not change much over time. This is not surprising as geographical distance and economic importance of countries should remain fairly stable across time.

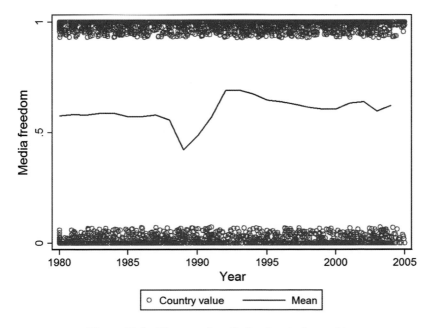

Figure 13.4   The rate of media freedom in the world

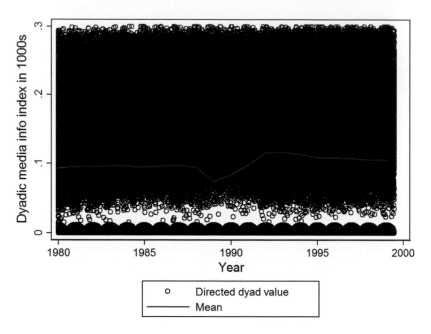

Figure 13.5   The measure on information flows in the world

13.3   The Effect of Information Flows on Armed Civil Conflict

In the previous section, we have seen how information flows via mass media can be approximated using theoretical considerations and empirical data. This section introduces research on how information flows from even distant other countries can affect a state's likelihood of armed civil conflict.

There may be many ways in which information from elsewhere can affect the likelihood of civil conflict, and exploring them is an important subject for researchers, especially as global information flows are likely to increase further in the future (Beiser, 2012). One way in which news from foreign countries may affect the security situation for a state's citizens is via a phenomenon that is sometimes referred to as conflict contagion (e.g. Braithwaite, 2010). This concept compares armed civil conflict with a disease that can infect subsequent states (Braithwaite, 2010).

Some researchers have analysed this phenomenon and found that armed conflicts can indeed increase the likelihood of conflict breaking out in other states that are geographically close (e.g. Buhaug and Gleditsch, 2008; Forsberg, 2008).

How exactly could conflict contagion via information flows take place? There are a number of theories that can provide insight into that question. Firstly, it is possible that once information about a conflict becomes known elsewhere, people in other states learn strategies about how to fight a government from the observed foreign conflict[3].

As a result, they expect to be more successful at fighting the government than they previously thought and some of them might take the chance and rebel. Similarly, it could be that

---

[3] See for example Hill and Rothchild (1986) as well as Hill et al., (1998) in the context of peaceful protest.

people become fearful that conflict breaks out in their state, and as a result increasing suspicion and mobilisation eventually trigger conflict (see Kuran, 1998, in the context of ethnic dissimilation).

If this is the case, we can expect that foreign civil conflicts increase the likelihood of conflict in countries where the population can be expected to have gained information about them (also see Hill and Rothchild, 1986). Whether armed conflicts can also spread via the information network – even to countries that are not close-by in physical space – has to date only been analysed in Weidmann (2015) who finds that direct communication between populations triggers the spread of civil conflicts. This chapter argues that if above arguments hold, foreign conflicts should also be contagious to countries that are not close-by but likely to gain information about them via mass media networks. This expectation constitutes proposition 1 of this chapter.

Secondly, it could be that specific characteristics of observed conflicts as well as of the potentially infected country increase the likelihood of information-based conflict contagion even further. Groups that experience ethnic discrimination have an incentive to challenge the government, but for that they need to be clearly aware of their unfair treatment (Beiser, 2012 drawing on others such as Buhaug et al., 2014). Foreign conflict of groups that are similarly discriminated can raise this awareness among domestic discriminated groups and as a result increase their likelihood of rebellion (Kuran, 1998; Weidmann, 2015; Beiser, 2012). If this mechanism holds, we can expect that foreign civil conflicts that are fought by discriminated groups increase the likelihood of civil conflict in other countries where ethnic groups experience discrimination as well – but again, only if the population there has gained information about them. This empirical expectation has been first formulated in Beiser (2012) and constitutes proposition 2 in this chapter.

The expectations derived from these arguments can be tested empirically. Here, we investigate the outbreak of conflicts involving ethnic groups in particular. Quantitative data on whether an ethnic conflict has started in a state in a year from the Growup database is used[4]. This database combines data on ethnic conflict and the status of ethnic groups in all countries in the world between 1946 and 2009.

In the empirical tests, following Beiser (2012), the overall measure on information flows is modified slightly in order to allow correct estimation. Now, we measure the information index $ind_{i,j}$ as

$$ind_{i,j} = (p(\text{Radio/TV}_i) + p(\text{GDP}_j) + p(1/\text{Distance}_{i,j})) \tag{13.2}$$

This index includes the three factors that were expected to increase the likelihood of the population in a country i gaining information on events in each country j. To test proposition 1, the number of all high intensity civil conflicts in each country j based on Gleditsch et al., (2002) will be multiplied by this measure on the likelihood of information transmission from country j to country i[5]. For each country i, all these weighted foreign events will be summed to create the overall foreign information about relevant conflicts that can be expected to reach country i. This measure is subsequently multiplied by the measure on media freedom in i in

---

[4] This database includes data from the Correlates of War State System Membership List v2008.1 (Correlates of War Project, 2008), the Ethnic Power-Relations Dataset Version 2.0 (Cederman et al., 2010), the GeoEPR-ETH Dataset 2.0 (Wucherpfennig et al., 2011), the Non-State Actor Dataset Version 3.1 (Cunningham et al., 2009), the ACD2EPR Docking Dataset (Wucherpfennig et al., 2012) and the UCDP Armed Conflict Dataset Version 4-2010 (Gleditsch et al., 2002).

[5] In both models, weighted measures are created using the Stata package spmon (Neumayer and Plümper, 2010).

order to test whether media censorship indeed inhibits foreign conflicts to exert an effect via media. The interaction term and constitutive terms are lagged in time by 1 year.

More specifically, in the test of proposition 1, we expect the following relation:

$$y_i^* = \beta_0 + \beta_1 \sum_{J \neg i} (\text{ind}_{i,j} x_j) z_i + \beta_2 \sum_{J \neg i} (\text{ind}_{i,j} x_j) + \beta_3 z_i + \beta_c c_i + \epsilon \tag{13.3}$$

where y_i^* is a latent variable that translates into the dichotomous outcome y, that is, whether a new ethnic conflict starts in a country or not. $\sum_{J \neg i} (\text{ind}_{i,j} x_j)$ are foreign civil conflicts weighted by the index and summed for each country i and z_i is the measure on media freedom. $\beta_c c_i$ is a vector of control variables with the associated coefficients. A list of control variables can be found in Appendix.

The second proposition on the contagious effect of conflicts when foreign and domestic groups are mutually discriminated is tested similarly, only that here the number of foreign ethnically discriminated groups' involvements in high-intensity armed civil conflict constitutes x_j[6] that is weighted by the index, $\text{ind}_{i,j}$, and subsequently multiplied by the measure on media freedom, z_j. In this case, the measure is also multiplied by the number of discriminated groups in a state over 5%, u_i , as only those groups are expected to react to information about this particular type of foreign conflict in Beiser (2012). This empirical test of proposition 2 has first been shown in Beiser (2012). Again, the interaction term and constitutive terms are lagged in time by 1 year and control variables are discussed in Appendix.

Thus, in the test of proposition 2, we expect[7]

$$y_i^* = \beta_0 + \sum_{J \neg i} (\text{ind}_{i,j} x_j) z_i u_i + \beta_2 \sum_{J \neg i} (\text{ind}_{i,j} x_j) + \beta_3 z_i + \beta_4 u_i + \beta_c c_i + \epsilon \tag{13.4}$$

In both models, coefficients are estimated using regression analysis. As ethnic conflict is a very rare phenomenon, estimates need to be corrected for this. Thus, following Beiser (2012), these relations are tested using a rare events logistic regression model as suggested by King and Zeng (2001) using a Stata package developed by Tomz et al. (2003). In addition, again following Beiser (2012), standard errors are clustered by country i.

So what do we find in statistical analyses of both propositions? Proposition 1 suggests that ethnic conflict becomes more likely if the domestic population gains information about intrastate conflicts in other countries. It was expected that foreign conflicts – weighted by domestic media availability, distance to the conflict country and economic prominence of the conflict country – can only exert an effect on countries that have free media. The model therefore tests whether the effect of weighted intrastate conflicts is different in countries with media freedom and in countries without. The proposition is not found to be supported by the empirical data.

Figure 13.6 shows results from the statistical test of assertion one. On the y-axis, we see the probability of an ethnic conflict breaking out in a country as predicted from the statistical model. On the x-axis, we see a scenario without media freedom on the left-hand side and a scenario with media freedom on the right-hand side. In each scenario, the grey dot stands for a case where the number of foreign civil conflicts, weighted by the information index, is low

---

[6] Moreover, extrasystemic conflicts of high intensity from the UCDP Armed Conflict Dataset Version 4-2010 (Gleditsch et al., 2002) until the year of independence (Gleditsch and Ward, 1999, 2013) are added.
[7] In addition, all twofold interactions between the constitutive terms are added as well.

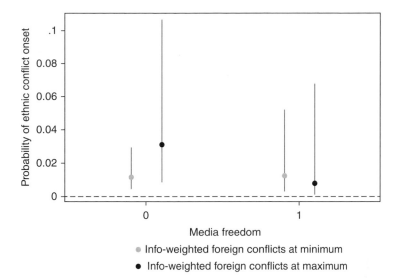

Figure 13.6   The effect of information about foreign civil wars

and the black dot for one where this number is high. Thus, we can understand the difference between the grey and the black dot in each scenario as the effect of an increase in information about foreign civil conflicts. When there is no media freedom in a country, an increase in information-weighted foreign civil conflicts actually has a positive effect on the probability of ethnic conflict in a state. When there is media freedom, on the other hand, the effect of information-weighted foreign conflicts becomes negative. In addition, the confidence intervals of conflict probabilities in all scenarios overlap. Thus, we can conclude that information flows about foreign events do not affect a country's security situation through a mechanism by which information about foreign civil conflicts affects the likelihood of ethnic conflicts. It could, however, be possible that these conflicts would have an effect on civil conflicts in general, ethnic as well as non-ethnic.

Proposition 2 suggests that information about ethnically discriminated groups fighting abroad can increase the likelihood of ethnic conflict, and that this effect is more pronounced the more ethnic groups are discriminated in a potentially affected country.

Figure 13.7 shows results from the regression model testing this proposition from Beiser (2012). Here, we see again the probability of ethnic conflict onset in a country on the y-axis. The left-hand side shows scenarios of a country with no media freedom and the right hand side shows the scenarios of a country with media freedom. Here, the grey dot stands for a case where the number of foreign discriminated groups in conflict, weighted by the information index, is low and the black dot for one where this number is high. The difference between the grey and the black dot can here be understood as the effect of an increase in information about discriminated groups fighting in other countries. In the scenario on the far left, where media is not free and there are no ethnic groups discriminated in a country, information about foreign discriminated groups fighting has a negative effect on ethnic conflict. In the scenario on the far right, on the other hand, when media is free and many groups are discriminated in a country, an increase in this information has a substantive positive effect on the likelihood of an

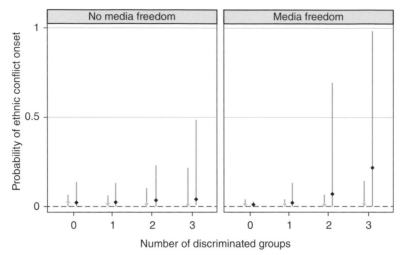

Figure 13.7   The effect of information about foreign civil wars moderated by mutual discrimination

ethnic civil conflict breaking out. This supports proposition 2 substantively, albeit neither the simulated interactions between discrimination and conflict nor the interaction between media freedom and the moderating effect of discrimination is significant here (Beiser, 2012).

We can unpack this result further by looking at the individual contributions of each factor that was expected to play a role for information transmission via mass media. Additional statistical tests in Beiser (2012) suggest that it is the factor on the availability of radios and TVs that is driving this result. Figure 13.8 from Beiser (2012) illustrates this. Here, the grey dot shows a scenario where the sum of discriminated groups fighting in other countries is at its lowest and the black dot shows a scenario where this number is at its highest. The difference between the grey and the black dot in each scenario can be interpreted as the effect of an increase in the number of such groups fighting in the world. We see that the effect of increases in groups fighting is largest when the availability of radio and TV in a country is high and when there are many domestic discriminated groups[8]. In this model, the simulated interaction between discrimination and foreign conflicts is significant in a scenario with high media availability and the effect of media availability on the moderating effect of discrimination is also significant (Beiser, 2012). This finding is interesting and suggests that discriminated groups in different countries may inspire each other (Beiser, 2012). The availability of media sources seems to be an important transmission channel for armed conflict between discriminated groups (Beiser, 2012). Of course, as this test is conducted on the country level as opposed to looking explicitly into the actions of groups, it is also possible that other actors instigate the conflict (Beiser, 2012). Interestingly, this effect of media source availability as such is barely present when all foreign high-intensity civil conflicts are considered under proposition 1.

---

[8] However, as noted in Beiser (2012), the variable on media freedom covers a shorter time period and the hypothesis on media availability would not receive support in this smaller sample either. Thus, results need to be interpreted with some caution.

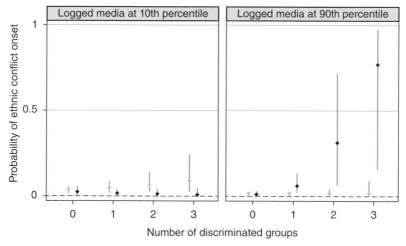

Figure 13.8   The effect of the availability of radio and TV on conflict contagion between discriminated groups

13.4   The Effect of Information Flows on Government Repression

In the previous section we have seen that foreign conflicts are likely to spread if foreign groups in conflict share the grievance of ethnic discrimination with domestic groups. But how does information from other countries affect a government's decision on whether and in how far to use physical repression against the population?

One potential explanation could be that repression itself is contagious between states. It could be that governments observe that other states that repress their citizens are not sanctioned by the international community despite threats to the opposite. This might encourage them to suppress their own population as well. Similarly, if states observe a high number of other states repressing their population, they might think that they are less likely to be punished because of a strength in numbers logic (also see Elkins and Simmons, 2005). If these mechanisms held, information about foreign governments' repression might have an effect on domestic repression. To date, to the best of the author's knowledge, this assertion has not yet been tested empirically.

Another way in which information about foreign events might affect governments' likelihood of repression may be as a result of their fear of conflict contagion. It is possible that governments fear that domestic groups start rebelling as a result of information about foreign conflicts and repress them in order to pre-empt this altogether (Beiser, 2013; Danneman and Ritter, 2014). Danneman and Ritter (2014) find that governments use high levels of repression – measured on the country level – when close-by countries experience armed civil conflict. Governments may also expect discriminated groups to take up arms as a result of information about foreign discriminated groups rebelling (Beiser, 2013) as evidence shown above suggests. If that was the case, we could expect that information about foreign conflicts of discriminated groups increases a government's likelihood of repression against domestic discriminated groups. This has been tested empirically in Beiser (2013).

Beiser (2013) finds mixed evidence with regard to the role of media sources in this process. However, Beiser (2013) finds some evidence suggesting that governments that censor media react with repression against discriminated groups as a result of an increase in foreign conflicts involving similar groups. This suggests that governments may indeed expect discriminated groups to rebel as a result of observing foreign groups engaging in similar conflicts abroad (Beiser, 2013). It may also suggest that governments that censor media are generally more likely to use repressive measures against groups (Beiser, 2013). Finally, the finding also suggests that governments that censor media do not seem to expect that censoring media can prevent information from entering the country (Beiser, 2013).

13.5   Conclusion

In many aspects, the world becomes increasingly interdependent. This is particularly the case with respect to global information flows. This chapter shows research on the effect of increased global information flows via mass media on armed civil conflict and government repression. It shows that increasing global information interdependence can have effects on security-related issues within countries. In particular, the availability of media sources has been found to play a role for conflict contagion in previous research. However, this is only the case if conflict groups address grievances that are also experienced by parts of the domestic population.

It is likely that other types of global interdependencies as well as other types of security issues are concerned by such processes as well. For example, and as noted earlier, governments may mimic other states' observed repression levels. Moreover, news about foreign repression may lead domestic groups to mobilise against the government out of fear of being subjected to a similar treatment such as foreign groups. Investigating these effects of global information flows is important, particularly as information flows increase.

The findings presented here also suggest that additional data is necessary to measure global information flows more accurately. In previous research, only media availability has been found to play a role for conflict contagion processes. Does that mean only devices are important or other factors have so far been overlooked? Answering this question is important to help investigating consequences of global information flows further.

References

Banks, A. (2011) *Cross-National Time-Series Data Archive*, Databanks International, Jerusalem.

Beiser, J. (2012) Looking beyond borders: inspiration, information, and ethnic conflict contagion. University College London working paper.

Beiser, J. (2013) Trampling out the spark? Governments' strategic reaction to the threat of ethnic conflict contagion. University College London working paper.

Beiser, J. (2016) Contagion processes of ethnic violence: group inspiration and government strategic reaction. PhD thesis. University College London.

Braithwaite, A. (2010) Resisting infection: how state capacity conditions conflict contagion. *Journal of Peace Research*, **47** (3), 311–319.

Buhaug, H., Cederman, L.E., and Gleditsch, K.S. (2014) Square pegs in round holes: inequalities, grievances, and civil war. *International Studies Quarterly*, **58** (2), 418–431.

Buhaug, H. and Gleditsch, K.S. (2008) Contagion or confusion? Why conflicts cluster in space. *International Studies Quarterly*, **52** (2), 215–233.

Carter, D.B. and Signorino, C.S. (2010) Back to the future: modeling time dependence in binary data. *Political Analysis*, **18** (3), 271–292.

Cederman, L.E., Wimmer, A., and Min, B. (2010) Why do ethnic groups rebel? New data and analysis. *World Politics*, **62** (1), 87–119, http://www.icr.ethz.ch/data.

Cingranelli, D.L., Richards, D.L., and Clay, C.K. (2014a) Short Variable Descriptions for Indicators in the Cingranelli-Richards (CIRI) Human Rights Dataset, http://www.humanrightsdata.com.

Cingranelli, D.L., Richards, D.L., and Clay, C.K. (2014b) The CIRI Human Rights Dataset. Dataset Version: 2014.04.14, http://www.humanrightsdata.com.

Collier, P. and Hoeffler, A. (2004) Greed and grievance in civil war. *Oxford Economic Papers*, **56** (4), 563–595.

Correlates of War Project (2008) State System Membership List, v2008.1, http://correlatesofwar.org.

Cunningham, D.E., Gleditsch, K.S., and Salehyan, I. (2009) It takes two: a dyadic analysis of civil war duration and outcome. *Journal of Conflict Resolution*, **53** (4), 570–597.

Danneman, N. and Ritter, E. (2014) Contagious rebellion and preemptive repression. *Journal of Conflict Resolution*, **58** (2), 254–279.

Davenport, C. (2007) State repression and political order. *Annual Review of Political Science*, **10**, 1–23.

Elkins, Z. and Simmons, B. (2005) On waves, clusters, and diffusion: a conceptual framework. *Annals of the American Academy of Political and Social Science*, **598** (1), 33–51.

Forsberg, E. (2008) Polarization and ethnic conflict in a widened strategic setting. *Journal of Peace Research*, **45** (2), 283–300.

Freedom House (2015) Freedom of the Press Historical Data. Detailed Data and Subscores 1980–2008. Global Data, http://www.freedomhouse.org/template.cfm?page=274.

Gleditsch, K.S. (2002) Expanded trade and GDP data. *Journal of Conflict Resolution*, **46** (5), 712–724, http://privatewww.esscx.ac.uk/ ksg/exptradegdp.html.

Gleditsch, K.S. and Ward, M.D. (1999) A revised list of independent states since the congress of Vienna. *International Interactions*, **25** (4), 393–413.

Gleditsch, K.S. and Ward, M.D. (2013) System Membership Case Description List: Release 5.1, http://privatewww.essex.ac.uk/ ksg/data/iisyst_casedesc.pdf.

Gleditsch, N.P., Wallensteen, P., Eriksson, M., Sollenberg, M., and Strand, H. (2002) Armed conflict 1946–2001: a new dataset. *Journal of Peace Research*, **39** (5), 615–637.

Goldstein, R. (1978) *Political Repression in Modern America. From 1870 to the Present*, Schenkman, Cambridge, MA.

Harbom, L. (2010) UCDP/PRIO Armed Conflict Dataset Codebook. Version 4-2010, Uppsala Conflict Data Program (UCDP), Centre for the Study of Civil Wars, International Peace Research Institute, Oslo (PRIO), http://www.pcr.uu.se/digitalAssets/124/124920_1codebook_ucdp_prio_armed_conflict_dataset_v4_2010.pdf.

Hill, S. and Rothchild, D. (1986) The contagion of political conflict in Africa and the world. *Journal of Conflict Resolution*, **30** (4), 716–735.

Hill, S., Rothchild, D., and Cameron, C. (1998) Tactical information and the diffusion of peaceful protest, in *The International Spread of Ethnic Conflict: Fear, Diffusion, and Escalation* (eds D.A. Lake and D. Rothchild), Princeton University Press, Princeton, NJ.

Katz, E. and Wedell, G. (1977) *Broadcasting in the Third World*, Harvard University Press, Cambridge, MA.

King, G. and Zeng, L. (2001) Logistic regression in rare events data. *Political Analysis*, **9** (2), 137–163.

Kuran, T. (1998) Ethnic dissimilation and its international diffusion, in *The International Spread of Ethnic Conflict: Fear, Diffusion, and Escalation* (eds D.A. Lake and D. Rothchild), Princeton University Press, Princeton, NJ.

Mowlana, H. (1997) *Global Information and World Communication*, New Frontiers in International Relations, Sage Publications, London.

Neumayer, E. and Plümper, T. (2010) Making spatial analysis operational: commands for generating spatial-effect variables in monadic and dyadic data. *Stata Journal*, **10** (4), 585–605.

Polity IV Project (2009) Polity IV Data Set, http://www.systemicpeace.org/inscr/inscr.htm.

Themnér, L. and Wallensteen, P. (2014) Armed conflict, 1946–2013. *Journal of Peace Research*, **51** (4), 541–554, http://www.pcr.uu.se/research/ucdp/datasets/ucdp_prio_armed_conflict_dataset/.

Tomz, M., King, G., and Zeng, L. (2003) ReLogit: rare events logistic regression. *Journal of Statistical Software*, **8** (2), 1–27.

UN General Assembly (1948) The Universal Declaration of Human Rights, http://www.un.org/en/documents/udhr/.

Uppsala Conflict Data Program (2014) *About UCDP*, Uppsala University Department of Peace and Conflict Research, Uppsala, http://www.pcr.uu.se/research/UCDP/.

Vreeland, J.R. (2008) The effect of political regime on civil war: unpacking anocracy. *Journal of Conflict Resolution*, **52** (3), 401–425.

Weidmann, N.B. (2015) Communication networks and the transnational spread of ethnic conflict. *Journal of Peace Research*, **52** (3).

Weidmann, N.B., Kuse, D., and Gleditsch, K.S. (2010) The geography of the international system: the CShapes dataset. *International Interactions*, **36** (1), 86–106.

Wucherpfennig, J., Metternich, N.W., Cederman, L.E., and Gleditsch, K.S. (2012) Ethnicity, the state and the duration of civil war. *World Politics*, **64** (1), 79–115, http://www.icr.ethz.ch/data.

Wucherpfennig, J., Weidmann, N.B., Girardin, L., Cederman, L.E., and Wimmer, A. (2011) Politically relevant ethnic groups across space and time: introducing the GeoEPR dataset. *Conflict Management and Peace Science*, **28** (5), 423–437, http://www.icr.ethz.ch/data.

Appendix

The following variables are controlled for in model 1: The number of ethnically discriminated groups over 5 % population share in t-1 introduced above, the natural log of a country' GDP per capita again using data from Gleditsch (2002), a country's population from Gleditsch (2002), a measure of the regime type from Polity IV Project (2009) coded following a suggestion of Vreeland (2008), the year of observation, the number of years since the last ethnic conflict in a country plus a squared and cubed term as suggested in Carter and Signorino (2010) and the dependent variable in t-1.

The following variables are controlled for in model 2: The natural log of a country's GDP per capita again using data from Gleditsch (2002); a country's population from Gleditsch (2002), a measure of the regime type from Polity IV Project (2009) coded following a suggestion of Vreeland (2008), the year of observation, the number of years since the last ethnic conflict in a country plus a squared and cubed term as suggested in Carter and Signorino (2010) and the dependent variable in $t - 1$ and an additional time-lagged dummy control variable on additional conflicts in neighbours that are not included in the main variable on foreign conflicts of discriminated groups. Neighbours are defined using data from Weidmann et al., (2010).

Part Five

Aid and Development

14

International Development Aid: A Complex System

Belinda Wu

14.1 Introduction: A Complex Systems' Perspective

International development aid plays an important role in human progress, as it is crucial to help developing nations grow out of poverty. As the gap between the developed and developing countries continues to grow, aid is provided in the belief that it will close such gaps and ultimately help improve the lives of those who really need it (Andrews, 2009; Radelet, 2008). To achieve this goal, it requires effective aid allocation. However, the aid system demonstrates a great degree of complexity that originates from various sources including the organisation of the aid system, the interwoven relationship of different actors and factors in the system, the data and the research method. Such complexity makes the goal of achieving aid effectiveness very difficult, and this has always fascinated researchers, government officials and practitioners in the field.

International development aid is part of a complex system where multiple organisations are providing aid over multiple issues through multiple relationships, driven by multiple motivations. There are so many intertwined factors playing important roles in this system that it makes the study of it very difficult, as it will be impossible to capture all the factors or describe them accurately in any model. The system also demonstrates the characteristics of a complex system such as connectedness/network, self-adapt/self-organising and emergence. Such features are challenging for traditional macroeconomic methods (Ramalingam, 2013), especially with heterogeneous donor and recipient behaviours changing the dynamics within the system. The interactions with other global systems such as international trade, security and migration also further complicate the situation.

While the debates on whether aid is an effective way to help improve the life of those in need, a better approach, or how to measure the effectiveness of aid, no generally accepted agreement has been reached. Both the policy need and the complexity within modern aid are

Global Dynamics: Approaches from Complexity Science, First Edition. Edited by Alan Wilson.
© 2016 John Wiley & Sons, Ltd. Published 2016 by John Wiley & Sons, Ltd.

now demanding a more holistic approach that can take into consideration various aspects of this complex system.

With the surfacing of the new forms of aid organisations and new issues, there is a pressing need to evaluate progress to date and shed new light on emerging issues and agendas against the shifting aid landscape. With the advance in data collection and research method, we are now in a better position to take various aspects of the issue into consideration using a complexity science approach. This is a promising approach that can enable researchers to explore the non-linearity, connectedness/network, self-adapt/self-organising and emergence features of the aid system (Ramalingam, 2013; Mavrotas, 2010; Barder, 2012; Porter, 2011).

14.2   The International Development Aid System: Definitions

In the 1940s, the great divergence between the incomes of countries in the global economy gave rise to international development aid issues (Pritchett, 1997; Madisson, 2008). However, from the beginning, the definitions (and names) of international development aid have reflected the complexity of this system.

International development aid has also been called development assistance, technical assistance, international aid, overseas aid, the formally named Official Development Assistance (ODA), or foreign aid. Other terms such as development co-operation are used, for example, by the World Health Organisation (WHO) to emphasise the partnership between donor and recipient countries. Generally speaking, aid is a financial aid given by governments and other organisations to support the economic, environmental, social and political development of developing countries (Riddell, 2014; Coscia et al., 2004; Radelet, 2008; WHO, 2008).

From the aspect of aid purposes, there is a range of aid from humanitarian emergency assistance, food aid, military assistance to longer-term development aid that aims to transform the lives of the nations and populations in need, contributing to their development in a sustainable way. Traditionally, a distinction has also been made between humanitarian aid and development aid, the former being a response to emergencies and short term in nature, the latter being the contribution of aid to support longer-term development processes, but recently there is a growing trend of less distinctive differentiation between the two (Bauer and Yamey, 1982; Moyo, 2009; Wikipedia, 2015).

From the aspect of donor types, aid may be bilateral: given from one country directly to another; or it may be multilateral: given by the donor country to an international organisation such as the World Bank or the United Nations Agencies (UNDP, UNICEF, UNAIDS, etc.) which then distributes it among the developing countries; or it can also come from private donors (such as the Bill and Melinda Gates Foundations) and other sources (OECD, 2015a).

The most important and the majority form of the international development aid is ODA. About 80–85% of aid comes from government sources as ODA. The remaining 15–20% comes from private organisations such as non-governmental organisations (NGOs), foundations and other development charities (e.g. Oxfam). In addition, remittances received from migrants working or living in diasporas form a significant amount of international transfer.

The Organisation for Economic Co-operation and Development (OECD) (2015a) categorises aid as bilateral, multilateral and private donors' aid (ODA) and other resource flows to developing countries. The aid flows include ODA, OOF (Other Official Flows) and private. ODA is the largest and the most important, as it is the key measure used in practically all aid targets and assessments of aid performance. Due to its importance in international development aid and data availability, our study will focus on ODA for the first instance and here we explore further the definition of ODA.

ODA is a measure of government-contributed aid, compiled by the Development Assistance Committee (DAC) of OECD since 1969. The DAC consists of 29 of the largest aid-donating countries. The DAC first defined ODA in 1969 and revised the definition in 1972. The DAC defines ODA as 'those flows to countries and territories on the DAC List of ODA Recipients and to multilateral institutions which are:

- provided by official agencies, including state and local governments, or by their executive agencies; and
- each transaction of which:
  - is administered with the promotion of the economic development and welfare of *developing* countries as its main objective; and
  - is *concessional in character* and conveys a grant element of at least 25 per cent (calculated at a rate of discount of 10 per cent) (OECD, 2015a)'

14.3   Features of International Development Aid as a Complex System

14.3.1   Introduction

The aid system has developed incrementally rather than systematically over decades. The aid system here mainly includes the international development aid organisations, their political owners and managers, as well as their sources and uses of funds. In its about 60 years of aid history, no major institution has exited the system, while new institutions have been created due to the economic divergence, decolonisation, decentralisation and the changing conceptual framework. This has resulted in an increasing number of donors, recipients and issues, further complicated by heterogeneous donor and recipient behaviours, emergence of new features and dynamic changes, making international development system a truly complex undertaking. In the following sections, we will discuss four features of the complex system of international development aid that demonstrate the characteristics of a complex system (Rogerson et al., 2004; Andrews, 2009; Riddell, 2014).

14.3.2   Non-linearity

Issues related to aid heterogeneity and disaggregating aid flows in order to better to understand aid effectiveness have been long recognised (Mavrotas and Nunnenkamp, 2007; Mavrotas, 2010). OECD just published the financial flows to developing countries. Here we list the ODA commitments from DAC countries, World Bank, UN Agencies and Regional development Banks in Figures 14.1–14.4. We can see that not only the geographical distribution patterns are non-linear but also each type of donor behaviours are also heterogeneous (OECD, 2015b).

Figures 14.1–14.4 demonstrate that there are clear distinctions between behaviours of different organisations towards different regions in the world. For instance, DAC countries consistently give the lowest amount of donations to Oceania of all geographies. In contrast with DAC countries, the World Bank, UN agencies and regional development banks all consistently donate the largest proportions to Oceania of all geographies. Although DAC's largest

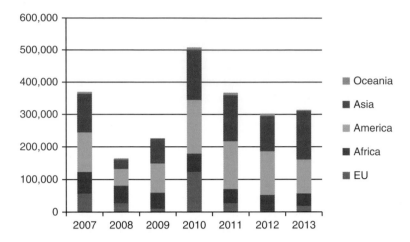

Figure 14.1   Total net disbursements by DAC countries (USD million). *Source*: OECD (2015b)

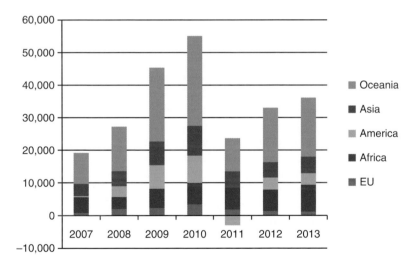

Figure 14.2   Total net disbursements by World Bank (USD million). *Source*: OECD (2015b)

donations went to America during the period 2007–2013, the donations to Asia is not far behind and there is an increase of donations to Asia recently and in 2013 the amount even got higher than that to America, while it seems that World Bank's donations are more focused on Asia and Africa than America. The analyses also suggest that United Nations is more likely to give more donations to Africa than the rest three organisations and the regional development banks also increased their assistance to Africa, while reducing their donations to Asia more recently (Figures 14.1–14.4).

Even in the multilateral donations, we can see the non-linear behaviours from the donor. Needless to say that there is more non-linearity in the disaggregate behaviours among

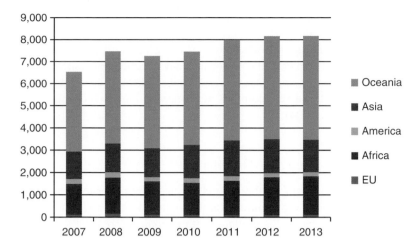

Figure 14.3   Total net disbursements by UN agencies (USD million). *Source*: OECD (2015b)

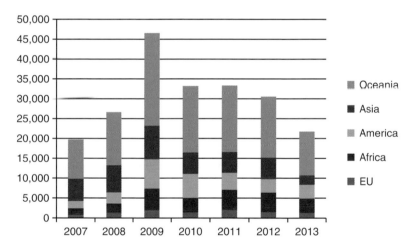

Figure 14.4   Total net disbursements by regional development banks (USD million). *Source*: OECD (2015b)

individual countries. Figures 14.1–14.4 demonstrate that there are clear distinctions between behaviours of different organisations towards different regions in the world.

14.3.3   Connectedness

The system of international development aid has changed rapidly in recent decades. Twenty years ago there were fewer than 100 official aid donors, while today there are over 70 bilateral and more than 230 multilateral aid agencies. In the mid-1960s, the 22 largest DAC donors each provided aid to fewer than 40 countries, but by the late 2000s they each provide to over 120 countries. The number of donors each recipient country has to interact with has also increased dramatically from an average of 12 major donors in the 1960s to 33 by 2005 (Rogerson et al., 2004; Riddell, 2014). According to the Directory of Development Organisation, there are more than 70,000 related organisations around the world (DEVDIR, 2015). Meanwhile with the environmental, social and economical changes in today's world, new issues have emerged in aid, such as climate change, international migration and conflicts. The multiplication of donors, recipients and issues implies a very complex network with hundreds of millions of possible connections (Coscia et al., 2004).

14.3.4 Self-Adapting and Self-Organising

Due to the human 'components' of the aid system, it is the opposite of a mechanical system where there is a stable context and operating environment to perform a straightforward task to produce identical, duplicable products with compliant, predictable and reliable parts (Ramalingam, 2013). In the complex adaptive system of aid, there are many interacting agents and organisations of agents that each has distinctive behaviours and act and react to situations according to their decisions. As the environment changes, they adapt and their strategies change, too. As a result, the designs or organisations of the international development aid system are also constantly self-adapting and self-organising.

14.3.5 Emergence

As different agents within the aid system interact dynamically with each other over various issues, such micro behaviours at the bottom level of the system can lead to complex and quite often, unexpected macro patterns at the global level of the system. This is an example of 'emergence' in complexity science terms. A simple change of action from some individual agents can lead to changes to the relationship to other agents that it interacts with. In turn, the impacted agents further influence the rest of the agents within the system through their interactions. It can trigger complex chains of changes that eventually lead to unintentional outcomes and change the course of the evolution of the system. The fact that the property of emergence arises from the individual level, but only observable at the system level, demands a holistic system approach that enables the study of the macro–micro linkage within the international development aid system (Barder, 2012; Porter, 2011). For example, the individual impact from efforts of the aid at the disaggregate level may be driven by various political economic strategies and cultural influences, but together they seldom reflect the global pattern of the international development aid at the aggregate level. This is a common property of complex systems where the 'emergence' or aggregate patterns are not intentional at the disaggregate level and they are only observable at the global level. The total is more than the sum of all components.

14.4 Complexity and Approaches to Research

14.4.1 Organisations

There are many sources of the complexity in the international development aid system. In this section, we examine four such sources: organisations, topics/issues, data and research methods.

The current international development aid system can be roughly divided into three major categories: multilateral organisations, bilateral organisations and other organisations. Multilateral organisations are normally developed through either the United Nations or the Bretton Woods organisations (e.g. the World Bank, the IMF and the GATT), while bilateral organisations are created by developed countries. Many bilateral organisations adopted the acronym ID to stand for International Development and the letter A for Agency, such as USAID (for the United States), DFID (United Kingdom), CIDA (Canada) and SIDA (Sweden). Other nations preferred to use the term 'cooperation' instead of 'development' for their bilateral efforts, such as Japan, Korea, Spain and the Netherlands. In addition to official multilateral and bilateral organisations, private philanthropic organisations have emerged from Church groups (e.g. CARE), wealthy individuals (e.g. Rockefeller, Ford, Gates), action-oriented groups (e.g. Medecins Sans Frontier, ACCION International) and other organisations (e.g. Human Rights Watch). Regional multilateral development banks modelled on the World Bank, such as AfDB, ADB and IADB, and EBRD, represent another important addition to multilateral and bilateral development organisations (Rogerson et al., 2004; Coscia et al., 2004).

The nature of such a system where many major actors are nation-states or their collective agents makes the demand and supply of the aid often not easy to interpret as in a free market (Raffer and Singer, 1996). Indeed, neither competitive forces nor a grand collective of owners drives the shape of the aid system today. Major governmental shareholders generally behave as if the aid system of cross-holdings as a whole is beyond even their collective grasp. Few have tried to develop comprehensive strategies or to compare the value for money of their investments across the whole architecture of the aid system (Rogerson et al., 2004).

Today's international aid system has been criticised as being too complex to operate as an effective collective. For instance, with the continued expansion and increased complexity from the emergence of new donors, modalities and instruments, the multilateral development finance system has been criticised as being in danger of becoming a 'non-system' with far-reaching implications for both donor and recipient communities (World Bank, 2007). The main concern has been with the 'architecture' of the aid system, especially the proliferation of bilateral agencies, international bodies and special-purpose 'vertical' funds. There are too many in total and too many in each country, with overlapping mandates, complex funding arrangements and conflicting requirements for accounting and reporting.' In addition, duplication, overlap and fragmentation are creating substantial transaction costs for donors and recipient governments (Burall and Maxwell, 2006; Roodman, 2006; Knack and Rahman, 2007; Mavrotas and Reisen, 2007).

On the other hand, new institutions for aid, such as the US Millennium Challenge Account and the Global Fund to fight AIDS, Tuberculosis and Malaria, are built on radically different premises of what constitutes effective aid delivery in the 21st century and are arguably incompatible with the aid system that preceded them. With the increasing impact from such institutions, they increase the complexity in studying the aid system (Rogerson et al., 2004).The organisation of the aid system introduces the first dimension of complexity into the study of international development aid.

14.4.2 The Range of Issues

The second dimension of the aid complexity originates from the vast range of aid issues. From the focus on the availability of capital and infrastructure as in the Marshall Plan in

Europe (1948–1952), the interpretation of the obstacles to development has now moved towards a much broader range of issues including education, public health, demographic pressures, industrial bottlenecks, macroeconomic stability, appropriate market regulation, environmental sustainability, human rights, institutional quality, governance, inequality and gender gaps (Coscia et al., 2004). There are numerous forms of aid to address different topics, ranging from short-term aid such as humanitarian emergency assistance to food aid, military assistance to long-term development aid that aims to transform the lives of the population in need, contributing to their development in a sustainable way. Allocation of aid is now influenced by multiple factors and it has expanded from donors' political, security, commercial interests and recipient's needs (McKinlay and Little, 1978) to concerns about democracy, human rights and trade openness, as well as new emerging topics such as climate change, sustainability and global governance issues (Coscia et al., 2004).

The changes in aid issues have in turn resulted in structural changes, too. Dedicated organisations and programmes have been formed to support specific purposes as discussed earlier. For instance, the UN system responded by creating specialised bodies for different purposes such as the WHO (for health), UNESCO (education), UNIDO (industrial development), FAO (for agriculture and food security), UNDP (technical assistance for development), UNCTAD (trade), ILO (labour) and UNHCR (human rights and refugees). Concomitantly, the regional organisations created similar structures (e.g. the Organisation of American States, the Pan-American Health Organisation, ASEAN, Organisation for African Unity). Within the developed countries, governments also started to involve ministries and departments to share expertise with their development agencies in international co-operations in areas such as financial regulation, tax administration and health (Coscia et al., 2004).

14.4.3   Research Approaches

Although over the decades, great efforts have been made to collect, store and publicise aid data, the development aid in most countries is still predominantly provided in the form of discrete projects. Despite a steady expansion in the numbers of databases assessed and an increase in analytic rigour, most data are not evaluated and only a small proportion of data are the focus of any in-depth evaluation. Few official agencies and no NGOs undertake the assessment in a systematic way. The usage of the data is typically hampered by a number of common problems, including data inadequacies (insufficient or non-existent baseline and monitoring data); problems of attribution (not knowing what specific contribution aid inputs make to outputs, and especially to wider outcomes) and problems of the counterfactual (not knowing what would have happened to those assisted if the aid inputs had not been provided) (Riddell, 2014). Sometimes, data from different sources are contradictory. For instance, in many countries, the OECD figures for ODA differ sharply from nationally generated data on aid levels, and both suffer from serious gaps and inaccuracies, for example, Ugandan national statistics showed a 50% lower total ODA disbursements in the year 2005 than the OECD records (MoFPED, 2005, p. 22; MoFPED, 2006, pp. 22–23 and OECD QWIDS, 2015). Due to such reasons, we are still far from having reliable, consistent and robust aid project information and it adds to the complexity of understanding the international development aid system.

Foreign aid is intended to have beneficial macroeconomic effects, most notably to raise a country's economic growth rate. Therefore, traditionally, the macroeconometric approach has been used in studying international development aid (Burnside and Dollar, 2000; Collier and Dollar, 2002; Easterly and Pfutze, 2008; Rajan and Subramanian, 2008). Some studies have carried out using the statistical, especially regression, analyses. For instance, based on ordinary least squares regression analyses, Dollar and Easterly (1999) argue that aid has a significantly positive effect on investment in 8 of 34 country cases. International development aid researches will benefit from Alesina and Dollar's (2000) investigation of the relationship between bilateral aid and a number of variables to extract the main drivers behind them using a series of linear regression analyses, while Morrissey (2001) and Gomanee et al. (2005) focus on the long-run impact of aid on GDP and its main macroeconomic determinants (including gross investment, and private and government consumption). Recently, Juselius et al. (2011) offer a unique perspective in coverage by studying a total of 36 sub-Saharan countries to provide a broad and statistically well-founded picture of the effect of aid on the macroeconomy of 36 sub-Saharan countries from the mid-1960s to 2007.

As Kanbur (2006) points out 'the macro-econometric investigation of aid-growth regressions will no doubt continue into the next century. There are sufficient issues of data (how exactly is "aid" defined?), of econometrics (how can the truly independent effects of aid be identified from a mix of independent relationships?) and of development doctrine (what is "good policy"?) to keep the debate alive'.

However, there has been some criticism on such methods. Over the years, development aid topics have been associated with many perceived paradoxes and dilemmas. One example is the micro–macro paradox pointed by Mosley (1980, 1987), who believes that aid is ineffective at the macro level. Mosley offers three explanations for the macro–micro paradox. Firstly, inaccurate measurement in either micro or macro studies; secondly, fungibility of aid within the public sector; thirdly, backwash effects from aid-financed activities that adversely affect the private sector. Mavrotas (2010) adds two more explanations: One is that the paradox may be a result of overaggregation on a country level, the macro and micro results are not so inconsistent. The other is that microeconomic evaluations use economic (social) data, whereas macro studies use the national financial (private) data, so the studies are not comparable.

More recently, the main aid donors are ready to acknowledge that aid-giving has become so complex and fragmented that a series of systemic problems have developed. With such a complex aid context, the study of the aid is now more difficult. The ever expanding list of each country's own programmes, multilateral agencies, special programmes and projects including technical assistance, debt relief or through vertical funds makes the choice even harder. The complex international aid architecture raises some serious issues and promotes new approaches to study the aid system (Riddell, 2014).

It is widely recognised that the research on the foreign aid system is in need of a revolution to adapt to the drastic changes that have happened in the field. But there are conflicting opinions as to what is needed. Recently among scientists, there has been a growing movement to expand the tools and techniques available, precisely because of a longstanding appreciation that the tools and assumptions are not appropriate for all contexts, let alone a radically changed aid environment. It is quite encouraging to see authors in the development field considering the principles from complexity to move away from more traditional econometric methodologies. In that sense, Ramalingam (2013) from the Active Learning Network for Accountability and Performance in Humanitarian Action, Barder (2012) from the Center for Global development in Washington, DC, and Porter (2011) from the Centre for Complexity Science in the University of Warwick have written on the issue. These researchers show that the linear, mechanistic models and assumptions on which foreign aid is built are inadequate in the dynamic, complex world we face today. Instead, they argue that a new approach embracing the 'new science' of complex systems can make foreign aid more relevant, more appropriate, more innovative and more catalytic.

14.4.4   The Complexity Science Approach

Just as there is no clear definition of science (Gross et al., 1996; Park, 2000; Shermer, 1997; Sokal and Bricmont, 1998), there is no simple definition for complexity science, either. Complexity science adopts an interdisciplinary approach and introduces a new way to study regularity uses developed in the natural sciences, utilising techniques such as network analysis, agent-based models and power laws to study simple causes for complex effects. At the core of complexity science is the assumption that complexity in the world arises from simple rules. Actions of individuals guided by simple rules at the bottom level of the system can lead to complex and often unexpected patterns at the system level. Such a phenomenon is called emergence. However, these rules are unlike the rules of traditional science. They do not predict an outcome for every state of the world, but instead, generative rules can use feedback and learning algorithms to enable the agents to adapt to their environment. Complexity science also takes a holistic view and believes that the connectedness and interactions are the key to understanding such systems (Wolfram, 2002; Phelan, 2001; Richardson et al., 2000).

Ramalingam (2013) calls for the application of complexity theory, offering approaches for solving development problems more effectively. He argues that there is a pervasive and long-standing bias in aid: aid agencies are incentivised to treat the world as a simple, predictable system where aid can be delivered to bring about positive changes. Current aid is often too static for the complex environment that is operates in. Development is an organic process, as it arises out of the interaction between complex social, political, economic and environmental systems. Instead of operating like a machine turning out ready-made solutions, aid must be guided by a holistic management and adapt to a dynamic environment. Also, it is recognised that aid rarely provides external momentum for development, but must rather act as a catalyst for change.

Ramalingam believes that learning the complexity approach can help aid align with its complex environment. He listed some of the potential benefits that complexity science can bring to development studies in terms of systems, behaviours, relations and dynamics (Table 14.1).

14.5   The Assessment of the Effectiveness of International Development Aid

14.5.1   Whether Aid Can Be Effective

In order to address development issues, we need to allocate aid effectively. However, current assessments return a mixed picture: in terms of overall aid impact, in most countries aid has had a positive overall impact in some time periods, but in some countries negative. At the country level, aid has worked in some countries at certain periods of time. Research suggests that although aid projects tend to work better in countries with a supportive policy and institutional environment, project success seems to vary more within than between countries (Denizer et al., 2013).

Table 14.1 Comparison between science of simplicity and complexity science approaches to development studies

	Systems	Behaviours	Relations	Dynamics
Sciences of simplicity	Systems and problems are closed, static, linear systems; reductionist	Individuals use rational deduction; behaviour and action can be specified from top-down	Actors can be treated as independent and atomized	Change is direct result of actions; proportional, additive and predictable
What sciences of complexity bring	Systems are open, dynamic, interconnected and interdependent. Macro patterns emerge from micro behaviours and interactions	Humans are adaptive tinkerers; subject to errors and biases; self-organise and co-evolve with system and each other	Relationships, flows, ties, values, beliefs are vital. Are path dependent and historical	Change is non-linear, unpredictable, with phase transitions, characterised by power laws and discontinuities

Source: Ramalingam (2013). Reproduced with permission of OUP.

Discussions about aid have been extremely polarised: those in favour of aid have been reluctant to draw attention to its problems and inadequacies, those against aid have been reluctant to acknowledge that it has done any good (Radelet, 2008). Critics such as Friedman (1958), Bauer and Yamey (1982) and Easterly (2001) have argued that aid has enlarged government bureaucracies, perpetuated bad governments, enriched the elite in poor countries or just been wasted. They cite widespread poverty in Africa and South Asia despite decades of aid and point to countries that have received substantial aid yet have had disastrous records such as the Democratic Republic of the Congo, Haiti, Papua New Guinea and Somalia. These authors suggest that aid programmes should be dramatically reformed, substantially curtailed or even eliminated altogether.

Supporters counter that these arguments are overstated. They argued that although aid sometimes failed, it has supported poverty reduction and growth in some countries and prevented worse performance in others (Sachs, 2005, Stiglitz, 2002, Stern, 2002; Arndt et al., 2010). They believe that many of the weaknesses of aid have more to do with donors than recipients, and point to a range of successful countries that have received significant aid such as Botswana, Indonesia, Korea and, more recently, Tanzania and Mozambique, along with successful initiatives such as the Green Revolution (Mavrotas, 2010).

In his influential study 'Does Aid Work?', Cassen et al. (1994) argued that 'most aid does succeed in terms of its own objectives and obtains a reasonable rate of return; but a significant proportion does not'. Twenty years after the publication, aid issues are still dominated by politics and ideology in many cases and Kanbur (2006), in his comprehensive review of the economics of international aid, points out that 'if the historical evolution of the aid literature experience is anything to go by, it is unlikely that a survey of international aid in ten years time will have entirely and dramatically new policy issues from the ones highlighted here'. Last year

Riddell (2014) also supported the opinion that aid had not worked as well as it could, or as well as it should. Although aid has had a positive impact, sustaining its benefits has often proved challenging. He believes that there is an important role for aid to play, but donors need to apply more of this knowledge, as well as to learn more about how best to address some of the more systemic challenges of giving, receiving and using aid (Riddell, 2014).

Apart from the different perspectives of the supporters and critics of aid, there is also a mix of complexity that makes the assessment of aid effectiveness a challenging exercise and we will discuss some of the issues in the next section.

14.5.2 Complexity in the Measurement of Aid Effectiveness

Recently, political scientists have also started to co-operate more with economists to facilitate more practical decision-makings in governments in relation to development finance and measurement of the effectiveness of aid (Schneider and Tobin, 2013). Impact evaluation and policy-related research in this area have again become a focus due to a lack of indication how effective the aid has been. For instance, WHO found that there was 'not enough information to draw conclusions about the WHO's development effectiveness' from 2007 to 2010 (Freeman, 2013, p. viii). Similarly, a 2013 study that looked at the evaluation reports of six of the largest Norwegian NGOs raised questions about the methodologies used and concluded that the evaluations did not say much that was reliable about impact in large measure because of incomplete baseline and monitoring data. There are also similar findings reported from various aid databases (Forss et al., 2013; Riddell, 2014).

The majority of the aid effectiveness literature is concerned with understanding the macroeconomic relationship of aid inflows on economic growth in the recipient countries (Beynon, 2002; Hansen and Tarp, 2001; Collier and Dollar, 2002; Addison et al. (2005); Easterly, 2006; Easterly et al., 2004; Rajan and Subramanian, 2008 and Antipin and Mavrotas, 2006). The findings have been mixed. Some scholars find a positive relationship between aid and economic growth (Hansen and Tarp, 2001; Sachs, 2005), others find no relationship (Rajan and Subramanian, 2008) or even a negative relationship (Svensson, 2000; Easterly, 2006; Djankov et al., 2008; Moyo, 2009).

There are many reasons contributing to the difficulties in measuring the effectiveness of aid. On the one hand, the allocation of official aid is still influenced by the short-term political, security and commercial interests of donor governments. On the other hand, what is central to contemporary discourse about aid is that the justification for providing aid is intimately linked to its impact. Researchers believe that the complex context in which aid operates needs to be inserted more centrally into the debate about aid, to the extent of readjusting our expectations of what effectiveness might mean: the failure of aid to 'work', especially in the short term, does not provide a sufficient basis for not giving it (Riddell, 2014).

Traditionally, a distinction has been made between humanitarian aid and development aid, the former being a response to emergencies and short term in nature, the latter being the support for longer-term developments. Most effectiveness critiques are focused on development aid. However, such distinction between the two is now growing narrower. Increasingly emergency aid has been deployed not only to save lives and respond to the immediate aftermath of a disaster but also to help rebuild the lives and restore the livelihoods. In doing so, it has been used to fund projects identical to those supported with development aid funds. On the other hand, a growing number of 'development aid' projects have been channelled into immediate life-saving initiatives and immunisation against deadly killer diseases. Such fading distinction between the emergency and development aid also complicates the measurement of the effectiveness of the aid.

Although the aid effectiveness literature has been focused on aid impact on economic growth, in reality donors have repeatedly stressed that they pursue multiple and often contradictory objectives in development aid (Isenman and Ehrenpreis, 2003). Although primarily donors have to base their aid allocation on the measurable policy of recipients and their institutional readiness to use aid for growth and poverty reduction, often there are other political obstacles. For instance, although massive aid redirection towards South Asia would sharply increase both EC and global aid effectiveness, this does not meet the requirement of geographical 'balance'. Even IDA, for example, is required to spend at least half its resources on Africa, regardless of need or effectiveness. This is inconsistent with the institution's poverty-reduction mandate (Beynon, 2002). On the other hand, the overoptimism when chronically weak countries show fresh signs of hope and oversell of the sustainability of early success in more stable countries make it more difficult to periodical release from these recipients to fund new priorities. Added to this mix, there are the distortions of aid flows resulting from historic and colonial/cultural ties. Also aid selectivity and aid absorption concerns can be mutually reinforcing. The idea of diminishing returns to aid at levels of ODA to GDP of 20%, 30% or 40% can seem quite unlikely if aid is widely distributed. However, these thresholds are more likely to be passed if aid is increased considerably and awarded to only a few countries. The resulting messy aid landscape produces donor favourites and donor orphans, with only transitory links to sustained development performance (Rogerson et al., 2004).

14.5.3   Complexity in Methods/Standards of Measurement of Aid Effectiveness

Such complexity in the aid effectiveness research context discussed earlier leads to the question of how we should measure the effectiveness of aid? What factors would be appropriate for measuring the effectiveness of aid? We need to address the complexity of evaluating whether an improvement in a country's growth rate or GDP per capita is attributable to aid inflow or not. Or, whether there is a link between aid, poverty and bad policies, so that we can safely say aid is ineffective (Andrews, 2009). Some call this intricacy the 'chicken and egg' problem (Alesina and Dollar, 2000).

It is difficult to determine if aid is effective, partly also because of the different perspectives of the assessors. Even for the independent reviews and assessments, judgements can be influenced by the overall perspectives of those undertaking the assessments and the specific questions raised by the organisations that commissioned the assessment. Those undertaking these studies who are more critical of aid tend to emphasise the problems while those more favourably inclined emphasise aid's strengths. Furthermore, such assessments are often conducted by the donor countries, and it was only in 2008 that a recipient country, Uganda, first commissioned a comprehensive evaluation of its overall development efforts that focused explicitly on the contribution that aid had played in helping to achieve its poverty-reducing targets (Matheson et al., 2008).

Also as discussed previously, the aid information and databases are often weak and not easily accessible due to the fact that aid in most countries is still predominantly provided in the form of discrete projects.[11] Few official agencies and no NGOs undertake aid impact assessment in a systematic way. Assessments of their impact are typically hampered by a number of common problems, including data inadequacies (insufficient or non-existent baseline and monitoring data); problems of attribution (not knowing what specific contribution aid inputs make to outputs, especially to wider outcomes); problems of the counterfactual (not knowing what would have happened to those assisted if the aid inputs had not been provided) (Riddell, 2014).

A further problem is that it is challenging to track down aid impact information that does exist. Notwithstanding improvements in the transparency of most donor agencies in recent years, it still remains difficult to locate and access data on aid impact, especially assessments of aid that are particularly critical. Beyond the official aid sector, only a handful of NGOs (almost certainly the largest) give prominence to evaluations on their websites, and few place any critical evaluations in the public domain (Gomanee et al., 2005; Juselius et al., 2011).

Although most studies use data from the same publicly available databases, such as aid and macro data from the DAC of the OECD, the Penn World Tables (PWT) and the World development Indicators (WDI), differences in results are used to support polarised opinions towards the effectiveness of aid that are embedded in the use of different econometric models and methods; different exogeneity/endogeneity assumptions; different choices of data transformations. For example, the literature reports different assumptions about exogeneity and endogeneity of aid as well as different measurements of variables (logs, levels, ratios, growth rates, etc.). Unfortunately, such choices regularly change the empirical results, sometimes crucially so, and can, therefore, be problematic (Juselius et al., 2011).

14.5.4   Standardising Aid Effectiveness

As described earlier, aid is characterised by an ever-increasing number of donors overseeing a growing number of discrete projects, creating an ever-more complex web of transactions and parallel management systems. This causes great difficulties in governance and measurement of the effectiveness of aid. To a certain extent, the lack of consensus stems from the existence of several factors that limit the evaluation of aid effectiveness. Recent aid-effectiveness studies continue to test the existence of different aid impact determinants, not all of them in relation to the recipients' characteristics but also in relation to the donors' managing procedures. On the one hand, these studies suggest that aid may be especially effective for circumstances related to the characteristics of the recipient economies:

1. if countries have sound institutions, in rule of law and respect civil and political rights (Burnside and Dollar, 2000); stability of the political system (Chauvet and Guillaumont, 2009); democracy (Svensson, 1999; Kosack, 2003); government effectiveness and control of corruption (Tezanos et al., 2009);
2. if countries suffer from adverse shocks, such as climate and trade shocks (Chauvet and Guillaumont, 2009; Collier and Hoeffler, 2007);
3. if countries suffer from structural disadvantages, such as their geographic location within the tropics (Dalgaard et al., 2004);
4. if countries are within post-conflict periods (Collier and Hoeffler, 2007).

On the other hand, some studies point out that donors' managing procedures also determine the impact of aid on growth:

1. aid volatility (Lensink and Morrissey, 2000; Bulir and Hamann, 2005; Hudson and Mosley, 2008; Tezanos et al., 2009);
2. donors' insufficient coordination, which generates problems of 'aid-fragmentation' (Djankov et al., 2008; Tezanos et al., 2009);
3. the preponderance of foreign interest – not always in accord with development goals – in the geographical allocation of aid (Minoiu and Reddy, 2010).

Based on such findings, initiatives designed to improve coordination among the donor community have emerged. The most important ones include the Declaration on Aid Harmonisation by donors in Rome in 2003 and the Paris Declaration on Aid Effectiveness in 2005 which stressed the importance of ownership, harmonisation, alignment and managing for results for accelerating progress on aid effectiveness and MDG attainment. The year 2008 was also very special for development aid and development finance in general, in view of the Third High-Level Forum on Aid Effectiveness in September 2008, where heads of governments, multilateral institutions and civil society organisations gathered in Accra to review the progress made in implementing the Paris Declaration. The Accra Forum was closely followed by the Follow-up International Conference on Financing for development in Doha at the end of the year, which reviewed the implementation of the Monterrey Consensus. The Accra Forum dealt with some key problems of the aid business, namely non-transparency, too many donor missions overstretching recipient capacities, the mix of (often incoherent) donor reporting standards and so on. These initiatives helped to reduce strategic donor behaviour of using their influence through foreign aid to achieve their own goals, but instead to standardise the measurement of aid effectiveness and improve the effectiveness (OECD, 2015c).

Multi-Dimensional Country Reviews (MDCRs) OECD (2015d) are a new tool for tailoring broad OECD expertise to the realities of developing economies. Policy-makers need to reconcile economic, social and environmental objectives to ensure that their country's development path is sustainable and that the lives of citizens improve. To support this need, the MDCRs aim to design policies and strategies which do not simply promote growth but rather development in this more comprehensive sense, and the area/indicators should be assessed are illustrated in Figure 14.5.

Under such frameworks, various attempts have been made to develop indicators for monitoring aid effectiveness. For instance, in the aid for trade area, the International Finance Corporation (IFC) paper *Doing Business Project* has played a major role in promoting the culture of results by monitoring selected indicators and benchmarking countries against each other (World Bank, 2013). The OECD's trade facilitation indicators measure a country's trade facilitation capabilities that identify areas for action and enable the potential impact of reforms to be assessed. Estimates based on the indicators provide a basis for governments to prioritise trade facilitation actions and mobilise technical assistance and capacity building efforts for developing countries in a more targeted way. In addition, *Doing Business* contains a *Trading Across Borders* indicators series that specifically measures a country's trade facilitation capabilities. Other initiatives have followed, which attempt to provide a more or less comprehensive list of trade-related indicators, sometimes aggregated in synthetic indexes and country fact-sheets or global rankings. These have included the *World Trade Indicators* collected by the World Bank

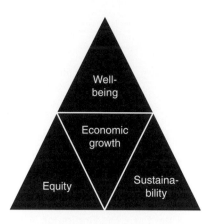

Figure 14.5   OECD's multi-dimensional country review framework for aid effectiveness

(2009), which contains a broad set (about 500 variables) of trade policy and outcome indicators for 211 countries and territories, and the *Global Competitiveness* and *Enabling Trade* indexes (WEF, 2014), which contain over 100 indicators (based on available statistics and on surveys) of relevance to trade, supply chain management and competitiveness issues. Some more specific indexes have also been developed, for example, by the World Bank in the field of logistics. Those efforts have been relayed by more specialised agencies that have long collected data in their field of competence which appear to be directly relevant to measuring trade performance. They include *World Communication/ICT Indicators Database* on telecommunications (ITU, 2014), data on air transport collected by the International Air Transport Association (IATA) and International Civil Aviation Organization (ICAO), and *Travel and Tourism Satellite Accounts* data on tourism collected by the UN World Tourism Organisation (UNWTO, 2010; OECD, 2013).

As the above-mentioned discussion revealed, to be able to facilitate effective aid, studies need to adopt a more holistic approach that includes a comprehensive set of indicators with careful use of official databases and research method. We also need to broaden the research area to other non-economic aspects of the aid.

14.6   Relationships and Interactions

14.6.1   Relationships between Donor and Recipient Countries

Apart from the complexity within the architecture of the aid system, various relationships/interactions inside and outside the aid system are also making an important impact on the aid. The relationship between the donor and recipient countries is the relationship that carries predominant weight in the system. One reason for the limited effectiveness of foreign aid is the variation in donor characteristics. A large body of research demonstrates a general inclination of donors to provide more foreign aid to countries that are of economic, military, geopolitical interest to the donor or to buy votes in international organisations such as United Nations (Alesina and Dollar, 2000; Neumayer, 2003; Stone, 2011; Easterly and Pfutze, 2008; Kilby, 2009; Vreeland, 2011). Aid tends to be less effective when it is provided for strategic purposes (Rajan and Subramanian, 2008; Dreher et al., 2014; Minoiu and Reddy, 2010). Although multilateral aid institutions are generally considered to be less strategic in their foreign aid allocation (Neumayer, 2003; Headey, 2007; Dietrich, 2013), scholars point out that multilateral aid institutions can also exhibit biases due to the influence that powerful member states or coalitions with strategic interests (Vreeland, 2011; Stone, 2011; Cruz and Schneider, 2014).

Another reason for the limited effectiveness of foreign aid is seen in recipient country politics. One important challenge is that some recipient countries do not have the capacity that is necessary to absorb foreign aid effectively (Herbst and Mills, 2009; Prunier, 2009). Most research on recipient characteristics focuses on perverse political incentives when receiving a large influx of foreign aid. Politicians often exploit foreign aid in order to maximise their survival in power (Collier and Hoeffler, 2007; Kono and Montiola, 2009; Licht, 2010; Labonne and Chase, 2011; Cruz and Schneider, 2014). The economic conditions in the recipient country also have an important effect on the success of foreign aid projects. The World Bank induced a paradigmatic shift in research, building on an influential study by Burnside and Dollar (2000), which found that aid could foster economic growth under the condition that recipient countries pursued 'good' economic policies.

The heterogeneous donor and recipient characteristics, behaviours and their interactions between each other further complicate the study of the system and create dynamics that need to be captured in the models of aid system.

14.6.2   Relationships between Aid and Other Systems

Globalisation has transformed our planet and created a more integrated world. As discussed in previous sections, the international development aid system is connected by multiple organisations around multiple issues that are important for the development of human societies. Therefore, interactions with other global system potentially can affect the aid system. In this section, we try to identify the relationships between the aid system and other three global systems studied in Enfolding project, namely Trade, Security and Migration.

Trade is a powerful engine for economic success and for improving living standards. By connecting local producers to global markets, trade helps developing countries to rapidly integrate into world markets through export-led industrialisation and thereby share in the prosperity generated by globalisation. Aid and trade have a close relationship and the interactions between the two can have a substantial effect on the dynamics within the aid system. Especially in recent years, there is a trend of aid for trade, where developing and industrial countries are bound together through trade, which serves as a transmission mechanism, with an increase in the value for trade (measured in terms of jobs, income, socio-economic upgrading, etc.) as the final objective of aid projects (OECD, 2013).

Aid flows may affect trade flows because of the general economic effects they induce in the recipient country through aid supplements in domestic saving that leads to increased investment which contributes to higher rates of economic growth than without aid (White, 1992). On the other hand, in the same macroeconomic tradition, the most celebrated argument of a relationship between aid and trade flows is the so-called 'Dutch disease' and 'aid dependency' effects in the recipient economy, illustrating the potential conflict between aid flows and the export competitiveness of the recipient country. The most obvious way that aid can influence trade is through direct ties to trade (Morrissey, 1991; Jepma, 1991), but it can reinforce bilateral economic and political links (or a combination of all three reasons earlier).

As is widely recognised, the causality between aid flows and trade flows can obviously also go the other way around (Morrissey et al., 1992; McGillivray and Oczkowski, 1992; McGillivray and White, 1993). Aid allocation in the donor country can be influenced by various lobby and business groups, which may be associated with particular trade interests; trade can lead to further aid if donors give preference in the allocation of their aid to countries with which they have the greatest commercial ties. The donor might want to reward the recipient for the purchase of its export goods or it might want to consolidate or expand its market in the country in which the expectations of an aid-induced trade dependency is highest. Such relationship and interactions introduce important dynamics into the aid system and therefore makes it necessary to study such interactions between the two systems to understand the aid dynamics.

Security and peace are essential to livelihood and a stable domestic order goes to the heart of the contract between states and their citizens that is the basis of the modern state. The discourse of statehood revolves around three narratives that encapsulate its core functions of providing security, welfare and representation, where economic, social and political life can then evolve normally. The provision of basic security is a precondition for political, social, economic development and well-being (Milliken and Krause, 2002; Schwarz, 2005; Krause and Jütersonke, 2005).

Therefore, it is a reasonable assumption that if two countries are at war with each other, one may not want to deliver aid to the other, but it is a different situation if they are two allied states in a conflict. Conflict is frequently portrayed as a product of low levels of development as well as political and social marginalisation. More recently, there is a changing emphasis on complex mandates of security-driven aid, spanning immediate stabilisation and protection of civilians to supporting humanitarian assistance, organising elections, assisting the development of new political structures, engaging in security sector reform, disarming, demobilising and reintegrating former combatants and laying the foundations of a lasting peace. This integration of diplomatic, human rights, military and development responses has been driven primarily by the requirement to effectively consolidate fragile peace agreements and make the delicate transition from war to a lasting peace. Such requirement indicates a lot of interactions between security and aid (Gordon, 2010; Eide et al., 2005; Fishstein and Wilder, 2012).

On the one hand, security-driven aid can provide the vital peaceful environment for the development of societies. On the other hand, the focus on integrated and comprehensive approaches to secure the support of the population has led humanitarian and development actors to criticise both stabilisation and counterinsurgency doctrines for leaving little room for the fundamental humanitarian principles of independence, impartiality and neutrality (Collier and Hoeffler, 2007; Duffield, 2001). Assistance is increasingly instrumentalised behind security and political objectives. Patrick and Brown (2007) argue that these trends are potentially damaging and raise concerns that 'foreign and development policies may become subordinated to a narrow, short-term security agenda at the expense of broader, longer-term diplomatic goals and institution-building efforts in the developing world'.

Globalisation effectively removes national boundaries and opens the way not only to free mobility of capital and goods but also, in effect, to free movement of vast labour pools from regions of rapid population growth and hence **migration**. The latter movement will have much more substantial impact on development in both origin and destination countries in areas such as migration labour markets, human capital, productivity, production, entrepreneurship, economic growth and public finances and goods (Castles and Wise, 2008; Keeley, 2009).

Immigration can affect per capita GDP and GDP growth through aggregate supply and aggregate demand effects (Facchini and Mayda, 2009) or increase entrepreneurship and boost productivity and technological change, with the mechanisms including specialisation and productivity-enhancing characteristics of immigrant workers (Ortega and Tanaka, 2015). While the inflow of migrants directly affects the human capital stock of the country of destination, there could be secondary effects that have the potential to affect the development trajectory of the country of destination in the long-run such as education and unemployment issues (Fauvelle-Aymar, 2014). The question whether immigrants 'pay their way' is also a frequent topic of public debate in many immigrant-receiving societies (Edward et al., 1997). On the other hand, migration helps shape the economic and social direction of some of the world's poorer countries. Sometimes, its role is mainly positive, such as when it produces a flood of remittances. However, it can be negative, especially when it is causing 'brain drain' (Keeley, 2009).

However, there have been some new trends in the recent international migration and they may have a deep impact on development. Edward et al. (1997) and Kapur (2004) suggest a 'new mantra' of international migration and believe that benefits that migration brings for development can be extended to

- migrants also transfer home skills and attitudes – known as 'social remittances' – which support development;
- 'brain drain' is being replaced by 'brain circulation', which benefits both sending and receiving countries;
- temporary (or circular) labour migration can stimulate development and should therefore be encouraged;
- migrant diasporas can be a powerful force for development through transfer of resources and ideas back to sending countries;
- economic development will reduce outmigration, encourage return migration and create the conditions necessary to utilise the capital and know-how provided by diasporas.

With the benefits that migration introduces into the development, this may in turn change the shape of the aid landscape as characteristics of recipient countries. Migration sometimes is also linked to climate change, economic shocks and conflicts, causing even more chained reactions to the aid system. Recently, there has been a surge of interest in examinations of some of these complex links between migration and development (UN, 2008; Newland, 2007; Castles and Wise, 2008).

14.7 Conclusions

Development aid has received a lot of attention since the implementation of the best known and probably most successful aid programme, the Marshall Plan, in Western Europe following the Second World War. Since then the aid landscape has changed many times and the discussion about whether aid can be effective has never stopped. Discussions about aid have been extremely polarised: those in favour of aid have been reluctant to draw attention to its problems and inadequacies, those against aid have been reluctant to acknowledge that it has done any good. Controversies about aid effectiveness go back decades. This is mainly due to the degree of complexity within this system.

There are many sources of the complexity causing the complexity of studying the international development aid system. In this paper, we examined four sources of such complexities among them, namely organisations, topics/issues, data and research methods. As the aid system has not been able to develop in a systematic way, now the architecture of the aid system has been recognised as the main concern for the ineffectiveness of the aid delivery. However, it is not only the system structure itself but also the relationships/interactions between the components within the system and potentially between the system and other global systems that have important impact on shaping the dynamics within aid system, too.

References

Addison, T., Mavrotas, G. and McGillivray, M. (2005) Aid, Debt Relief and New Sources of Finance for Meeting the Millennium development Goals. *Journal of International Affairs*, **58**, 113–127.

Alesina, A. and Dollar, D. (2000) Who Gives Foreign aid to Whom and Why? *Journal of Economic Growth*, **5**, 33–63.

Andrews, N. (2009) Foreign aid and development in Africa: What the literature says and what the reality is. *Journal of African Studies and development*, **1** (1), 008–015, Available online http://www.acadjourn.org (accessed 08 January 2016).

Antipin, J.E. and Mavrotas, G. (2006) *On the Empirics of aid and Growth: A Fresh Look* UNU-WIDER Research Paper 2006/05, UNU-WIDER, Helsinki.

Arndt, C., Jones, S. and Tarp, F. (2010) Aid, Growth, and development: Have We Come Full Circle? *Journal of Globalization and development*, **1** (2), 1–29.

Barder, O. (2012) *Development and Complexity, Presentation from his Kapuscinski Lecture*. http://www.cgdev.org/content/multimedia/detail/1426397 (accessed 08 January 2016).

Bauer, P. and Yamey, B. (1982) Foreign aid: What is at stake. *The Public Interest*, **86**, 69.

Beynon, J. (2002) Policy Implications for aid Allocations of Recent Research on aid Effectiveness and Selectivity, in *New Perspectives on Foreign Aid and Economic Development* (ed B. Mak Arvin), Praeger, Westport, Conn.

Bulir, A. and Hamann, J. (2005) *Volatility of development aid: from the frying pan into the fire?*, IMF, Washington, DC, paper submitted to this symposium.

Burall, S. and Maxwell, S. (2006) *Reforming the International Aid Architecture: Options and Way Forward Overseas development Institute Working Paper 278*, ODI, London.

Burnside, C. and Dollar, D. (2000) Aid, policies and growth. *American Economic Review*, **90**, 847–868.

Cassen, R. et al. (1994) *Does Aid Work? Report to an Intergovernmental Task Force*, Oxford University Press, Oxford.

Castles, S. and Wise, R. (eds) (2008) *Migration and Development: Perspectives from the South*, IOM (international Organization for Migration).

Chauvet, L. and Guillaumont, P. (2009) Aid, Volatility, and Growth Again: When aid Volatility Matters and When it Does Not. *Review of Development Economics*, **13** (s1), 452–463.

Collier, P. and Dollar, D. (2002) Aid allocation and poverty reduction. *European Economic Review*, **46** (8), 1475–1500.

Collier, P. and Hoeffler, A. (2007) Unintended consequences: Does aid promote arms races? *Oxford Bulletin of Economics and Statistics*, **69** (1), 1–27.

Coscia, M., Hausmann, R. and Hidalgo, C. (2004) The Structure and Dynamics of International Development Assistance. *Journal of Globalization and development*, **3** (2 (March 2013)), 1–42.

Cruz, C. and Schneider, C. (2014) *Foreign Aid and the Politics of Undeserved Credit-Claiming*. Working Paper.

Dalgaard, C.-J.L., Hansen, H. and Tarp, F. (2004) On the empirics of foreign aid and growth. *The Economic Journal*, **114** (496), 191–216.

Denizer, C., Kaufmann, D. and Kraay, A. (2013) Good countries or good projects? Macro and micro correlates of World Bank project performance. *Journal of Development Economics*, **105**, 288–302.

DEVDIR (2015) Source Guide to Development Organizations and the Internet, http://www.devdir.org/index.html (accessed 08 January 2016).

Dietrich, S. (2013) Bypass or engage? Explaining donor delivery tactics in aid allocation. *International Studies Quarterly*, **57** (4), 698–712.

Djankov, S., Montalvo, J. and Reynal-Querol, M. (2008) The curse of aid. *Journal of Economic Growth*, **13** (3), 169–194.

Dollar, D. and Easterly, W. (1999) The Search for the Key: aid, Investment and Policies in Africa. *Journal of African Economies*, **8** (4), 546–577.

Dreher, A., Vera, E. and Kai, G. (2014) Geopolitics, Aid and Growth. CEPR Discussion Paper No. 9904.

Duffield, M. (2001) Governing the borderlands: Decoding the power of aid. *Disasters*, **25** (4), 308–320.

Easterly, W. (2001) *The Elusive Quest for Growth: Economists' Adventures and Misadventures in the Tropics*, MIT Press, Cambridge, MA.

Easterly, W. (2006) *The White Man's Burden: Why the West's Efforts to aid the Rest Have Done So Much Ill and So Little Good*, Penguin Press, New York, pp. 113–127.

Easterly, W., Levine, R. and Roodman, D. (2004) New data, new doubts: a comment on Burnside and Dollar's aid, policies, and growth. *American Economic Review*, **94**, 774–780.

Easterly, W. and Pfutze, T. (2008) Where Does the Money Go? Best and Worst Practices in Foreign aid. *Journal of Economic Perspectives*, **22** (2), 29–52. doi: 10.2139/ssrn.1156890

Edward, T., Martin, P. and Fix, M. (1997) *Poverty Amid Prosperity: Immigration and the Changing Face of Rural California*, Urban Institute Press, Washington, DC.

Eide, E., Kaspersen, A., Kent, R. and Hippel, K. (2005) *Report on Integrated Missions: Practical Perspectives and Recommendations*, Independent Study for the Expanded UN ECHA Core Group.

Facchini, G. and Mayda, A. (2009) *The Political Economy of Immigration Policy*, United Nations Development Programme Human Development Reports Research Paper.

Fauvelle-Aymar, C. (2014) The welfare state, migration, and voting rights. *Public Choice*, **159** (1), 105–120.

Fishstein, P. and Wilder, A. (2012) *Winning Hearts and Minds? Examining the Relationship between aid and Security in Afghanistan*, Fishstein International Centre Publication.

Forss, K., Befani, B., Kruse, S.-E. et al. (2013) *A study of monitoring and evaluation in six Norwegian civil society organisations* Report 7/2012, Norwegian Agency for Development Cooperation (NORAD), Oslo, https://www.norad.no/en/toolspublications/publications/2013/a-study-of-monitoring-and-evaluation-in-six-norwegian-civil-society-organisations/ (accessed 27 January 2016).

Freeman, T. (2013) *Development effectiveness review of the World Health Organization (WHO 2007–2010)*, Strategic Planning and Evaluation Directorate, Canadian international development Agency, Ottawa, pp. viii–ix.

Friedman, M. (1958) Foreign Economic Aid. *Yale Review*, **47** (4), 501–516.

Gomanee, K., Girma, S. and Morrissey, O. (2005) Aid and Growth in Sub-Saharan Africa: Accounting for Transmission Mechanisms. *Journal of International Development*, **17** (8), 1055–1075.

Gordon, S. (2010) *Aid and Conflict Drivers in Southern Afghanistan: Assessing Stabilisation Strategies*, paper presented at Wilton Park conference: Winning Hearts and Minds in Afghanistan: Assessing the Effectiveness of development aid in COIN operations. February 2010.

Gross, P., Levitt, N. and Lewis, M. (eds) (1996) *The Flight from Science and Reason*, Johns Hopkins University Press, Baltimore, MD.

Hansen, H. and Tarp, F. (2001) Aid and growth regressions. *Journal of Development Economics*, **64** (2), 547–570.

Headey, D.D. (2007) Geopolitics and the effect of foreign aid on economic growth: 1970–2001. *Journal of International Development*, **20** (2), 161–180.

Herbst J, Mills G. (2009) *There Is No Congo*. Available at http://www.foreignpolicy.com/story/cms.php?story id=4763 (accessed 08 January 2016).

Hudson, J. and Mosley, P. (2008) Aid volatility, policy and development. *World Development*, **36**, 2082–2102.

Isenman, P. and Ehrenpreis, D. (2003) Results of the OECD-DAC development Centre Experts' Seminar on "aid Effectiveness and Selectivity: Integrating Multiple Objectives into aid Allocations". *DAC Journal*, **4**, 7–25.

ITU (2014) *World Communication/ICT Indicators Database 2014*, http://www.itu.int/en/ITU-D/Statistics/Pages/publications/wtid.aspx (accessed 08 January 2016).

Jepma, C. (1991) *E-wide Untying*, International Foundation for development Economics and Department of Economics, University of Groningen.

Juselius, K; Møller, N and Tarp, F (2011) *The Long-Run Impact of Foreign aid in 36 African Countries: Insights from Multivariate Time Series Analysis,* UNU-WIDER ReCom Working Paper No. 2011/51, ISSN 1798–7237, ISBN 978-92-9230-418-8

Kanbur, R. (2006) The Economics of international aid, in *The Handbook on The Economics of Giving, Reciprocity and Altruism*, Vol. 2 (eds S. Christophe-Kolm and J. Mercier-Ythier), North-Holland, ISBN: 978-0-444-52145-3.

Kapur, D. (2004) *Remittances: the New development Mantra? Discussion Paper*, World Bank, Washington, DC.

Keeley, B. (2009) *OECD Insights of International Migration: The Human Face of Globalisation*, OECD, DOI: 10.1787/9789264055780-en, available from: http://www.oecd-ilibrary.org/social-issues-migration-health/international-migration_9789264055780-en.

Kilby, C. (2009) The political economy of conditionality: An empirical analysis of world bank loan disbursements. *Journal of Development Economics*, **89** (1), 51–61.

Knack, S. and Rahman, A. (2007) Donor Fragmentation and Bureaucratic Quality in aid Recipients. *Journal of Development Economics*, **83**, 176–197.

Kono, D. and Montinola, G. (2009) Does foreign aid support autocrats, democrats, or both? *Journal of Politics*, **71**, 704–718.

Kosack, S. (2003) Effective aid: How democracy allows development aid to improve the quality of life. *World Development*, **31** (1), 1–22.

Krause, K. and Jütersonke, O. (2005) Peace, security and development in post-conflict environments. *Security Dialogue*, **36** (4), 447–462.

Labonne, J. and Chase, R.S. (2011) Do community driven development projects enhance social capital? Evidence from The Philippines. *Journal of Development Economics*, **96**, 348–358.

Lensink, R. and Morrissey, O. (2000) Aid instability as a measure of uncertainty and the positive impact of aid on growth. *Journal of Development Studies*, **36**, 31–49.

Licht, A. (2010) Coming Into Money: The Impact of Foreign aid on Leader Survival. *Journal of Conflict Resolution*, **54** (1), 58–87.

Madisson, A. (2008) Shares of the Rich and the Rest in the World Economy: Income Divergence between Nations, 1820–2030. *Asian Economic Policy Review*, **3** (1), 67–82.

Matheson, A., Scott, C., Thomson, A. et al. (2008) *Independent Evaluation of Uganda's Poverty Eradication Action Plan (PEAP): Final Synthesis Report*, Oxford Policy Management, pp. 22–33.

Mavrotas, G. (ed) (2010) *Foreign Aid for Development: Issues, Challenges, and the New Agenda*, WIDER Studies in development Economics, ISBN: 978-0-19-958093-4.

Mavrotas, G. and Nunnenkamp, P. (2007) Foreign aid Heterogeneity: Issues and Agenda. *Review of World Economics*, **143**, 585–595.

Mavrotas, G. and Reisen, H. (2007) *The Multilateral Development Finance "Non-System"*, paper presented at the Experts' Workshop on Performance and Coherence in Multilateral development Finance, Berlin, January.

McGillivray, M. and Oczkowski, E. (1992) A two-part sample selection model of British bilateral foreign aid allocation. *Applied Economics*, **24**, 1311–1319.

McGillivray, M. and White, H. (1993) *Explanatory Studies of Aid Allocation among Developing Countries*, The Hague: ISS Working Paper No. 148.

McKinlay, R.D. and Little, R. (1978) A Foreign-Policy Model of the Distribution of British Bilateral aid, 1960–70. *British Journal of Political Science*, **8**, 313–332.

Milliken, J. and Krause, K. (2002) State Failure, State Collapse and State Reconstruction: Concepts, Lessons and Strategies. *Development and Change*, **33** (5), 753–774.

Minoiu, C. and Reddy, S. (2010) Development aid and economic growth: A positive long-run relation. *Quarterly Review of Economics and Finance*, **50** (1), 27–39.

MoFPED (2006) *Development cooperation Uganda, 2005/06 report*, MFPED, Kampala, pp. 22–23.

MoFPED (Ministry of Finance, Planning and Economic development, Uganda) (2005) *Development cooperation Uganda 2004/05 report*, MFPED, Kampala, p. 22.

Morrissey, O. (1991) An evaluation of the economic effects of aid and trade provision. *Journal of Economic Studies*, **28**, 104–129.

Morrissey, O. (2001) Does aid Increase Growth? *Progress in Development Studies*, **1** (1), 37–50.

Morrissey, O., Smith, B. and Horesh, E. (1992) *British aid and international Trade*, Open University Press, Buckingham.

Mosley, P. (1980) aid, Savings and Growth Revisited. *Oxford Bulletin of Economics and Statistics*, **42** (2), 79–95.

Mosley, P. (1987) *Overseas aid: Its Defence and Reform*, Wheatsheaf Books, Brighton.

Moyo, D. (2009) *Dead Aid: Why Aid Is Not Working and How There Is Another Way for Africa*, Penguin Books, London, 188 pages. ISBN-13: 978–0374139568.

Neumayer, E. (2003) *The Pattern of Aid Giving – The Impact of Good Governance on Aid Giving*, Routledge, London.

Newland, K. (2007) *A New Surge of Interest in Migration and Development*, Migration Information Source, Washington, DC, www.migrationinformation.org (accessed 08 January 2016)..

OECD (2013) *aid for Trade and development Results: A Management Framework, The development Dimension*, OECD Publishing. 10.1787/9789264112537-en (accessed 08 January 2016).

OECD (2015a) *Official Development Assistance – Definition and Coverage*: http://www.oecd.org/dac/stats/officialdevelopmentassistancedefinitionandcoverage.htm (accessed 08 January 2016).

OECD (2015b) *Geographical-Distribution-of-Financial-Flows-to-Developing-Countries*: http://www.keepeek.com/Digital-Asset-Management/oecd/development/geographical-distribution-of-financial-flows-to-developing-countries-2015_fin_flows_dev-2015-en-fr#page106 (accessed 08 January 2016).

OECD (2015c) *The High Level Fora on Aid Effectiveness: A History*, http://www.oecd.org/dac/effectiveness/thehighlevelforaonaideffectivenessahistory.htm (accessed 08 January 2016).

OECD (2015d) *Multi-Dimensional Country Reviews*, http://www.oecd.org/dev/mdcr-countries.htm (accessed 08 January 2016).

OECD (QWIDS) (2015) *Query Wizard for international development Statistics (QWIDS)*, http://stats.oecd.org/qwids/ (accessed 08 January 2016).

Ortega, F. and Tanaka, R. (2015) *Immigration and the Political Economy of Public Education: Recent Perspectives*, IZA (The Institute for the Study of Labour in Bonn) Discussion Paper No. 8778, January 2015.

Park, R. (2000) *Voodoo Science: The Road from Foolishness to Fraud*, Oxford University Press, New York.

Patrick, S. and Brown, K. (2007) *The Pentagon and Global Development: Making Sense of the DOD's Expanding Role*, Working Paper Number 131 (November 2007).

Phelan, S. (2001) What is complexity science really? *EMERGENCE*, **3** (1), 120–136.

Porter, J. (2011) *UK Collaborative on Development Sciences Report: Complexity Science and International Development*, http://www.ukcds.org.uk/_assets/file/publications/Complexity%20and%20international%20development%20-%20%20Final%20report%20(2).pdf (accessed 09 January 2016).

Pritchett, L. (1997) Divergence, big time. *Journal of Economic Perspectives*, **11** (3), 3–17.

Prunier, G. (2009) *Africa's World War: Congo, the Rwandan Genocide, and the Making of a Continental Catastrophe*, Oxford Univ. Press, New York.

Radelet, S. (2008) Foreign aid, in *International Handbook Of Development Economics*, Vol. 2 (eds A. Dutt and J. Ros), Edward Elgar publishing, London.

Raffer, K. and Singer, H. (1996) *The Foreign aid Business: Economic Assistance and Development Co-operation*, Edward Elgar Publishing Company, Brookfield, Vermont.

Rajan, R. and Subramanian, A. (2008) Aid and growth: what does the cross-country evidence really show? *The Review of Economics and Statistics*, **90** (4), 643–665.

Ramalingam, B. (2013) *Aid on the Edge of Chaos*, Oxford University Press, NY, USA, ISBN:9780199578023..

Richardson, K A., Mathieson, G. and Cilliers, P. (2000) The theory and practice of complexity science: Epistemological considerations for military operational analysis. *SysteMexico*, **1** (1), 25–66.

Riddell, R. (2014) *Does Foreign Aid Really Work? An Updated Assessment*, development Policy Centre Discussion Paper 33, Crawford School of Public Policy, The Australian National University, Canberra.

Rogerson, A., Hewitt, A. and Waldenberg, D. (2004) *The International Aid System 2005–2010: Forces For and Against Change*, Overseas development Institute Working paper, http://citeseerx.ist.psu.edu/viewdoc/download?doi=10.1.1.195.4233&rep=rep1&type=pdf

Roodman, D. (2006) *Aid Project Proliferation and Absorptive Capacity. UNU-WIDER Research Paper*, 2006/04.

Sachs, J.D. (2005) *The End of Poverty: Economic Possibilities for Our Time*, Penguin, London.

Schneider, C. and Tobin, J. (2013) Interest Coalitions and Multilateral Aid Allocation in the European Union. *International Studies Quarterly*, **57** (1), 103–114.

Schwarz, R. (2005) Post-Conflict Peacebuilding: The Challenges of Security, Welfare and Representation. *Security Dialogue*, **36** (4), 429–446.

Shermer, M. (1997) *Why People Believe Weird Things: Pseudo-science, Superstition, and Bogus Notions of Our Time*, MJF Books, New York.

Sokal, A. and Bricmont, P. (1998) *Fashionable Nonsense: Postmodern Intellectuals' Abuse of Science*, St Martin's Press, New York.

Stern, N. (2002) Making the Case for aidin, (in *A Case for Aid: Building a Consensus for Development Assistance* (ed The World Bank) ed.), The World Bank.

Stiglitz, J. (2002) Overseas aid is money well spent. *Financial Times,* April 14.

Stone, R. (2011) *Controlling Institutions: international Organizations and the Global Economy*, Cambridge University Press, Cambridge, ISBN: 9780521183062.

Svensson, J. (1999) Aid, growth and democracy. *Economics and Politics*, **11**, 275–297.

Svensson, J. (2000) Foreign aid and rent seeking. *Journal of International Economics*, **51**, 433–461.

Tezanos, V., Madrueño, S. and Guijarro, M. (2009) The impact of aid on economic growth: The case of Latin America and the Caribbean. *Cuadernos Económicos del ICE*, **78**, http://hdl.handle.net/10902/3379 (accessed 27 January 2016).

UN (Department of Economic and Social Affairs: Office for ECOSOC Support and Coordination) (2008) *Achieving Sustainable development and Promoting development Cooperation*, Dialogues at the Economic and Social Council, United Nations Publications, ISBN: 978-92-1-104587-1.

UNWTO (2010) *SA Data Around the World - Worldwide summary*, http://statistics.unwto.org/sites/all/files/pdf/tsa_data.pdf (accessed 09 January 2016).

Vreeland, J.R. (2011) Foreign Aid and Global Governance: Buying Bretton Woods - The Swiss-bloc Case. *Review of international Organizations*, **6** (3–4), 369–391.

WEF (2014) *The Global Enabling Trade Report 2014*, http://www3.weforum.org/docs/WEF_GlobalEnablingTrade_Report_2014.pdf (accessed 09 January 2016).

White, H. (1992) The Macroeconomic Impact of development aid: A Critical Survey. *Journal of Development Studies*, **28**, 163–240.

WHO (2008) *WHO Glossary of Terms*, http://www.who.int/trade/glossary/story016/en/ (accessed 09 January 2016).

Wikipedia (2015) *Development Aid*, http://en.wikipedia.org/wiki/development_aid (accessed 09 January 2016).

Wolfram, S. (2002) *A New Kind of Science*, Wolfram Media Inc. ISBN 1-57955-008-8, book website: http://www.wolframscience.com/nksonline/toc.html (09 January 2016).

World Bank (2007) *Aid Architecture: An Overview of the Main Trends in Official development Assistance Flows*. http://documents.worldbank.org/curated/en/2007/02/7411698/aid-architecture-overview-main-trends-official-development-assistance-flows (accessed 09 January 2016).

World Bank (2009) *World Trade Indicators 2009/10*, http://web.worldbank.org/WBSITE/EXTERNAL/TOPICS/TRADE/0,,contentMDK:22421950~pagePK:148956~piPK:216618~theSitePK:239071,00.html (accessed 09 January 2016).

World Bank (2013) *Doing Business 2013: Smarter Regulations for Small and Medium-Size Enterprises*, World Bank Group, Washington, DC. doi: 10.1596/978-0-8213-9615-5

15

Model Building for the Complex System of International Development Aid

Belinda Wu, Sean Hanna and Alan Wilson

15.1 Introduction

The international development aid system is complex. Since its formation more than 60 years ago, it has developed incrementally rather than systematically over decades. No major institution has since exited from the system, while there has been a continued expansion that has resulted in increased complexity from the emergence of new donors, modalities and instruments. As multiple organisations provide aid over multiple topics through multiple relationships driven by multiple motivations, the aid system has now become very complex with a great deal of dynamic interaction between different agents. Recent issues arising from such complexities have fascinated researchers and decision-makers.

Studying such a system is not easy and there are many debates on how we should do so. With the progress of globalisation, our planet has become more and more connected. Thus, it is not surprising that there are many interwoven factors from both donor and recipient sides and even other global systems such as international trade, security and migration also interact with the international development aid system. Given the changing modern aid context, many researchers argue that the linear, mechanistic models and the assumptions on which aid is built are no longer inadequate in the dynamic, complex world we face today.

Although macroeconomic methods will still have their value in the foreseeable future, there has been an increasing need to incorporate complexity thinking into the process

Global Dynamics: Approaches from Complexity Science, First Edition. Edited by Alan Wilson.
© 2016 John Wiley & Sons, Ltd. Published 2016 by John Wiley & Sons, Ltd.

of understanding and management of the system of aid. Researchers argue that a new approach embracing the 'new science' of complex systems can make aid more relevant, more appropriate, more innovative and more catalytic, where an interdisciplinary approach provides a new way to study regularities that differs from traditional science. Complexity science takes a holistic view of complex system and makes connectedness and interactions the key to understand such systems (Ramalingam, 2013; Mavrotas, 2010; Barder, 2012; Porter, 2011).

We propose to use a complexity approach to explore the connectedness and interactions in our model after a series of experiments based on the influential study of Alesina and Dollar (2000). Compared to their study, where certain recipient countries' characteristics are of interest to certain donors to explain who donates to who and why, we are more interested in the general pattern in the global aid system that affect aid flows.

In our model, we will look at the characteristics of both donors and recipients, as well as the relationships and interactions between them. As our planet has become more integrated, we also find that other global systems, such as international trade, security and migration, can also have important impact on aid dynamics. Therefore, we propose to also investigate such interactions with those systems.

In the following sections, we will discuss the study method in detail from the aspects of both data collection and model building.

15.2   Data Collection

15.2.1   Introduction

Various data has been collected to facilitate the understanding and improvement of international development aid. The United Nations, World Bank and OECD are the major sources of development aid data and various bilateral and other organisations also provide their own data.

We have explored various data sources and focused on five main collections of data for this analysis: aid, trade, migration, security and associated geographical data. In the following sections, we will describe the five data sets in more detail.

15.2.2   Aid Data

15.2.2.1   World Bank WDI Data[1]

World Development Indicators (WDI) is the primary World Bank database for development data from officially recognised international sources. This data set contains vast information on various development indicators. The data captured includes 25 DAC donors and EU institutions and 214 recipients during the period of 1960–2013 with 1,346 indicators detailed. For instance, with GDP alone, there is information on:

---

[1] Source: http://data.worldbank.org/data-catalog/world-development-indicators.

GDP (constant 2005 US$)
GDP (constant LCU)
GDP (current LCU)
GDP (current US$)
GDP deflator (base year varies by country)
GDP growth (annual %)
GDP per capita (constant 2005 US$)
GDP per capita (constant LCU)
GDP per capita (current LCU)
GDP per capita (current US$)
GDP per capita growth (annual %)
GDP per capita, PPP (constant 2005 international $)
GDP per capita, PPP (current international $)
GDP per person employed (constant 1990 PPP $)
GDP per unit of energy use (constant 2005 PPP $ per kilogram of oil equivalent)
GDP per unit of energy use (PPP $ per kilogram of oil equivalent)
GDP, PPP (constant 2005 international $)
GDP, PPP (current international $)

15.2.2.2   Data from Alesina and Dollar (2000)[2]

Alesina and Dollar are distinguished researchers in the field of international development aid and their works have been well cited. These particular data sets have been used in various works that try to test and replicate their model (Roodman, 2006). This data set includes a wide range of information; examples include the measurement of political democracy, economical openness, net ODA by individual donor countries, GDP per capita, population, percentage of recipient votes the same as donor in United Nations, being a colony of the donor since 1900, religion and particular country factors (e.g. Israel and Egypt).

There are 180 recipient countries whose information is recorded in this data, during the period 1960–1990 in 5-year intervals, each with 149 variables.

15.2.2.3   OECD Data[3]

OECD development data include total flow of ODA (Official Development Assistance), OOF (Other Official Flows) and private donations during the period of 1960–2009. For ODA definition, please refer Chapter 14; OOF refers to transactions by the official sector with countries on the List of Aid Recipients which do not meet the conditions for eligibility as ODA or Official Aid because either they are not primarily aimed at development or they have a grant element of less than 25%. Private donations refer to donations from individuals or private

---

[2] Source: requested directly from the authors.
[3] Source: http://stats.oecd.org/.

organisations such as the Bill and Melinda Gates Foundation. Information on 287 countries has been recorded in this data, including 73 donors (including multilateral and private organisations) and 220 recipients.

15.2.2.4 World Bank Migration and Remittance Data[4]

Increasingly remittances have played an important part in international development aid, as some studies have noted. Our remittance data is from the World Bank, where data is collected in conjunction with various organisations. In this data, migrant remittances are defined as the sum of workers' remittances, compensation of employees and migrants' transfers. Workers' remittances, as defined by the International Monetary Fund (IMF, 2010) in the *Balance of Payments Manual, 6th edition*, are 'current private transfers from migrant workers who are considered residents of the host country to recipients in the workers' country of origin. If the migrants live in the host country for 1 year or longer, they are considered residents, regardless of their immigration status. If the migrants have lived in the host country for less than 1 year, their entire income in the host country should be classified as compensation of employees' (World Bank, 2011). Remittance flows between 216 countries are recorded between 2010 and 2014.

15.2.3   Trade Data

15.2.3.1   COMTRADE[5]

United Nations provides free access to detailed global trade data and UN COMTRADE is a repository of official trade statistics and relevant analytical tables. The database contains annual trade flows between 162 economies between 1962 and 2014. The trade flows include export, import and re-export and re-import between the reporter and partner countries.

15.2.3.2   Input–Output Data[6]

Other trade data we obtained is from the World Input Output Database. This data covers 27 EU countries and 13 other major countries in the world, excluding Africa, between 1995 and 2008 and includes 35 sectors of product output/input from each country.

15.2.3.3   Dyadic Trade Data[7]

This trade data is downloaded from the Correlates of War (COW) project website and covers the annual trade between 243 countries during the period 1920–2009. However, due to the focus on war, the entry of each country varies hugely. This may be connected to the randomness of the country and year that the conflicts occurred.

---

[4] Source: http://databank.worldbank.org/data/home.aspx.
[5] Source: http://comtrade.un.org/.
[6] Source: http://www.wiod.org/.
[7] Source: http://www.correlatesofwar.org/.

15.2.3.4 Enfolding Trade Data[8]

Unlike from the previous data, this data set is the output of the Enfolding global trade model, which is based on an input–output approach. The modelled data set currently covers a collection of national input–output tables (NIOTs) for 40 (mostly OECD) countries across 17 years, 1995–2011. However, the authors intend to extend the model to include data of over 200 countries (Levy et al., 2014).

15.2.4 Security Data[9]

As with the dyadic trade data earlier, this data is also from the COW website. There are 243 countries recorded sporadically as wars occurred during the period 1823–2003. In general, there is one record for each state war participant, unless the state changed sides during the war, in which case there are two records.

15.2.5 Migration Data

15.2.5.1 World Bank Migration and Remittances Data[10]

From the same data source as the remittance data, we also collect the migration data. The Global Bilateral Migration Database, constructed by the World Bank and its collaborating organisations, covers the bilateral migration data between 226 economies. The data spans the period 1960–2000 and can be disaggregated by gender. The migrants are selected primarily on the basis of the foreign-born statues in over 1,000 census and population register records that are combined to construct this database. The data sets are available in 10-year intervals.

15.2.5.2 Enfolding Migration Data[11]

Again, this is a data set with bilateral migration flow estimates from a model. The Enfolding migration model is based on a spatial interaction model, and the resulting output covers annual migration flows between 231 countries during the period 1960–2010.

15.2.6 Geographical Data

Due to our interest in spatial patterns, some geographical data is also collected, including the coordinates of capitals and country and region boundary shape files. Such data is used to produce maps to illustrate the spatial patterns of the aid distributions.

---

[8] Source: Enfolding trade workstream.
[9] Source: http://www.correlatesofwar.org/.
[10] Source: http://databank.worldbank.org/data/home.aspx.
[11] Source: Enfolding trade workstream.

15.2.7   Data Selected

15.2.7.1   Aid Data Selected

After careful examination and consideration in terms of both the data itself and the possibility of linking to other data sets in our collection, we have selected a set of data to be used in our model. We decided to use the Alesina and Dollar (2000) data as a base with updated information from the World Bank WDI, such as income and population. This is because the Alesina and Dollar data covers a broad range of variables that have an important impact on the aid flows and their data, and this reported study has been widely cited and studied. We also want to compare our model to theirs in the following sections.

Another advantage of using two data sets is that both the Alesina and Dollar data and WDI data have the ISO codes, which are useful to link to other data. In addition, it provides the opportunity to explore the spatial patterns, geography is useful in providing a bridge to integrate data from various data sources. This is an important factor, as using a complexity thinking, we want to explore the connections and interactions between different factors and systems.

15.2.7.2   Trade Data Selected

We chose the COMTRADE data from United Nations as it provides a comprehensive list of imports and exports from a selection of countries that are more in line with our aid data than other actual data. COMTRADE data also use ISO codes for individual countries, and this makes the linkage to the rest of the data sets easier.

Although the modelled data from Enfolding project can potentially provide more detailed information, at present we are focusing on the general trend and patterns and we also want to explore such patterns with some actual data before we move to the modelled data. Using the modelled data may be our next step in future work.

15.2.7.3   Security Data Selected

The COW data is selected, as this is the only reliable data that we have found so far. This data also uses the standard ISO3 codes for individual countries.

15.2.7.4   Migration Data Selected

For similar reasons as described in the selected trade data section, we chose the Global Bilateral Migration Database by the World Bank as we consider it is necessary to focus on general patterns first and may look into more details with the modelled data from Enfolding project at a later date.

15.2.7.5   Geography Data Selected

The coordinates of capitals and country and region boundary shape files are used in our study to produce maps to illustrate the spatial patterns of international development aid.

15.3 Model Building

15.3.1 Modelling Approach

The main aim of the Enfolding project focuses on explaining, modelling and forecasting global dynamics of various complex systems such as international development aid, international trade, security and migration. Therefore, our aid model will focus on the general pattern of the bilateral ODA only. This is because ODA is the most important part of the aid flows and bilateral ODA forms the major part of the ODA, as described in the previous chapter. Given that the ultimate purpose of international development aid is intended to have beneficial macroeconomic effect, most notably to raise a country's economic growth rate, traditionally the macro-econometric approaches have been used to study aid. Through the literature review in the previous chapter, we recognised some of the most influential models in the field. We believe that it would be good practice to build on the strength of those important models, while attempting to extend them by introducing some complexity thinking into the model to reflect the complex and dynamic international development aid system we are facing today.

Among other studies, the study by Alesina and Dollar (2000) is a major contribution to the studies of international development aid, where the authors investigated the drives behind certain donors giving foreign aid to certain recipients. In this section, the patterns of foreign aid flow from donor countries to recipients are studied through the examinations of the heterogeneous characteristics of individual recipient countries. The analyses cover the period 1970–1994 at 5-year intervals. A series of multiple regression analyses have been carried out between the natural log of bilateral aid and a selection of independent variables to explore various drives of the aid donations through the examination of recipient characteristics at aggregate level of all recipient countries, as well as at the disaggregate level of the individuals. The coefficient values on the independent variables and their t-statistics are used to indicate which variables have more impact on aid.

In this study, we try to extend Alesina and Dollar's models, by introducing some complexity into it. The interactions between donors and recipients as well as the interconnections with other global systems will be explored. In the following sections, we will describe both Alesina and Dollar's model and their findings and then compare them to our models and findings.

15.3.2 Alesina and Dollar Model

From the aggregate model, Alesina and Dollar found that the characteristics of the recipient countries, such as income, population, economic openness, political democracy, strategic political alliances (represented as UN votes in alliance with the country studied), colonial history, religious preferences and being certain countries, all have an important impact on bilateral aid from the donors (with an R^2 of 0.63 in the base model). The authors also pointed out that the political goals are still playing an important role in aid allocation and political strategic variables have more significant correlations than measures of poverty, democracy and policy with the bilateral ODA. For instance, the authors explored how much of a difference one standard deviation over the average makes on country openness (20% more aid), democratic state (39% more aid), colonial past (87% more aid) and UN voting with Japan (172% more aid). While all countries demonstrate some variances in their individual characteristics, being Egypt itself alone, this characteristic will enable Egypt to receive 481% more and being Israel more than

600%. However, such special cases are not generally applicable. In fact, it is not applicable for any of the rest countries. On the other hand, the authors also found that some variables included in the aggregate model do not have such significant influence on aid. For instance, the religious preferences of recipient countries do not have much influence on the pattern of aid flows.

In the disaggregate model, multi-regression analysis is carried out by a selection of individual donors, using a selection of variables to represent recipient countries' characteristics from the aggregate model. Some heterogeneity is found in the donor behaviour, for instance the Nordic countries tend to send aid to low-income economies and they also seem to acknowledge and reward economic openness and political democracy of the recipients. On the other hand, although United States follows similar patterns, it is also affected by UN voting patterns and its interests in the Middle East. Similarly for the United Kingdom, aid is driven by low levels of income, openness and democracy levels of the recipients; however, UN friendship and colonial past are also significant factors affecting the aid allocation patterns. To a large extent, the United Kingdom demonstrates a similar pattern to the United States. In contrast, France and Japan are the two most representative countries whose aid is mainly driven by colonial relationships and UN voting alliance.

15.3.3   Our Models

15.3.3.1   Data

We used the Alesina and Dollar data as the base of our model, but made some modifications in order to benefit the model in reflecting the more up-to-date information available now.

15.3.3.2   Replaced Variables: Population and Income

Two variables were replaced using the WDI data on population and income, the original data originates from Summers and Heston (1988), but the WDI data is more up to date and it is one of the major sources for international development aid indicators. More importantly, they have data on population and income information on more countries, including all of the donor countries. Similarly, we also updated the economic openness and political democracy records from the source that the authors referred to. Although Alesina and Dollar's study only focuses on the recipients' characteristics, we think that it is important to capture the interaction between the donors and recipients, as complexity thinking points out that the connectedness and interactions between the components of complex systems are just as important as the components of the systems themselves, if not more.

15.3.3.3   Excluded Variables

As discussed in the previous section, Alesina and Dollar in their influential study found that some factors pose a greater impact on the allocation of aid than others. As our interest in this study is to look at the general patterns of the complex system of international aid, we decided to exclude those factors that are only applicable to certain countries, but less influential to the rest countries such as the 'Egypt' and 'Israel' factors. As religious influence is also found to be less significant to most countries, those factors are also excluded from our model.

15.3.3.4   Removed and Added Records

As the Egypt and Israel dummy variables are removed from the model, we also removed the records for these two countries, as they are special cases that could obscure the general pattern in our model. However, we added the donor records to the data, as we believe that the interaction between the donors and recipients is also important and should be captured in the model.

15.3.3.5   Model A: Assess the General Patterns

Based on the above-mentioned discussion, we then decided to use the Alesina and Dollar data for an exploration, but we excluded the less influential variables from our model, including religion variables, Therefore, our first attempt at the aid model in effect has transformed the Alesina and Dollar aggregate model from the original form:

$$
\begin{aligned}
\ln(\text{Aid}) = {} & \beta_0 + \beta_1 \ln(\text{Income}) + \beta_2 \ln (\text{Income})^2 + \beta_3 \ln(\text{Population}) + \beta_4 \ln (\text{Population})^2 \\
& + \beta_5 \text{Openness} + \beta_6 \text{Democracy} + \beta_7 \text{USAFriend} + \beta_8 \text{JapanFriend} \\
& + \beta_9 \ln(\text{Colony}) + \beta_{10}(\text{Egypt}) + \beta_{11}(\text{Israel}) + \beta_{12}(\text{Muslim}) + \beta_{13}(\text{RomCatholic}) \\
& + \beta_{14}(\text{OtherReligion}) + \varepsilon
\end{aligned}
\tag{15.1}
$$

to:

$$
\begin{aligned}
\ln(\text{Aid}) = {} & \beta_0 + \beta_1 \ln(\text{Income}) + \beta_2 \ln (\text{Income})^2 + \beta_3 \ln(\text{Population}) + \beta_4 \ln (\text{Population})^2 \\
& + \beta_5 \text{Openness} + \beta_6 \text{Democracy} + \beta_7 \text{USAFriend} + \beta_8 \text{JapanFriend} \\
& + \beta_9 \ln(\text{Colony}) + \varepsilon
\end{aligned}
\tag{15.2}
$$

and using US aid as an example, the Alesina and Dollar disaggregate model has been transformed from

$$
\begin{aligned}
\ln(\text{USAAid}) = {} & \beta_0 + \beta_1 \text{Income} + \beta_2 \text{Openness} + \beta_3 \text{Democracy} + \beta_4 \text{USAfriend} + \beta_5 \text{Colony} \\
& + \beta_6 \text{OtherColony} + \beta_7 (\text{Egypt}) + \beta_8 (\text{Israel}) + \beta_9 (\text{Muslim}) \\
& + \beta_{10}(\text{RomCatholic}) + \beta_{11}(\text{OtherReligion}) + \varepsilon
\end{aligned}
\tag{15.3}
$$

to

$$
\begin{aligned}
\ln(\text{USAAid}) = {} & \beta_0 + \beta_1 \text{Income} + \beta_2 \text{Openness} + \beta_3 \text{Democracy} + \beta_4 \text{USAfriend} + \beta_5 \text{Colony} \\
& + \beta_6 \text{OtherColony} + \varepsilon
\end{aligned}
\tag{15.4}
$$

15.3.3.6   Findings from Model A

The initial experiments have revealed a mixed picture. The aggregate model has resulted in a slightly lower, but very close R^2 value of 0.62 in comparison with Alesina and Dollar's R^2 value of 0.63. However, the disaggregate results have seen some departures from Alesina and Dollar's model results.

   The fact our aggregate model is very close to what Alesina and Dollar reported is encouraging, as our focus is more concerned with a generic model for the global system of aid. Therefore, we do not focus on special cases such as Egypt and Israel whose characteristics are of particular interest to the United States, but not to the rest of the world in general. Our purpose is to try to understand the relationship/correlations that are generically applicable.

   First, we apply the original model as described in Alesina and Dollar's paper to our data. The model output revealed lower R^2s than what reported by Alesina and Dollar. However, it is worth noting that we are interested in a generic model for a global system, while they are interested in certain aspects instead of the whole system of aid. Although our initial attempt resulted in lower R^2s, we do not include special cases such as Egypt and Israel in Alesina and Dollar's study and specific dummy variables that are also used in combination in regression. As the focus of our study is different from Alesina and Dollar's, we also exclude some variables that are not significant for the majority of bilateral ODAs. We also updated some variables using data from different sources such as WDI. However, we are confident with the data and believe that they are the most up to date and reliable data that are suitable for our study (Table 15.1).

Table 15.1   Comparison of the results of AD model and our three models

	AD model	Model A	Model B	Model C
All	0.63	0.6174	0.688	0.6946
AUS	0.55	0.1356	0.1741	0.1742
AUT	0.47	0.1383	0.179	0.1804
BEL	0.54	0.1186	0.2468	NA
CAN	0.41	0.02375	0.2427	0.2493
CHE	NA	0.02923	0.113	0.1148
DEU	0.51	0.1519	0.594	0.6012
DNK	NA	0.02651	0.1118	0.1118
ESP	NA	0.03341	0.03257	0.03381
FRA	0.6	0.4093	0.5818	0.582
FIN	NA	0.1479	0.1104	0.1105
IRL	NA	0.15	0.1905	0.1906
ITA	0.56	0.03373	0.1736	0.176
JPN	0.62	0.07609	0.3476	0.3482
LUX	NA	0.09609	0.2468	NA
NLD	0.48	0.06775	0.4039	0.4041
NZL	NA	0.09503	0.06594	0.07174
NOR	NA	0.05236	0.1292	0.1294
PRT	NA	0.05693	0.06623	0.06624
SWE	NA	0.01497	0.05483	0.05529
GBR	NA	0.2163	0.5448	0.5468
USA	0.5	0.2156	0.3151	0.3152

15.3.4   Model B: Introducing Donor Interactions and Modification of the Model

The results from our aggregate model are better than the disaggregate model in Model A experiments. Therefore, based on the assumption that the aggregate model may be scaleable to a disaggregate model to a degree and, more importantly, as the purpose of aid is to elevate the lower income economies and their populations out of poverty, we decided to include the variables of population and income in the next model, for they may be just as important as in the disaggregate model and in the aggregate model. We decided to keep the aggregate model as it is in Alesina and Dollar (2000). In our disaggregate model, we have added in the natural log of income and population to the study's original disaggregate model, as well as using the log value of the years being a colony of the studied country. Those variables are also in common with those in the Alesina and Dollar's original aggregate model. As a result, such modifications have transformed our disaggregate model as in Equation (15.4) into

$$\ln(\text{USAAid}) = \beta_0 + \beta_1 \ln(\text{Income}) + \beta_2 \ln(\text{Income})^2 + \beta_3 \ln(\text{Population}) + \beta_4 \ln(\text{Population})^2$$

$$+ \beta_5 \text{Openness} + \beta_6 \text{Democracy} + \beta_7 \text{USAfriend} + \beta_8 \ln(\text{Colony})$$

$$+ \beta_9 \text{OtherColony} + \varepsilon \tag{15.5}$$

15.3.5   Findings from Model B

Our second attempt, by including the donor information and some variables from the aggregate model, have resulted in much better results than the previous simplified version of Alesina and Dollar's model. The aggregate model has improved considerably and produced a much higher R^2. In fact with a value of 0.688, the R^2 of this improved model is even higher than that of Alesina and Dollar's model of 0.63. In the disaggregate model, we have also seen an improvement for all countries. However, the R^2s are still lower than those reported from Alesina and Dollar's disaggregate model (Table 15.1).

As described earlier, we included the donors' records instead of just using recipients' in the base data as we want to explore the interactions between the donors and recipients. For instance, whether the characteristics of the donor countries, such as the income and population, also have an impact on their decisions of aid donation to recipient countries. Based on such assumptions, we applied this modified version of our model and saw that both our aggregate and disaggregate models have produced a much better result. The aggregate model even achieved a higher R^2 than the Alesina and Dollar's. It suggests that the donor characteristics are also playing an important role in aid allocation and should be included in the model to enable consideration of the interactions between donors and recipients. This may also indicate that this is a better model to study the general global patterns of the aid flows than that of Alesina and Dollar. In the next section, we will introduce the interaction with other complex global systems, such as international trade, into the aid model in an attempt to reflect the interactions and connectedness between the complex systems.

15.3.6   Model C: Introducing Interactions with Trade System and Further Modification of the Model

Encouraged by the results from our second attempt by improved results from introducing donor information and incorporation of more variables, we carry on exploring the interactions between aid and other global systems. International trade was selected to demon-strate the impact from other complex systems interacting with aid. Many studies have pointed out the extensive interactions between the two complex systems and these interactions can have a substantial effect on the dynamics within the aid system. In recent years, in particular, there has been a trend of aid for trade, where developing and industrial countries are bound together through trade, which serves as a transmission mechanism, with an increase in the value for trade (measured in terms of jobs, income, socio-economic upgrading, etc.) as the final objective of aid projects. Aid can help trade growth, but it is possible to introduce the so-called 'Dutch disease' and 'aid dependency' effects in the recipient economy and aid can also reinforce bilateral economic and political links (OECD, 2013; Morrissey et al., 1992; McGillivray and Oczkowski, 1992; McGillivray and White, 1993).

For this purpose, we use the UN COMTRADE data. First, we explored the correlations between the existing variables in the model with the bilateral export and import flows and found that the strongest correlation is between the economic openness and logarithm of export volume, which is perhaps not so surprising. Therefore we decided to add the bilateral export in our model. Thus, the aggregate model (Equation (15.3)) has been transformed into the following form:

$$\ln(\text{Aid}) = \beta_0 + \beta_1 \ln(\text{Income}) + \beta_2 \ln(\text{Income})^2 + \beta_3 \ln(\text{Population}) + \beta_4 \ln(\text{Population})^2$$
$$+ \beta_5 \text{Openness} + \beta_6 \ln(\text{Export}) + \beta_7 \text{Democracy} + \beta_8 \text{USAfriend}$$
$$+ \beta_9 \text{JapanFriend} + \beta_9 \ln(\text{Colony}) + \varepsilon \tag{15.6}$$

And the disaggregate donor model (Equation (15.5)) has been transformed into the following:

$$\ln(\text{USAAid}) = \beta_0 + \beta_1 \ln(\text{Income}) + \beta_2 \ln(\text{Income})^2 + \beta_3 \ln(\text{Population}) + \beta_4 \ln(\text{Population})^2$$
$$+ \beta_5 \text{Openness} + \beta_6 \ln(\text{Export}) + \beta_7 \text{Democracy} + \beta_8 \text{USAfriend}$$
$$+ \beta_9 \ln(\text{Colony}) + \beta_{10} \text{OtherColony} + \varepsilon \tag{15.7}$$

15.3.7   Findings from Model C

Linking aid to trade model has produced some encouraging results, with improvements in both aggregate and disaggregate model. There has been a slight improvement in the aggregate model with an elevated R^2 of 0.69, as well as slight improvement on all countries in the disaggregate results (as shown in Table 15.1). The findings from Model C indicate that trade does have a significant impact on the global patterns of aid allocation.

The model results also suggest that by taking into consideration the interactions with other factors that are outside the complex system of aid, the classical research method can be extended to better capture the generic pattern of aid flows. In the next section, we will discuss this further.

15.4   Discussion and Future Work

A series of regression models have been built to try to improve the correlations between the aid allocation and the important factors. As the purpose of our research is to produce a generic model that allows us to explore the general patterns of aid allocation in the context of the complex global systems, we decided to exclude special cases that are used in the Alesina and Dollar studies. Our first attempt to improve the model was to take into account of the interactions between the recipients and the donors, using the updated data with donor information and modifying the Alesina and Dollar model for the generic modelling purpose. Such modifications have improved results from both our aggregate and disaggregate models, although there is more improvement in the aggregate model. The improved results suggest that there are stronger correlations between aid and income and population (especially the logarithm value of them), economic openness and political democracy, UN friends and colony histories, but religion and special countries such as Israel and Egypt are less important for the general global patterns. Our model may also suggest that the disaggregate models actually share more in common with the aggregate model in terms of the general patterns, as the disaggregate correlations actually improved with more variables added in common with the aggregate model.

Encouraged by the findings, we then tried to explore the interactions between aid and other complex global systems. International trade was selected due to its strong relationship with the international development aid system. The second version of the model has led to further improvement in the model results. Such results indicate that the complexity science approach can be useful in studying the system of aid. Combined with the classic regression models that are widely applied in the study of aid, our model can be extended to enable the study of interactions and connectedness between a variety of factors within the complex system or even with other complex global systems such as international trade.

Due to the time limit of this project, we have to focus on the trade system only. The next step would be to incorporate such data as international security and international migration systems into our model, using the data sets identified in the data description sections. Ideally, we also would like to incorporate various data sets produced from other Enfolding workstreams, which have more details than the actual data. Other future work that we would also like to carry out includes the exploration of time and space patterns and the application of network approach to the international aid flow. These will be our next steps in the future to understand the important property of any complex system: the interactions and connectedness. We believe that complexity science is a new approach in the study of international development aid and has great potential to help us understand the interconnectedness of various factors, both within and outside the complex system of aid itself. It broadens the horizon and provides a more holistic way to study the system. Findings using this new approach may help to shed new light on the studies of this challenging complex system.

References

Alesina, A. and Dollar, D. (2000) Who gives foreign aid to whom and why? *Journal of Economic Growth*, **5** (1), 33–63.

Barder, O. (2012) *Development and Complexity, Presentation from His Kapuscinski Lecture*. http://cf.owen.org/wp-content/uploads/Development-and-Complexity-Slides.pdf (accessed 25 June 2015)

IMF (2010) *Balance of Payments Manual*, 6th edn, IMF, Washington, DC, p. 210, para12.22 https://www.imf.org/external/pubs/ft/bop/2007/pdf/bpm6.pdf (accessed 11 January 2016).

Levy, R., Oléron-Evans, T. and Wilson, A. (2014) *A Global Inter-Country Economic Model Based on Linked Input-Output Models*, Centre for Advanced Spatial Analysis, University College London, London, CASA WORKING PAPERS SERIES 198, ISSN 1467-1298.

Mavrotas, G. (ed) (2010) *Foreign Aid for Development: Issues, Challenges, and the New Agenda*, WIDER Studies in Development Economics, ISBN: 9780199580934.

McGillivray, M. and Oczkowski, E. (1992) A two-part sample selection model of British bilateral foreign aid allocation. *Applied Economics*, **24**, 1311–1319.

McGillivray, M. and White, H. (1993) *Explanatory Studies of Aid Allocation among Developing Countries*, The Hague: ISS Working Paper No. 148.

Morrissey, O., Smith, B. and Horesh, E. (1992) *British Aid and International Trade*, Open University Press, Buckingham, ISBN 0335156525.

OECD (2013) *Aid for Trade and Development Results: A Management Framework*, The Development Dimension, OECD Publishing. 10.1787/9789264112537-en.

Porter, J. (2011) *UK Collaborative on Development Sciences Report: Complexity Science and International Development*, http://www2.warwick.ac.uk/fac/cross_fac/complexity/people/students/dtc/students2007/porter/international_development_and_complexity_science.pdf (accessed 25 July 2015).

Ramalingam, B. (2013) *Aid on the Edge of Chaos*, Oxford University Press, NY, USA, ISBN:9780199578023.

Roodman, D. (2006) *Aid Project Proliferation and Absorptive Capacity*. UNU-WIDER Research Paper, 2006/04

World Bank (2011) *Migration and Remittances Factbook 2011*, 2nd edn, World Bank. https://openknowledge.worldbank.org/handle/10986/2522 (accessed 11 January 2016).

Summers, R. and Heston, A. (1988) A new set of international comparisons of real product and price levels estimates for 130 countries, 1950–1985. *Review of Income and Wealth*, **XXXIV**, 1–25, (updated to 1992).

16

Aid Allocation: A Complex Perspective

Robert J. Downes and Steven R. Bishop

16.1 Aid Allocation Networks

16.1.1 Introduction

While much has been written on foreign aid allocation, relatively little work has considered mathematical models beyond regression analysis. Modelling aid allocation is a complex issue. Empirical findings on the allocation of foreign aid indicate that donor countries pursue a wide range of objectives, achieving complex outcomes sometimes with unintended consequences. Poverty alleviation is frequently cited as a key factor in the disbursement of aid, see United Kingdom Government (2002), United States Government (1961), Collier and Dollar (2002); donor countries often engage in less than altruistic behaviour, see Harrigan and Wang (2011); highly heterogeneous behaviour is the norm, see Collier and Dollar (2002), Harrigan and Wang (2011), Alesina and Dollar (2000), Bermeo (2008), Berthélemy (2006), and Balla and Reinhardt (2008). Recent research indicates that decision-making in the donor community also impacts the allocation of aid (see Riddell (2007) and Frot and Santiso (2011)). Such 'bandwagon' behaviour is widely recognised in financial markets, see Schiller (2000) and Hommes (2006), but has only recently been considered as a component of aid allocation.

The development of mathematical models can support thinking on aid allocation and effectiveness. Most mathematical models follow a statistical trajectory, postulating a range of variables upon which aid allocation depends, followed by a regression-based analysis of available data; McGillivray (2003), Berthélemy (2006), Alesina and Dollar (2000) and Tarp *et al.* (1999) give an excellent introduction to this approach. Other models take an econometric route, considering a utility maximisation process, often based on recipient need; Collier and Dollar

Global Dynamics: Approaches from Complexity Science, First Edition. Edited by Alan Wilson.
© 2016 John Wiley & Sons, Ltd. Published 2016 by John Wiley & Sons, Ltd.

(2002), Chong and Gradstein (2008) and Tarp *et al.* (1999) all cover such procedures and McGillivray (2004) gives a particularly delicate consideration in the context of so-called 'prescriptive and descriptive' analyses of aid allocation.

This chapter presents a novel formulation of aid allocation. Rather than adopting a statistical or data-driven approach, we develop a framework allowing the model user to explore the allocation of foreign aid through the mathematical theory of networks. We aim to allow users of this model to explore possible policy choices in aid allocation. Statistical analysis indicates how donor countries have allocated aid in the past. Looking forward, our model allows users to explore responses to this analysis. Highlighting the behaviour of donor countries, we focus on the complexity of aid allocation from a mathematical perspective. According to McBurney (2012), alternative perspectives can function as a 'locus of discussion' which

> ...provide[s] a means to tame the complexity of the domain. Modelling thus enables stakeholders to jointly explore relevant concepts, data, system dynamics, policy options, and the assessment of potential consequences of policy options, in a structured and shared way.

This is our intention throughout this chapter.

16.1.2   Why Networks?

The community of aid donors and recipients is complex in a mathematical sense. According to Wilson (2012):

> Complex systems are characterised by requiring many variables to describe them and having strong interdependencies between the elements of the system. When represented mathematically, these interdependencies will typically be nonlinear relationships.

In the aid allocation context, complex systems exhibit emergent behaviour: decisions made by individual system actors aggregate into global system states in an often unpredictable manner. Whether global states are desirable or otherwise is beyond the purview of any individual actor.

A motivating example from population studies is given in Schelling (1971): the population comprises two groups randomly distributed on a lattice (finite in extent); individuals have a preference for remaining close to members of their own group and will move accordingly in a series of discrete steps. The model demonstrates that each actor's relatively small preference for being close to members of its own group can lead to a global pattern of segregation and that, famously, 'inferences about individual motives can usually not be drawn from aggregate patterns'.

As currently available models of aid allocation are typically statistically oriented, only qualitative discussion of complex behaviour is possible, see for example Ramalingam (2011). Mathematical models allowing users to explore the complexity of aid allocation are therefore a valuable contribution to current discussion.

16.1.3   Donor Motivation in Aid Allocation

Overviews of donor behaviour in aid allocation are provided by the work of Alesina and Dollar (2000) and Fuchs *et al.* (2014): donor countries do not behave uniformly towards recipient countries [Berthélemy (2006) and Chong and Gradstein (2008)], and aid allocation generally depends upon the specific situation in recipient and donor nations (both in demographic and political terms).

A selection of common donor motivations includes poverty alleviation Collier and Dollar (2002), McGillivray (2003), Baulch (2003); colonial ties Alesina and Dollar (2000); commercial interests Harrigan and Wang (2011), Alesina and Dollar (2000), Younas (2008); strategic interests Harrigan and Wang (2011); governance and policy environment quality Collier and Dollar (2002), Burnside and Dollar (2004), McGillivray (2006). Recent work suggests that donor community bandwagon effects Frot and Santiso (2011), conflict Balla and Reinhardt (2008) and [im]migration Bermeo (2008) also play a role. Contenting ourselves with these factors throughout this chapter, we emphasise that this is not an exhaustive list.

The factors motivating donors can be divided into two distinct classes:

1. Factors associated with *individual* countries. These include governance quality and demographic factors related to poverty;
2. Factors associated with relationships *between* countries. These include colonial ties and commercial interests in the form of trade flows.

This division is essential in this chapter as we relate donor behaviour to an underlying network of countries.

There are many ways to classify social phenomena in complex systems. For example, in the context of ethno-political conflict, Gallo (2013) classifies contributory factors as so-called 'state' or 'activity' variables: state variables define structural aspects of the system, while activity variables are used to affect change in this structure.

16.2   Quantifying Aid via a Mathematical Model

16.2.1   Overview of Approach

In this section, we write down the basic mathematical objects used in this chapter. We do not assume the reader to be familiar with network theory and so include illustrative diagrams wherever possible. The mathematical details are kept to a bare minimum. We direct the interested reader to Newman (2010) for an accessible introduction to the theory of networks.

We base our model on an underlying network comprising a set of nodes connected by a number of different quantifiable relationships: nodes are identified with countries, and links between nodes with international relationships, material or otherwise. By further identifying countries as either donors or recipients and associating demographic information to each country, we construct a model of the international network of nations.

To each donor–recipient pair, we assign a function determining the preference the donor has for awarding aid to the recipient. This function is dependent upon the following:

1. Demographic information of donor and recipient;

2. Relationships between donor and recipient;
3. Relationships between donor and *all other* recipients (explicitly).
4. Relationships between donor and *all other* donors via *all* recipients (implicitly).

Allocations are made on the basis of this preference along with a minimum donation value below which donations are zero.

Demographic and relational information encoded in the network is related to donor behaviours discussed in Section 16.1.3: the model then simulates donor behaviour and subsequent aid allocation. For example, we explore the complex interaction between donors allocating aid for poverty alleviation and commercial gain simultaneously.

The preference function is stimulated by the generalised Cobb–Douglas production function see Cobb and Douglas (1928). In its original guise, this provides a functional relationship between two inputs, capital and labour, with production as output; the parameters governing this interaction are estimated statistically in Cobb and Douglas (1928). While the specification of the function is arbitrary, see Simon and Levy (1963), its utility can be seen in its simple presentation, understandability and elucidation of the complex interaction between factors and output of production, see (Bhanumurthy, 2002).

The preference function also shares the form of the utility function found in weighted product method approaches to multi-objective optimisation (see Marler and Arora (2004)). However, we do not adopt an optimisation approach here.

16.2.2 Basic Set-Up

A graph or, equivalently, network G is a collection of nodes (or vertices), denoted $V(G)$, linked by ties (or edges), denoted $E(G)$. In a bipartite graph, the node set is decomposed into two disjoint subsets: there are edges between nodes in each disjoint set, but no edges connecting nodes within the same set. In a complete bipartite graph, each node is connected to every node of the opposing disjoint set; this is illustrated in Figure 16.1(a).

A weighted graph has a numerical quantity associated with each edge. We work with a more general object, an m-vector-weighted graph, in which m numbers are associated with each edge. Finally, we associate to each node an additional n numerical quantities which we call the n-vector node-specific information.

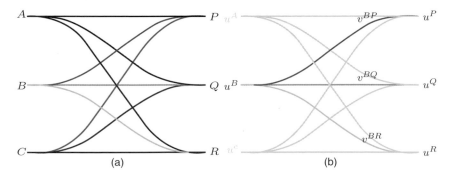

Figure 16.1 Complete bipartite graph (a) with vector weights and node-specific information detail (b)

16.2.3   The Network of Nations

Define the network of nations as a complete bipartite m-vector-weighted graph with n-vector node-specific information, with the following interpretation:

- Nodes represent countries, $A, B, C, \ldots$ donor nations and $P, Q, R, \ldots$ recipient nations. We index donors and recipients by i and j, respectively.
- The vertex set $V(G)$ is partitioned into donor and recipient country sets D and R. No donors are recipients, or vice versa, and the total number of countries is N.
- The m-vector edge weights v^{ij} represent m relationships between donor and recipient countries, material or otherwise. This will always be written so that the first superscript, i, is donor i and the second superscript, j, is recipient j. Then, v_α^{ij} is the αth element of the vector v^{ij}.
- The n-vector node-specific information u^k represents n quantities associated with each country k in our network. Then u_α^k is the αth element of vector u^k. Donors and recipients will generally have different node-specific information, denoted by u^i and u^j, respectively.

Figure 16.1(a) shows a network of nations with three recipients P, Q, R and three donors A, B, C; appropriate m-vector edge weights and n-vector node-specific information are given in Figure 16.1 (b).

16.2.4   Preference Functions

In this model, aid is allocated using a preference function representing the preference a donor has for awarding aid to a recipient. For a given donor i and recipient j, this function depends explicitly on v^{ij}, u^i, u^j and $v^{i\bullet}$, $u^\bullet$ where superscript $\bullet$ indicates 'all countries in R'.

For each donor i and recipient j, define the preference function as

$$P^{ij}(v^{ij}, u^j) := \prod_\alpha f_\alpha^i(v_\alpha^{ij}, v_\alpha^{i\bullet}) \cdot \prod_\beta g_\beta^i(u_\beta^j, u_\beta^\bullet). \tag{16.1}$$

The functions f_α^i, $\alpha = 1, \cdots, m$ and g_β^i, $\beta = 1, \cdots, n$ are to be specified for each donor D_i.

Note that the functions f_α^i and g_β^i are country specific: in general, two different donors, A and B, have two different sets of functions f_α^A and f_α^B. Although we could suppress this additional generality, enforcing the same set of functions for all donors, we exploit this at a later stage when exploring donor heterogeneity.

For clarity, consider the relationship defined by Equation (16.1): given donor i and recipient j, for each component of v^{ij}, we have a function f_α^i of v_α^{ij} and $v_\alpha^{i\bullet}$. Then f_α^i depends on both the relationship between donor i and recipient j *and* the relationship between donor i and all other recipients in R. The same holds for functions g_β^i.

This construction therefore emphasises the role of the network of nations: the preference function assigned to each donor–recipient dyad depends upon both the dyad and all other recipients.

16.2.5   Specifying the Preference Functions

In applying the model, we must specify numerical quantities associated with the behaviour under investigation and, crucially, functions f_α^i and g_α^i for each donor i. While there is an arbitrariness in this specification, it can nonetheless provide insight into aid allocation.

Here, the functions take the following form:

$$f_\alpha^i(v_\alpha^{ij}, v_\alpha^{i\bullet}) = \left[\frac{v_\alpha^{ij}}{\sum_{k \in R} v_\alpha^{ik}} \right]^{\mu_\alpha^i} \tag{16.2}$$

$$g_\alpha^i(u_\alpha^i, u_\alpha^\bullet) = \left[\frac{u_\alpha^i}{\sum_{k \in R} u_\alpha^k} \right]^{v_\alpha^i} \tag{16.3}$$

The role of the superscript $\bullet$ is seen in the denominator of the bracketed quantities in (16.2) and (16.3).

Equations (16.2) and (16.3) warrant the following explanation:

Input (positive correlation): in (16.2) and (16.3), indicators positively correlated with aid allocation have the functional input

$$\frac{v_\alpha^{ij}}{\sum_{k \in R} v_\alpha^{ik}} \quad \text{or} \quad \frac{u_\alpha^i}{\sum_{k \in R} u_\alpha^k},$$

that is, as a proportion of the total v_α^{ij} between i and all recipients, or the total quantity of u_α^i among all recipients, respectively. A country with a greater proportion of a given quantity receives a greater allocation of preference and, hence, aid than a country with a lesser proportion.

Input (negative correlation): in (16.2) and (16.3), indicators negatively correlated with aid allocation have the functional input

$$1 - \frac{v_\alpha^{ij}}{\sum_{k \in R} v_\alpha^{ik}} \quad \text{or} \quad 1 - \frac{u_\alpha^i}{\sum_{k \in R} u_\alpha^k}.$$

A country with a lesser proportion of a given quantity receives a greater allocation of preference and, hence, aid than a country with a greater proportion. In this case, we alter (16.2) and (16.3) accordingly.

Parameters: the parameters μ_α^i and v_α^i can take values in $\mathbb{R}_0^+$, the positive real numbers including zero. All other things being equal, if a parameter takes

- the value 0: allocation is uniform;
- the value 1: allocation is strictly proportional;
- values in $(0, 1)$: allocation is 'sub-proportional', tending towards ambivalence as the parameter approaches 0;
- values in $(1, \infty)$: allocation is 'super-proportional', increasingly favouring the most preferred recipient as the parameter tends to ∞.

16.2.6   Recipient Selection by Donors

In recipient selection, we assume there exists a minimum aid volume (a lower bound), Z^i, below which donor i does not allocate. This threshold could be set uniformly for all donor nations and can also be set to zero.

Each donor ranks all recipients by preference, then allocates aid to the largest subset of this ranking such that

- each recipient receives aid greater than or equal to the threshold Z^i;
- each recipient receiving aid has greater preference than all recipients not receiving aid.

If the threshold is set to zero, all donors receive some aid.

16.3 Application of the Model

16.3.1 Introduction

We illustrate the aid allocation model through three scenarios using the preference function (16.1), (16.2), (16.3) and the aforementioned model algorithm. All scenarios have the same set-up: three donors A, B, C and three recipients P, Q, R. This situation is described by Figure 16.1. By varying donor behaviour, these scenarios emphasise different aspects of the model, especially system feedbacks.

Figure 16.2 gives a schematic representation of our model (with only a single donor A present for clarity). Recipient states, donor states, donor–recipient relationships and donor community activity feed into A's preference, determining aid allocation. In turn, this allocation affects the recipient states and the donor community in two feedback loops. This feedback is a simple source of complexity in the model and will be explored explicitly in the sequel.

16.3.2 Scenario 1. No Feedback

This scenario examines the impact of heterogeneous donor motivation in aid allocation, introducing the model and its mode of operation. Donor motivation consists of recipient poverty p^j, colonial relationships c^{ij} and trade flows t^{ij} (see Appendix A.1 for functional definitions). The scenario set-up is described in Figure 16.3; a detailed explanation of this figure is provided in

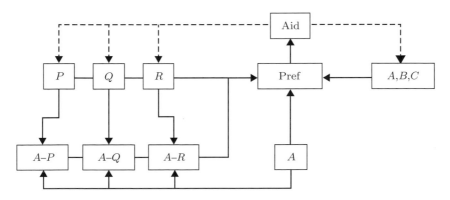

Figure 16.2 A schematic model representation with recipients P, Q, R and donor A. Arrows show information flows; dashed lines, feedback; hyphens, inter-country relationships

Data type	Name	Symbol
u_1^i	Aid volume	a^i
u_2^j	Poverty headcount	p^j
v_1^{ij}	Total trade volume	t^{ij}
v^{ij}	Colonial relationship	c^{ij}

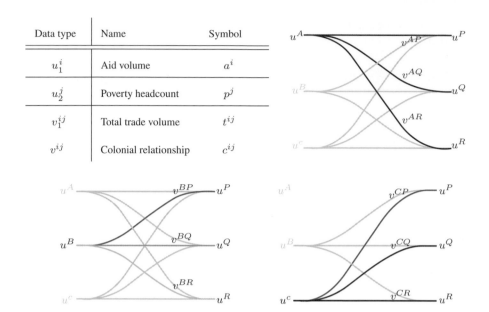

Figure 16.3   Scenario 1 set-up emphasising the role of node-specific information and inter-country relationships. The table notes the data included in this scenario, while each network diagram encodes relevant allocation data

Section 16.2.3. Note that all donors provide the same aid volume (normalised to 1); there is no minimum allocation threshold (see Section 16.2.6 for details).

The ratio of poverty headcount between recipient countries is $p^P : p^Q : p^R \equiv 1 : 2 : 4$; trade relationships are the same for each donor, with the ratio between recipients given by $t^{iP} : t^{iQ} : t^{iR} \equiv 1 : 4 : 2$, for $i = A, B, C$. Finally, for completeness, P is a former colony of A and B while Q is a former colony of B.

Using (16.1), (16.2) and (16.3), we write the preference function:

$$P^{ij} = f_1^i(c^{ij}, c^{i\bullet}) \cdot f_2^i(t^{ij}, t^{i\bullet}) \cdot g_1^i(p^j, p^\bullet)$$

where

$$f_1^i(c^{ij}, c^{i\bullet}) = \left[\frac{c^{ij}}{\sum_{k=1}^3 c^{ik}} \right]^{\mu_1^i} ; \quad f_2^i(t^{ij}, t^{i\bullet}) = \left[\frac{t^{ij}}{\sum_{k=1}^3 t^{ik}} \right]^{\mu_2^i} ; \quad g_1^i(p^j, p^\bullet) = \left[\frac{p^j}{\sum_{k=1}^3 p^k} \right]^{v_1^i} .$$

$$(16.4)$$

For ease, we take the matrix of parameters as

$$\Phi_{i\alpha} := \begin{pmatrix} \mu_1^1 & \mu_2^1 & v_1^1 \\ \mu_1^2 & \mu_2^2 & v_1^2 \\ \mu_1^3 & \mu_2^3 & v_1^3 \end{pmatrix} .$$

The choice of parameters governs the distribution of aid. We consider three different donor behaviours, showing the impact of (sub-/super-)proportional allocation (see Section 16.2.5).

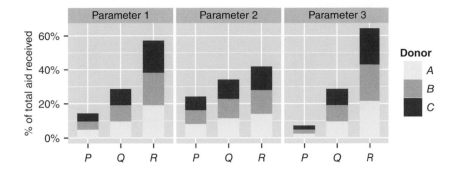

Figure 16.4   Model allocation for parameter choices 1, 2 and 3

Suppose alleviation of poverty is the only donor motive, and all donors value poverty identically (i.e. they have the same parameter choice). Let donors allocate proportionally, $\Phi^1_{i\alpha}$, sub-proportionally, $\Phi^2_{i\alpha}$ and super-proportionally, $\Phi^3_{i\alpha}$:

$$\Phi^1_{i\alpha} = \begin{pmatrix} 0 & 0 & 1 \\ 0 & 0 & 1 \\ 0 & 0 & 1 \end{pmatrix}, \quad \Phi^2_{i\alpha} = \begin{pmatrix} 0 & 0 & \frac{1}{2} \\ 0 & 0 & \frac{1}{2} \\ 0 & 0 & \frac{1}{2} \end{pmatrix}, \quad \Phi^3_{i\alpha} = \begin{pmatrix} 0 & 0 & 2 \\ 0 & 0 & 2 \\ 0 & 0 & 2 \end{pmatrix}.$$

For each parameter choice, the allocation is given by Figure 16.4.

The outcomes reflect recipient poverty headcount distribution, biased by donor motivation: when allocation is strictly proportional, Φ^1, recipient P has 14% of total poverty and therefore receives 14% of total aid. Parameter choices Φ^2 and Φ^3 place, respectively, lesser and greater emphasis on poverty as a motivator of allocation, relative to the proportional allocation Φ^1.

Suppose that countries now allocate based on the poverty and colonial relationships. Again, let donors allocate proportionally, $\Phi^4_{i\alpha}$, sub-proportionally, $\Phi^5_{i\alpha}$ and super-proportionally, $\Phi^6_{i\alpha}$:

$$\Phi^4_{i\alpha} = \begin{pmatrix} 1 & 0 & 1 \\ 1 & 0 & 1 \\ 1 & 0 & 1 \end{pmatrix}, \quad \Phi^5_{i\alpha} = \begin{pmatrix} \frac{1}{2} & 0 & \frac{1}{2} \\ \frac{1}{2} & 0 & \frac{1}{2} \\ \frac{1}{2} & 0 & \frac{1}{2} \end{pmatrix}, \quad \Phi^6_{i\alpha} = \begin{pmatrix} 2 & 0 & 2 \\ 2 & 0 & 2 \\ 2 & 0 & 2 \end{pmatrix}.$$

For each parameter choice, the allocation is given by Figure 16.5.

In the proportional case (Φ^4), including the positive aid allocation–colonial relationship, correlation in the preference function reduces by around 15% the allocation to R, the recipient allocated most aid when poverty alone comprises donor behaviour, compared with the proportional allocation from Figure 16.4. In turn, the allocation to both P and Q increases, especially to Q which is rewarded for its two colonial relationships with the donor community.

As earlier, parameter choices Φ^5 and Φ^6 place, respectively, lesser and greater emphasis on the overall combination of poverty and colonial history in aid allocation. Note also that C's parameter choice is the same as that from Figure 16.4 as it has no colonies.

These allocations are not obvious despite the simplicity of the situation (even assuming proportional allocation). In effect, identical signals from each recipient produce differing donor responses as a result of the unique set of donor–recipient relationships, even when donors share

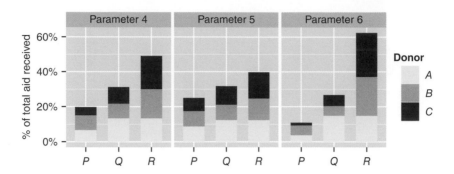

Figure 16.5   Model allocation for parameter choices 4, 5 and 6

the same 'values' (parameter choices). This hints at the complexity of donor heterogeneity coupled with aggregation of aid flows.

Suppose now that commercial ties influence allocation, in addition to colonial relationships and poverty: donors reward recipients with whom they enjoy a large trade volume. Suppose that only A and B are influenced in this manner, with C steadfastly continuing to allocate based on poverty alone. Again, let donors allocate proportionally, $\Phi^7_{i\alpha}$, sub-proportionally, $\Phi^8_{i\alpha}$ and super-proportionally, $\Phi^9_{i\alpha}$:

$$\Phi^7_{i\alpha} = \begin{pmatrix} 1 & 1 & 1 \\ 1 & 1 & 1 \\ 1 & 0 & 1 \end{pmatrix}, \quad \Phi^8_{i\alpha} = \begin{pmatrix} \frac{1}{2} & \frac{1}{2} & \frac{1}{2} \\ \frac{1}{2} & \frac{1}{2} & \frac{1}{2} \\ \frac{1}{2} & 0 & \frac{1}{2} \end{pmatrix}, \quad \Phi^9_{i\alpha} = \begin{pmatrix} 2 & 2 & 2 \\ 2 & 2 & 2 \\ 2 & 0 & 2 \end{pmatrix}.$$

For each parameter choice, the allocation is given by Figure 16.6.

While R has the largest trade volume and poverty in absolute terms, overall donor behaviour has drawn aid away from the 'most deserving' recipient: aid is shared so that Q, with middling poverty, strong colonial and modest trade relationships, is allocated approximately the same amount as R. This is in line with Alesina and Dollar (2000) who assert that former colonies are favoured over other nations; donors A and B act in this way, while C is 'Nordic' in that it

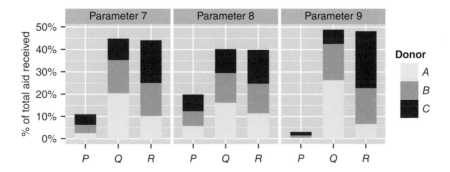

Figure 16.6   Model allocation for parameter choices 7, 8 and 9

allocates based only upon need (see Alesina and Dollar (2000)). This allocation with multiple criteria and heterogeneous donor motivation produces a radically different allocation to that given in Figure 16.4.

This scenario highlights both the operation of the model and the complex interactions at the heart of aid allocation, emphasising that aggregate flows are a poor indicator of donor behaviour when heterogeneous multi-objective behaviour is the norm in the donor community.

16.3.3   Scenario 2. Bandwagon Feedback

Bandwagon behaviour is the tendency of donors to act in a self-reinforcing collective manner. Certain recipient countries gain 'star' status relative to others despite seemingly little difference between such nations. While this may be attributed to increasing selectivity of aid, the tendency of donors to reward effective aid usage - see Dollar and Levin (2006), recent work suggests that herd behaviour can play a significant role, see Frot and Santiso (2011). Recipients previously awarded large aid volumes are then preferred by the donor community and are rewarded as such in subsequent allocation. (Although selectivity is not considered here, it could be incorporated as a positively correlated effectiveness input with a super-proportional parameter choice.)

Donors allocate based on recipient poverty p^j, commercial ties t^{ij} and previous success in receiving aid b_t^j (see Appendix A.1 for functional definitions). The model is dynamic: with fixed demographic factors, aid volume changes as the model is iterated forward in time from $t = 1$. Donors reward recipients which are successful in garnering aid; this 'community action' affects subsequent allocation and corresponds to the clockwise feedback loop in Figure 16.2.

The basic set-up of this scenario is as follows. Donors A and B allocate one unit of aid ($a^A = a^B = 1$) while donor C allocates two units ($a^C = 2$). As earlier, the ratio of poverty headcount between recipient countries is $p^P : p^Q : p^R \equiv 1 : 2 : 4$. Only donor C has significant trade relationships: the ratio between recipients is given by $t^{CP} : t^{CQ} : t^{CR} \equiv 10 : 1 : 1$.

Using (16.1), (16.2), and (16.3) we write:

$$P_t^{ij} = f_1^i(t^{ij}, t^{i\bullet}) \cdot g_1^i(p_t^j, p_t^\bullet) \cdot g_2^i(b_t^j, b_t^\bullet) \tag{16.5}$$

where

$$f_1^i(t^{ij}, t^{i\bullet}) = \left[\frac{t^{ij}}{\sum_{k=1}^3 t^{ik}} \right]^{\mu_1^i} ; \qquad g_1^i(p_t^j, p_t^\bullet) = \left[\frac{p_t^j}{\sum_{k=1}^3 p_t^k} \right]^{v_1^i} \tag{16.6}$$

$$g_2^i(b_t^j, b_t^\bullet) = \left[\frac{b_t^j}{\sum_{k=1}^3 b_t^k} \right]^{v_2^i} = \left[\frac{\sum_{m=1}^3 A_{t-1}^{mj}}{\sum_{k=1}^3 \sum_{n=1}^3 A_{t-1}^{nk}} \right]^{v_2^i} \tag{16.7}$$

We take the matrix of parameters as

$$\Phi_{i\alpha} = \begin{pmatrix} \mu_1^1 & v_1^1 & v_2^1 \\ \mu_1^2 & v_1^2 & v_2^2 \\ \mu_1^3 & v_1^3 & v_2^3 \end{pmatrix}.$$

We consider two different parameter choices, showing the impact of bandwagon feedback on the allocation of aid, driven by poverty and commercial interest, respectively.

Let all donors allocate according to recipient poverty and experience a sub-proportional tendency towards bandwagon behaviour (as this is a relatively subtle effect, see Frot and Santiso (2011)). Allocation based on poverty and bandwagon behaviour manifests in the following parameter choice:

$$\Phi_{i\alpha}^{10} = \begin{pmatrix} 0 & 1 & 0.4 \\ 0 & 1 & 0.4 \\ 0 & 1 & 0.4 \end{pmatrix} \tag{16.8}$$

For this parameter choice, the aid allocation from $t = 1$ to $t = 5$ is given by Figure 16.7.

As we move forwards in time, it is clear that R, the country allocated most aid based upon the poverty measure at $t = 1$, has increased its share of aid by approximately 50% at $t = 5$. Correspondingly, the country allocated least aid at $t = 1$, P, has seen its share of aid fall by a significant factor by $t = 5$.

This illustrates the bandwagon phenomenon driven by aid volume: R experiences an increase in aid allocation as a result of prior success in garnering aid. While a contrived example, this scenario demonstrates one approach to modelling bandwagon behaviour in the donor community.

Suppose now that A and B allocate based on poverty, but C allocates based on commercial interests; all donors are subject to a sub-proportional bandwagon influence.

Allocation based on this situation manifests in the following parameter choice:

$$\Phi_{i\alpha}^{11} = \begin{pmatrix} 0 & 1 & 0.4 \\ 0 & 1 & 0.4 \\ 1 & 0 & 0.4 \end{pmatrix} \tag{16.9}$$

For this parameter choice, the aid allocation from $t = 1$ to $t = 5$ is given by Figure 16.8.

In this case, we see that the conflicting motivations of donor nations have produced a complex result. A and B prioritise R as the 'neediest' nation, but have also been drawn towards P via a bandwagon effect resulting from C's commercially driven aid allocation. As neither A, B or C prioritise Q, the share of aid received by P and R increases. If the strength of the bandwagon

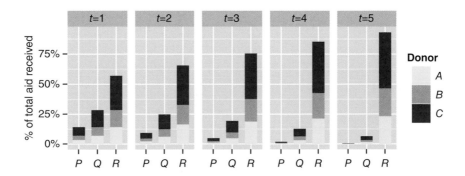

Figure 16.7   Model allocation for parameter choice 10

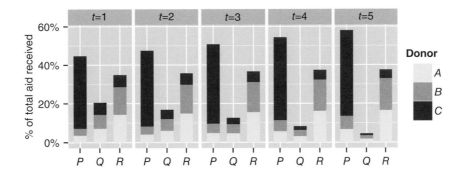

Figure 16.8   Model allocation for parameter choice 11

behaviour were stronger, the self-reinforcing behaviour would draw increasing volumes of aid from R toward P over time.

These two examples show the bandwagon allocation mechanism, albeit in an exaggerated manner. The first shows that, in our model, poverty itself can act as a driver of bandwagon behaviour: even though all nations prioritise poverty alleviation, they can be affected by the relative success of certain nations over time. The good intentions of donor nations can be subverted by their tendency towards 'groupthink'.

The second example shows how, in our model, varying motives and donated aid volumes can lead to a complex allocation outcome in the presence of bandwagon behaviour. A rich donor allocating a large quantity of aid for self-interested purposes can lead smaller but well-intentioned donors astray.

16.3.4   Scenario 3. Aid Effectiveness Feedback

This scenario presents a simple dynamic model of aid usage that allows us to explore how donors react to the changing fortunes of recipients, independent of the bandwagon effects discussed in Section 16.3.3. As earlier, we simplify the system under consideration for the purposes of elucidation.

Take recipient poverty p_t^j, trade relationships t_t^{ij} and governance quality q^j as the discriminators used by donors in aid allocation. We allow each recipient nation to use allocated aid to decrease poverty (in line with humanitarian concerns) and increase trade flows (as a proxy for economic development). Governance quality is constant throughout.

Using basic financial mathematics, aid usage decisions concerning trade expansion and poverty alleviation are treated as an investment portfolio for recipients: we assume that recipients 'invest' their allocated aid in a risk averse manner (where 'risk' is quantified as the variance of the time series of the corresponding 'asset' returns, see below).

This model is dynamic: starting from $t = 1$ the model is iterated forwards in time; underlying poverty and trade volumes change as a result of recipient investment decisions. This, in turn, impacts the way donor nations allocate their (fixed) aid volume at each time period.

The basic set-up of this scenario is as follows. All donors allocate one unit of aid ($a^A = a^B = a^C = 1$). As earlier, the ratio of poverty headcount between recipient countries is initially (at $t = 1$) $p_1^P : p_1^Q : p_1^R \equiv 1 : 2 : 4$; trade relationships are the same for each donor, with the ratio between recipients initially (at $t = 1$) given by $t_1^{iP} : t_1^{iQ} : t_1^{iR} \equiv 1 : 4 : 2$, for $i = A, B, C$; the ratio of governance quality between recipients is $q^P : q^Q : q^R \equiv 4 : 1 : 1$ and is constant throughout.

Using (16.1), (16.2) and (16.3) we write:

$$P_t^{ij} = f_1^i(t_t^{ij}, t_t^{i\bullet}) \cdot g_1^i(p_t^j, p_t^\bullet) \cdot g_1^i(q^j, q^\bullet) \tag{16.10}$$

where

$$f_1^i(t^{ij}, t^{i\bullet}) = \left[\frac{t^{ij}}{\sum_{k=1}^3 t^{ik}} \right]^{\mu_1^i} ; \quad g_1^i(p_t^j, p_t^\bullet) = \left[\frac{p_t^j}{\sum_{k=1}^3 p_t^k} \right]^{v_1^i} ; \quad g_2^i(q^j, q^\bullet) = \left[\frac{q^j}{\sum_{k=1}^3 q^k} \right]^{v_2^i} \tag{16.11}$$

Note that trade t_t^{ij} and poverty p_t^j are dynamic quantities, while governance quantity q^j is not. We take the matrix of parameters as

$$\Phi_{i\alpha}^{12} = \begin{pmatrix} \mu_1^1 & v_1^1 & v_2^1 \\ \mu_1^2 & v_1^2 & v_2^2 \\ \mu_1^3 & v_1^3 & v_2^3 \end{pmatrix} = \begin{pmatrix} 0.8 & 2 & 1 \\ 0.8 & 2 & 1 \\ 0.8 & 2 & 1 \end{pmatrix}.$$

Donors highly value a recipient country's level of poverty, allocate proportionally based on governance quality and have a sub-proportional interest in trade links.

16.3.5   Aid Usage Mechanism

We treat aid usage as an investment decision in the context of Modern Portfolio Theory (MPT) (see Adams *et al.* (2003) for an introduction to this topic). In essence, a portfolio in MPT consists of a number of possible assets each with normally distributed returns. An investment decision aims to minimise the risk of the total portfolio: risk is identified with the variance of each asset.

In the case of aid usage, each recipient country is allocated a certain aid volume. A fixed proportion of this, A_j, is then 'invested' in a portfolio consisting of a number of assets, here trade volume and poverty alleviation. The remaining proportion is lost to, for example, corruption; for ease, we allow 15% of allocated aid to be lost here; this quantity could be set for each country independently. An investment in poverty alleviation decreases poverty, and an investment in trade volume increases total exports.

When deciding how to invest, recipients aim to minimise risk across this portfolio, identified with the corresponding asset return variance: a large variance implies a riskier asset, a small variance a less risky asset. Assuming returns on each asset in the portfolio are normally distributed, with a specified mean and standard deviation, we can devise a risk-minimising investment decision.

MPT allows one to incorporate the correlation between each of the normally distributed variables in the portfolio as a whole. Note that this analysis can be extended to multiple correlated variables, assuming a multivariate normal distribution across the portfolio (see Adams *et al.* (2003)). We simplify matters here by assuming the asset returns are uncorrelated.

Denote by P^j recipient j's investment portfolio. This consists of two assets, poverty and trade volume:

$$P^j_t = \{p^j_t, t^j_t\}$$

We must determine the proportion of allocated aid a recipient invests in each possible asset. To this end, we determine the asset returns: for recipient j, define the asset return at time t as

$$R^{p^j}_t = \frac{p^j_t}{p^j_{t-1}}, \quad R^{t^j}_t = \frac{t^j_t}{t^j_{t-1}}$$

This then produces the time series of asset returns from $t = 1$:

$$R^{p^j}_1, R^{p^j}_2, R^{p^j}_3, \ldots$$

$$R^{t^j}_1, R^{t^j}_2, R^{t^j}_3, \ldots$$

Suppose these time series are normally distributed, as required by MPT. One can then calculate the means μ^{p^j} and μ^{t^j} and standard deviations σ^{p^j} and σ^{t^j}, respectively.

According to MPT, the expected return on the investment is

$$\mathbb{E}(R^{P^j}_t) = \alpha \mathbb{E}(R^{p^j}_t) + (1-\alpha)\mathbb{E}(R^{t^j}_t),$$

subject to the constraint

$$\sigma^2_{P^j_t} = \alpha^2 \sigma^2_{R^{p^j}_t} + (1-\alpha^2)\sigma^2_{R^{t^j}_t}$$

where $\alpha \in [0, 1]$ is a parameter to be found (see Adams $et\ al.$ (2003)): this parameter determines the proportion of A_j devoted to poverty alleviation (α) and the proportion devoted to trade volume expansion $(1 - \alpha)$.

Under the assumption of normally distributed asset returns

$$\mathbb{E}(R^{P^j}_t) = \alpha \mu^{p^j} + (1-\alpha)\mu^{t^j},$$

$$(\sigma^{P^j}_t)^2 = \alpha^2(\sigma^{p^j})^2 + (1-\alpha^2)(\sigma^{t^j})^2$$

as the means μ and standard deviations σ are constant in time.

This situation has an optimal 'risk-minimising' solution when

$$\alpha = \frac{(\sigma^{t^j})^2}{(\sigma^{p^j})^2 + (\sigma^{t^j})^2}.$$

This determines the 'risk-minimising' investment opportunity or, alternatively, the proportional investment in each possible asset which minimises the standard deviation of the total investment.

Then, aid usage is determined by investing αA^j in p^j_t and $(1 - \alpha)A^j$ in t^j_t, thus maximising the expected rate of return while minimising the overall portfolio risk. The model is re-evaluated at each time step t.

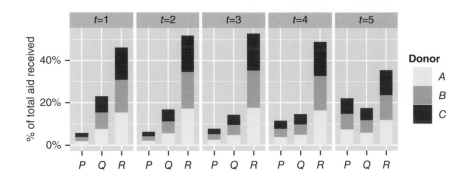

Figure 16.9   Aid allocation incorporating aid usage

Table 16.1   Parameters determining recipient investment

Country	μ^{p^j}	σ^{p^j}	μ^{t^j}	σ^{t^j}
P	2	0.2	2	0.2
Q	0.5	0.5	0.5	0.7
R	0.5	0.7	0.5	0.7

16.3.6   Application

Continuing with Scenario 3, we apply the mechanism presented in Section 16.3.5. The initial aid allocation is shown in Figure 16.9 ($t = 1$). We supplement this data with the mean and variance of each asset for each recipient. Guided by recipient governance quality, given in Section 16.3.4, we assign these according to Table 16.1.

Countries with lower governance quality q^j are likely to lack the necessary bureaucratic infrastructure to successfully utilise development aid or lose aid to corruption throughout the investment process. Therefore, we associate a greater risk to countries with lower governance quality.[1]

Iterating the model forward from $t = 1$, aid allocation is as shown in Figure 16.9.

We see that R is initially allocated the largest aid volume, reflecting its poverty headcount; R experiences its largest allocation at $t = 3$, with diminishing subsequent allocations. P has the smallest initial allocation, which grows as the model is iterated forwards in time. Q has a middling initial allocation, which diminishes initially, reaching a nadir at $t = 3$, but growing again thereafter.

We can explain this situation by considering the changes resulting from recipient investment of aid (see Figure 16.10).

Following the initial allocation, R draws most aid. The investment parameters indicate (see Table 16.1) that R invests equally in poverty reduction and trade expansion. However, having a significantly greater headcount than either P or Q, R continues to draw heavily from a donor community highly motivated by poverty.

---

[1] In a full model application, such quantities would be determined directly from available data, controlling for fluctuations of return produced by other factors.

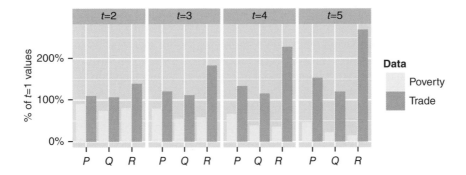

Figure 16.10   Poverty and trade levels of recipients following aid investment

P also splits its investment evenly between poverty reduction and trade expansion, but receives a relatively small aid volume compared to its comparator recipient nations. However, its fortunes begin to change after *R* has decreased its poverty headcount sufficiently for the governance advantage held by *P* to become a determining factor in aid allocation; aid volumes increase significantly after $t = 3$.

Q receives relatively little aid throughout, losing out initially to *R* and *P*. However, as *R*'s aid allocation drops after $t = 3$, *Q* begins to benefit. It is bias towards poverty reduction, as seen in investment parameters given in Table 16.1, mean its poverty headcount is reduced at a greater pace than trade expansion, which diminishes its likelihood of receiving aid in the subsequent time period.

16.3.7   Conclusions

Scenario 3 has shown how aid usage may be factored into a model of aid allocation, closing the counter-clockwise feedback loop of Figure 16.2 and feeding into donor preferences in allocation. Again, this process can produce complex outcomes.

While the approach sketched out earlier is clearly a heuristic, a stand-in for a full economic model encompassing the impact aid allocation on recipient economies, it does begin to encode behaviour identified in the literature.

16.4   Remarks

This Chapter presents a novel model of foreign aid allocation based on the mathematical theory of networks. Our intention is to capture the interconnected nature of this system, shedding light on the complex interactions between donors and recipients.

As we have shown, complex donor motivators in aid allocation can be described by our model. In particular, we have shown that heterogeneous donor motivation can lead to a wide variety of behaviours, even when donor nations share the same laudable intentions. Commercial and colonial relationships are shown to interact in an unpredictable manner with more altruistic, poverty-minded motivations.

Crucially, all allocations by our model are made in reference to the wider community of donors and recipients. This emphasises the increasingly networked nature of the global aid system which cannot be captured by more traditional dyadic analyses.

Our model allows for a simple characterisation of bandwagon behaviour (see Section 16.3.3). This characterisation is carried out in explicitly network theoretic terms and, hence, is an exploratory tool allowing connections not seen when using more traditional aid allocation models.

As we have shown, albeit in an exaggerated manner, bandwagon behaviour not only can skew the allocation of aid but can do so in an unexpected way in the context of our model. Over time, even a small tendency towards self-reinforcing behaviour can have a substantial impact upon the allocation of aid. This interaction becomes increasingly complex as the number of donor motivators increases.

This chapter joins an increasing chorus arguing that foreign aid should be viewed through the lens of complex systems analysis. Such a perspective suggests that aggregated data and assumptions in favour of homogeneity of motivation and behaviour are insufficient to describe global aid patterns. This chapter offers a new perspective on aid allocation and also suggests a new approach towards aid effectiveness using a conceptual-numerical modeling lens.

Acknowledgements

The authors thank R.G. Levy for a valuable tour of the development economics literature and the code used to generate the network images. A.G. Wilson and F.T. Smith provided helpful suggestions in developing this chapter. The authors acknowledge the financial support of the Engineering and Physical Sciences Research Council under the grant ENFOLDing—Explaining, Modelling, and Forecasting Global Dynamics, reference EP/H02185X/1.

References

Adams, A.T., Booth, P.M., Bowie, D.C., and Freeth, D.S. (2003) *Investment Mathematics*, John Wiley & Sons, Ltd, Chichester.

Alesina, A. and Dollar, D. (2000) Who gives foreign aid to whom and why? *Journal of Economic Growth*, **5** (1), 33–63.

Balla, E. and Reinhardt, G.Y. (2008) Giving and receiving foreign aid: does conflict count? *World Development*, **36** (12), 2566–2585.

Baulch, B. (2003) Aid for the poorest? The distribution and maldistribution of international development assistance. Chronic Poverty Research Centre (CPRC).

Bermeo, S.B. (2008) Aid strategies of bilateral donors. Department of Political Science, Yale University. Unpublished manuscript.

Berthélemy, J.C. (2006) Bilateral donors' interest vs. recipients' development motives in aid allocation: do all donors behave the same? *Review of Development Economics*, **10** (2), 179–194.

Bhanumurthy, K. (2002) Arguing a case for the Cobb-Douglas production function. *Review of Commerce Studies*, **20**, 21.

Burnside, A.C. and Dollar, D. (2004) Aid, policies, and growth: revisiting the evidence. World Bank, Washington DC.

Chong, A. and Gradstein, M. (2008) What determines foreign aid? The donors' perspective. *Journal of Development Economics*, **87** (1), 1–13.

Cobb, C.W. and Douglas, P.H. (1928) A theory of production. *The American Economic Review*, **18** (1), 139–165.

Collier, P. and Dollar, D. (2002) Aid allocation and poverty reduction. *European Economic Review*, **46** (8), 1475–1500.

Dollar, D. and Levin, V. (2006) The increasing selectivity of foreign aid, 1984–2003. *World Development*, **34** (12), 2034–2046.

Frot, E. and Santiso, J. (2011) Herding in aid allocation. *Kyklos*, **64** (1), 54–74.

Fuchs, A., Dreher, A., and Nunnenkamp, P. (2014) Determinants of donor generosity: a survey of the aid budget literature. *World Development*, **56**, 172–199.

Gallo, G. (2013) Conflict theory, complexity and systems approach. *Systems Research and Behavioral Science*, **30** (2), 156–175.

Harrigan, J. and Wang, C. (2011) A new approach to the allocation of aid among developing countries: is the USA different from the rest? *World Development*, **39** (8), 1281–1293.

Hommes, C.H. (2006) Heterogeneous agent models in economics and finance. *Handbook of Computational Economics*, **2**, 1109–1186.

Marler, R.T. and Arora, J.S. (2004) Survey of multi-objective optimization methods for engineering. *Structural and Multidisciplinary Optimization*, **26** (6), 369–395.

McBurney, P. (2012) What are models for? in *Post-Proceedings of the 19th European Workshop on Multi-Agent Systems (EUMAS 2011)*, Lecture Notes in Computer Science, vol. **7541** (eds M. Cossentino, K. Tuyls, and G. Weiss) Springer-Verlag, Berlin, pp. 175–188.

McGillivray, M. (2003) Modelling aid allocation: issues, approaches and results, 2003/49, WIDER Discussion Papers//World Institute for Development Economics (UNU-WIDER).

McGillivray, M. (2004) Descriptive and prescriptive analyses of aid allocation: approaches, issues, and consequences. *International Review of Economics & Finance*, **13** (3), 275–292.

McGillivray, M. (2006) Aid allocation and fragile states, in *Fragile States: Causes, Costs and Responses* (eds W. Naudé, A.U. Santos-Paulino, and M. McGillivray), Oxford University Press, Oxford, pp. 166–184.

Newman, M. (2010) *Networks: An Introduction*, Oxford University Press, Oxford.

Ramalingam, B. (2011) *Aid on the Edge of Chaos*, Oxford University Press, Oxford.

Riddell, R.C. (2007) *Does Foreign Aid Really Work?* Oxford University Press, Oxford.

Schelling, T.C. (1971) Dynamic models of segregation. *Journal of Mathematical Sociology*, **1** (2), 143–186.

Schiller, R.J. (2000) *The irrational exuberance*. Princeton University Press, Princeton.

Simon, H.A. and Levy, F.K. (1963) A note on the Cobb-Douglas function. *The Review of Economic Studies*, **30** (2), 93–94.

Tarp, F., Bach, C.F., Hansen, H., and Baunsgaard, S. (1999) *Danish Aid Policy: Theory and Empirical Evidence*, Springer-Verlag, Heidelberg.

United Kingdom Government (2002) International Development Act 2002 1(1)a.

United States Government (1961) Foreign Assistance Act of 1961 Sec. 101(a).

Wilson, A. (2012) *The Science of Cities and Regions*, Springer-Verlag, Berlin.

Younas, J. (2008) Motivation for bilateral aid allocation: altruism or trade benefits. *European Journal of Political Economy*, **24** (3), 661–674.

Appendix

A.1　Common Functional Definitions

Common functional definitions are collected in this appendix. Variables not discussed here may be assumed to be easily defined from Section 16.2.5.

Variable	Definition
c^{ij}	Colonial relationship between donor D_i and recipient R_j: $$c^{ij} = 1 + \begin{cases} 1 & \text{colonial relationship has existed between } D_i \text{ and } R_j \\ 0 & \text{no colonial relationship} \end{cases}$$ The 'null' condition 1 is arbitrary: the key point is that donors have greater preference for colonial over non-colonial recipients (see Alesina and Dollar (2000)). A more complex measure would be the fraction of the last century for which a colonial relationship existed (see Alesina and Dollar (2000)).
b_t^j	Prior success in receiving aid, the total volume of aid received in the previous year, b_t^j where $t = 1, 2, 3, \ldots$ indicates time. Assume that when $t = 1$, b_1^i is allocated identically for all recipients: $$b_t^j = \sum_i A_{t-1}^{ij}, \quad t = 2, 3, 4, \ldots, \quad b_1^j = 1$$

Part Six

Global Dynamics: An Integrated Model and Policy Challenges

17

An Integrated Model*

Robert G. Levy

17.1 Introduction

In Chapter 4 of this book, we introduced a model of linked input–output models using the 40 national input–output tables of the World Input–Output Database (WIOD). We will refer to this model throughout this chapter as "the global dynamics model". In Chapter 5 of *Geo-Mathematical Modelling*, we presented a method for extending the global dynamics model via mathematical estimation techniques to include a more complete set of countries.

In this final chapter, we draw these threads together and use the extended version of the global dynamics model to show how it might be used to address questions of global dynamics. Specifically, we use the global picture that our extended model gives us to look at the impact of three global *domains*: migration, aid and international security. In each case, we propose some simple assumptions about how the global dynamics model might interact with the phenomena described by the domain and examine the effects of these assumptions given either a set of domain data or a particular domain scenario. We will often assess the impact of a particular domain on the modelled countries by measuring the effect that domain has on modelled GDP. The rest of this chapter is structured as follows.

In Section 17.2, we take the migration flow estimates from the method outlined in Chapter 7 of this book as a starting point for examining how migration might affect the global economy. We first review the relevant parts of the literature, before proposing two mechanisms through which migration might affect the global economy: the *familiarity effect* whereby migration between two countries causes a commensurate increase in trade, and *consumption similarity* whereby migrants consume according to patterns which are similar to both their origin country and their destination country. We are then able to measure the impacts of GDP on the countries

*The author would like to acknowledge Peter Baudains who wrote the text of Sections 17.4.2–17.4.4; Thomas Evans for his work on the model and its 200-country extension; Adam Dennett for his migration estimates; Alan Wilson for his overall contribution to many of the ideas in this chapter.

in the model given the migration flows for a particular year, suggesting a set of countries most likely to benefit from migration, and a set of countries for whom the costs might be highest.

In Section 17.3, we turn to the domain of development aid, specifically so-called 'aid for trade', by assuming that aid is used to increase the export attractiveness of a recipient country. We model the effect of this increased export attractiveness using an estimation procedure first proposed in Chapter 4 of *Geo-Mathematical Modelling* for estimating missing bilateral services flows from flow totals. We are then able to assess the impact of an arbitrarily chosen increase in export attractiveness on each recipient country, thereby offering insights into which countries might make effective aid recipients and, more specifically, which recipients particular donors might choose to favour.

Finally, we look at the domain of security in Section 17.4. We employ a dyadic measure of threat developed in Chapter 12 of *Geo-Mathematical Modelling* and make an assumption about how a threatened country responds by increasing its military expenditure. The section begins with a short literature review, then introduces the dyadic threat measure and places it in the context of the global dynamics model, making additional assumptions about how military expenditure is reflected in the given input–output framework. We then introduce a simple model for how international trade responds to changes in security conditions and from there begin an analysis of the impact of particular threat relationships on the GDP of all countries in the global dynamics model.

17.2   Adding Migration

17.2.1   Introduction

A huge literature is dedicated to determining the economic factors which determine migration rates between countries.

Bertoli and Fernández-Huertas Moraga (2013) give a good overview and describe the recent literature as concerning itself with methodological and theoretical questions. They show how some authors Beine *et al.* (2011); Grogger and Hanson (2011); Ortega and Peri (2013) focus on discrete-choice micro-founded models, and others Clark *et al.* (2007); Mayda (2009); Pedersen *et al.* (2008) take a purely econometric approach.

Bertoli and Fernández-Huertas Moraga (2013) also describe how the 'traditional approach' of regressing only with variables relating to the origin and destination country contrasts with their 'multilateral resistance' approach (also sometimes described in the literature as a network effect) which emphasises the fact that migration decisions depend on the attractiveness/costs of all other destination options facing the migrant.

A large number of explanatory variables are employed in these papers, with some more common than others. Table 17.1 gives a summary of the variables used in each.

For our purposes what is significant here is that none of these recent papers attempts to describe the structure of the destination country's economy, something which the input–output tables used in the global dynamics model make available.

As the literature review in Felbermayr and Toubal (2012) demonstrates, the idea that migration between countries increases trade between them has a long history. Using the input–output approach of the global dynamics model, we can look further into what impact these increased trade flows might have had on both the source and destination countries.

Table 17.1   A list of the independent variables used by a selection of recent papers explaining migration flows. Each column represents one of the works cited in Section 17.2 with an 'x' showing that this variable is used by the author(s)

	Beine	Bertoli	Clark	Grogger	Mayda	Ortega	Pedersen
Linguistic distance	x		x	x	x	x	x
Immigration policy	x	x	x	x	x	x	
Geographic distance	x			x	x	x	x
Colonial relationships	x			x	x	x	x
GDP per capita		x			x	x	x
Average years of schooling			x	x			x
Schengen dummy	x			x		x	
Destination wages	x			x		x	
Relative population size	x						x
Social welfare spending	x						x
Shared land border				x	x		
Young population share			x		x		
Existing stocks	x		x				
Trade totals							x
Cultural similarity							x
Illiteracy rates							x
Political stability							x
Inequality ratio			x				
Source poverty rate			x				
Landlocked source			x				
Refugee/asylee share				x			
Common currency						x	
Common legislation						x	

Felbermayr and Toubal (2012) describe two ways in which migration can affect trade. First, in reducing trade costs by overcoming language, cultural and institutional barriers (we might term this the *familiarity effect*), and second in increasing international demand for goods produced in the source country (which could be termed *consumption similarity*).[1]

17.2.2   The Familiarity Effect

We can model the familiarity effect, whereby migration flows encourage trade between the source and destination countries, via two separate mechanisms in the global dynamics model. Firstly, the arrival of migrants will presumably boost demand simply for domestic goods in the destination country. We will name this as the *demand channel*. For now, we will model this as a simple scaling of the demand vector with no change in consumption patterns assumed. We will adjust this approach in section 17.2.3.

---

[1] The authors of the literature review name these terms the *trade cost channel* and the *preference channel*, respectively, but we find these terms less indicative of the underlying economics for our current purposes than our newly introduced terms.

Secondly, the source country might become a more preferred exporter for the destination. This we can name the *trade channel*. We can model this by adjusting the import propensities of the destination country. There is evidence in the literature for a pro-trade effect in both directions. For example, Girma and Yu (2002) find that, in the case of the United Kingdom and for migrants outside the Commonwealth, an increase of migrant flows by 10% increases the exports of the destination country to the origin country by 1.6%. Using a more international approach, Felbermayr and Jung (2009) find that a 10% increase in migration between a country pair increases trade by 1.1%. But the direction in which this trade increase happens is not specified since they take the geometric mean of the trade flows in the two directions.

These results and many more have been combined into a meta-analysis by Genc *et al.*(2012) who find that, although the picture is far from simple, on average, trade in both directions has an elasticity of 0.17 with migration. Thus, an increase of 10% in migration increases trade *in both directions* by 1.7%. It is this meta-analysis which we will use here, glossing over all the underlying detail behind the mean results reported in detail by the authors of the meta-analysis.

For modelling purposes, we must make an arbitrary decision about whether the state of the economy found in the global dynamics model in a particular year represents the state before or after the arrival of a particular year's immigrants. For notational simplicity, we will assume that the state of the economy (the level of final demand, etc.) is determined *before* the migration flow for that year occurs.

17.2.2.1   The Demand Channel

In modelling the demand channel, we use the concept of *final demand* discussed in Chapter 4. It is defined as the dollar value of the output of a particular sector which is consumed by households or governments and is in contrast to sector output which is used as intermediate demand, is invested or is exported. We represent this quantity as f, as is done in Equation (4.12) where no distinction is made between final demand for domestic and imported goods.

Final demand is assumed simply to scale with the associated population increase/decrease due to a particular migration flow. The migration flows themselves, denoted μ_{ij}, are taken from the migration estimation work of Chapter 7 of this book.

If we denote with a prime variables which are changed due to the effect of migration and denote the population in country i as $\bar{\pi}_i$, with the bar indicating a quantity taken from data, we can calculate the impacts on source and recipient country final demands of the flow of migrants as follows[2]:

$$\mathbf{f}'_i = \left(1 - \frac{\bar{\mu}_{ij}}{\bar{\pi}_i}\right)\mathbf{f}_i$$

$$\mathbf{f}'_j = \left(1 + \frac{\bar{\mu}_{ij}}{\bar{\pi}_j}\right)\mathbf{f}_j$$

(17.1)

where

$$\mathbf{f}_i = \begin{bmatrix} f_{i1} \\ \vdots \\ f_{iS} \end{bmatrix}$$

(17.2)

---

[2] Note that the notation here takes a slightly simplified form from that required by the formalism of Chapter 4 in, for example, Equations (17.1)–(17.3).

Table 17.2   The 10 biggest winners and 10 biggest losers from the final demand effect of global migration in GDP terms, measured in millions of $US. Also shown is the change in each country's balance of trade (BoT). Both model and migration flows are using 2010 data

	(a) 10 biggest increases				10 biggest decreases		
	BoT	ΔGDP	% GDP		BoT	ΔGDP	% GDP
United States	−5279	55226	0.42	Mexico	840	−2270	−0.29
Canada	−1409	7100	0.63	Turkey	265	−787	−0.12
Australia	−1224	7016	0.79	Poland	388	−723	−0.18
Germany	−983	6507	0.21	Portugal	302	−717	−0.31
France	−703	5313	0.22	Korea	362	−535	−0.07
Spain	−950	4480	0.36	Romania	204	−486	−0.35
Japan	266	2224	0.05	India	226	−233	−0.02
United Kingdom	−98	1978	0.09	Bulgaria	68	−124	−0.26
Italy	135	1166	0.06	Slovakia	101	−106	−0.13
Sweden	−191	1029	0.22	Finland	118	−95	−0.04

is the final demand vector in country i with elements associated with each sector from 1 to S. The term $\frac{\tilde{\mu}_{ij}}{\bar{\pi}_i}$ is the outflow *per capita* of the country of origin.

We will start by modelling the demand channel alone. Table 17.2 shows the overall effect of applying Equations (17.1)–(17.3) for every migration flow in the data set for 2010. What this highlights immediately is that gains in GDP (not GDP per capita, note) are broadly negatively correlated with changes to balance of trade (BoT), the difference between total exports and total imports. This makes intuitive sense since when a migrant moves from country i to country j, they stop demanding imported goods from j and start demanding imported goods from i. Generally speaking, countries with generally high import ratios (i.e. a reliance on imported goods to satisfy final demand) will do proportionally worse per migrant arriving than those with generally low import ratios.

The biggest winner in pure final demand terms is, by a huge margin, the United States with a gain in GDP of $US 55 billion (which represents almost half a percent of GDP). But in fraction of GDP terms, Canada and Australia are bigger winners still (with 0.63% and 0.79% of GDP, respectively.) This list should be contrasted with a list of overall migration destinations. Both lists would have the United States at the top, but Russia does not benefit as much as its position as the second most popular destination for immigrants suggests it should. In line with the above-mentioned reasoning, this suggests that Russia may have a greater reliance on imported goods to supply domestic consumption than does the United States. A similar comparison can be made between Canada, which does appear on the top 10 destinations list, and Australia which does not. This suggests that the structure of Australia's economy makes it especially able to benefit from the additional consumption of migrants or from the additional consumption of migrants to its trading partners.

Mexico is the biggest sufferer from migrant outflow in GDP terms, despite only being the seventh largest in emigration *per capita* terms. Its GDP fell by over $US 2 billion, almost a third of a percent of GDP. No other country has an impact on GDP of over a billion dollars, but Portugal, Romania and Bulgaria are hard hit in percentage-of-GDP terms with 0.31%, 0.35% and 0.26%, respectively.

17.2.2.2   The Trade Channel

To model the trade channel, we can calculate a new set of import propensities by first converting the set of propensities and import demands to a set of trade flows, then adjusting the relevant trade flows (in both directions) and recalculating the import propensities. We define the trade flows in terms of the import propensities and the import demands as

$$y_{ijs} = p_{ijs} m_{js} \qquad (17.3)$$

In order to apply the elasticity of 0.17 of Genc *et al.* (2012), we must first determine the change in migration flows between the year in question and the previous year, which we will denote $\Delta \bar{\mu}_{ij}$. Then from the definition of an elasticity:

$$\frac{\bar{\mu}_{ij}}{y_{ijs}} \frac{\Delta y_{ijs}}{\Delta \bar{\mu}_{ij}} = 0.17$$

$$\Rightarrow \Delta y_{ijs} = y'_{ijs} - y_{ijs} = 0.17 \frac{\Delta \bar{\mu}_{ij}}{\bar{\mu}_{ij}} y_{ijs}$$

$$\Rightarrow y'_{ijs} = \left(1 + 0.17 \frac{\Delta \bar{\mu}_{ij}}{\bar{\mu}_{ij}} \right) y_{ijs} \qquad (17.4)$$

When $\bar{\mu}_{ij} = 0$, that is, there is zero migration from i to j, a relatively common occurrence among the estimates, we simply set $y'_{ijs} = y_{ijs}$.

Notice that, for now at least, we assume the impact on trade flows is identical across all sectors. Improving on this assumption in a non-arbitrary way might be an interesting avenue for future research.

We can then use Equation (17.2) along with Equation (17.1) to calculate a new set of import propensities using the definition of an import propensity as a share of import coming from a particular country:

$$p_{ijs} = \frac{y'_{ijs}}{\sum_k y'_{kjs}} \qquad (17.5)$$

where all y' have been adjusted for the familiarity effect.

Thus, each migration flow affects the import propensities and, via Equations (17.1)–(17.3), the final demand in the source and destination countries. We can then use these changes to assess the impact of each migration flow on the GDP and GDP per capita of each country as well as the total production in the model as a whole.

Table 17.3 shows the biggest increases and decreases in emigration between 2009 and 2010 as a fraction of the country's 2010 population. Results are shown only for the 40 WIOD countries, although emigration figures were calculated across all countries in the estimate data set. These results will be important when we analyse the effect of the trade channel below. The biggest reduction in emigration in 2010 is in Ireland, where there were fewer migrants than in 2009 equal to 0.05% of the population. Estonia and Cyprus are next on the list with 0.030% and 0.027%, respectively. Luxembourg shows the biggest increase as a proportion of its population over the same period, followed by Portugal and Romania.

The effect of applying the trade channel to each country in the model is shown in Table 17.4. The largest winner from this effect by a wide margin is China, shown in Table 17.4a, which

Table 17.3   The 10 biggest increases and reductions in emigration between 2009 and 2010 across all destinations. Changes are shown as a percentage of the country's 2010 population. Results are only shown for the 40 countries of WIOD, but emigrations were calculated across all countries in the data set

(a) The 10 biggest increases		The 10 biggest reductions	
Country	Δ emigration (%)	Country	Δ emigration (%)
Luxembourg	0.037	Ireland	−0.051
Portugal	0.034	Estonia	−0.030
Romania	0.031	Cyprus	−0.027
Malta	0.030	Czech Republic	−0.026
Bulgaria	0.020	Poland	−0.025
Belgium	0.018	Lithuania	−0.022
Turkey	0.006	Latvia	−0.021
Slovenia	0.006	Finland	−0.008
Austria	0.005	Slovakia	−0.007
Mexico	0.005	Russia	−0.005

Table 17.4   The biggest gains and losses from applying the trade channel of the familiarity effect, governed by Equation (17.2). Only results for the 40 countries of WIOD are shown, and the trade channel was only calculated across these countries. GDP changes are shown in $US millions

(a) 10 biggest GDP gains			(b) 10 biggest GDP losses		
	ΔGDP	%		ΔGDP	%
China	2,245	0.06	Czech Republic	−3,057	−1.66
Belgium	1,124	0.22	Ireland	−2,420	−1.00
Netherlands	1,075	0.14	Russia	−1,864	−0.16
South Korea	792	0.11	Japan	−1,655	−0.04
Spain	647	0.05	Germany	−1,496	−0.05
United Kingdom	469	0.02	Poland	−1,125	−0.28
Italy	419	0.02	Indonesia	−765	−0.15
Mexico	353	0.04	Finland	−707	−0.33
Austria	171	0.04	Brazil	−498	−0.03
Sweden	94	0.02	Slovakia	−466	−0.58

gains $US 2 billion or 0.06% of its 2010 GDP. But far more significant from a percentage perspective are Belgium and the Netherlands, both with over a billion dollars of GDP increase from the trade channel, or 0.22% and 0.14% of GDP, respectively. But none of these gains is huge in percentage of GDP terms. The really significant results are among the GDP losses, shown in Table 17.4b. Two countries, the Czech Republic and Ireland, both suffer by more than 1% of GDP from the trade channel in 2010, with absolute reductions in GDP of $US 3 billion and $US 2.4 billion, respectively. Both these countries are high on the list of emigrant reductions in Table 17.3b. Russia, Japan, Germany and Poland all have GDP losses of over a billion dollars, but none of these is as significant in percentage terms.

17.2.2.3   A Combined Familiarity Effect

We now combine the demand channel and the trade channel into a single familiarity effect. We do this by simply applying both Equations (17.1)–(17.3) to adjust final demand and Equation (17.2) to adjust the import propensities, for every pair of countries in the model. The results of this experiment are shown in Table 17.5.

The three biggest GDP gains, shown in Table 17.5a, are experienced by the United States, Canada and Australia, the same three as at the top of the table for the demand effect in Table 17.2. The gains seen with the combined effect are very slightly increased for all three. It can be seen from the fact that none of these three countries is in the top 10 GDP gains from the trade channel, in Table 17.4a, that the combined effect on GDP of the two channels is not equal to the sum of the effects. This would be expected since the countries are interlinked in a complex manner via the trade model, and therefore changes in one economy affect all others in a way that is not immediately straightforward to predict (although increases in demand in one country can only ever lead to increases in others.) Spain, France and Germany then follow, as they do in the demand channel, but with Spain leapfrogging France and Germany into fourth place.

This reflects the fact that Spain appears in the biggest trade channel gains, where France and Germany do not. Japan, Italy and Sweden fail to appear in the combined table. Japan has one of the biggest losses from the trade channel, but both Italy and Sweden are among those benefiting from the trade channel. These latter two only fail to make the top 10 because Belgium, the Netherlands and China have such strong gains from the trade channel.

The two biggest sufferers from the combined familiarity effect, shown in Table 17.5b, are the same as those from the trade channel, shown in Table 17.4b. They are the Czech Republic and Ireland. Mexico is actually among the 10 biggest increases due to the trade channel, but is such a big sufferer from the demand channel that the costs outweigh the benefits. Poland suffers from both effects, as does Finland, explaining their positions among the worst combined sufferers. Turkey has the second-worst decrease in GDP due to the demand channel, so is therefore

Table 17.5   The biggest gains and losses from applying the both channels (demand and trade) of the familiarity effect. Only results for the 40 countries of WIOD are shown. GDP changes are shown in $US millions

(a) 10 biggest GDP gains				(b) 10 biggest GDP losses			
	BoT	ΔGDP	%		BoT	ΔGDP	%
United States	−5,588	54,917	0.42	Czech Rep.	−3,019	−2,991	−1.63
Canada	−1,667	6,842	0.60	Ireland	−2,397	−2,247	−0.92
Australia	−1,669	6,571	0.74	Mexico	1,194	−1,916	−0.24
Spain	−301	5,129	0.42	Poland	−741	−1,851	−0.46
France	−888	5,128	0.22	Russia	−1,795	−1,248	−0.11
Germany	−2,483	5,007	0.16	Turkey	176	−876	−0.14
China	3,536	2,821	0.08	Finland	−591	−804	−0.38
United Kingdom	372	2,448	0.12	Indonesia	−595	−769	−0.15
Netherlands	1,117	1,890	0.25	Portugal	390	−628	−0.27
Belgium	1,083	1,849	0.36	India	−120	−579	−0.05

among the worst affected by the combined effect, despite it having among the largest increases in emigration between 2009 and 2010. Together with Mexico and Portugal, which also had large increases in emigration in that year, Turkey shows a large improvement in its balance of trade, set against its large reduction in GDP. This shows that the various effects of migration tend to work in opposite directions, with gains (reductions) in GDP being accompanied by commensurate reductions (gains) in balance of trade. Notable exceptions to this are China, the Netherlands and Belgium, all of which see large gains in balance of trade accompanying their increases in GDP. On the other side, Czech Republic, Ireland and Russia all have large reductions in balance of trade as well as reductions in GDP.

This approach to modelling the effect of changing populations allows us some novel ways of thinking about how migration and economics interact. We have seen how migration has conflicting effects on GDP via the demand channel, whereby emigrants take their expenditure with them to their destination country, and via the trade channel, whereby a source country's trade is boosted bilaterally with a destination country. We have also seen how these opposing forces have varying effects on balance of trade and GDP, depending on a country's position in the global network, and its relationships in both trade and migration sense with the other countries in the model.

17.2.3   Consumption Similarity

In Section 17.2.2, we treated migrants' consumption patterns as being identical to those of the population of the destination country. This allowed us to simply scale the final demand vector according to migration as a proportion of the population.

A more interesting way of modelling this process might be to assume that migrants continue to consume in the same pattern either as the population of the source country or as some linear combination of the two patterns. By making these more subtle adjustments to the final demand vector as a result of migration, we can compare the effect on the economy of the destination countries and, by some measure, assess the extent to which a particular country's demand vector is changed by the immigrants it hosts. In this way, we are making a link to the vector rotation work of Chapter 25 of *Geo-Mathematical Modelling*.

Early work on this subject by Wallendorf and Reilly (1983) inspected the contents of the rubbish bins of Mexican Americans in an attempt to test the theory that migrants consume in a pattern mixed between the source and the destination countries. They in fact found that migrants consumed according to their own unique patterns. In the context of Greenlandic immigrants to Denmark, the work of Askegaard *et al.* (2005) find evidence both that consumption patterns matched the destination country, and that new patterns were developed. Karamba *et al.* (2011) find, in a food context, that migrants in Ghana did not change their consumption expenditure or consumption pattern from that of the source country. The picture in the literature is therefore a mixed one, between no consumption pattern adjustment and perfect consumption pattern assimilation. It therefore seems reasonable to take an approach some way between these two extremes.

We will therefore assess the impact on the destination country of migrants variously adjusting their consumption patterns. The patterns will vary between completely retaining the source country patterns, to complete consumption pattern assimilation. In order to simplify the analysis, we will focus on the three largest recipients of immigrants. Since adjusting consumption patterns requires knowing the patterns for the source country, only migrants from the 40 WIOD countries will be considered. With this restriction, the largest three are the United States, Germany and Canada with 0.9 million, 0.3 million and 0.2 million immigrants, respectively, in 2010.

To make the adjustment to the final demand vector, we will use the simplest possible extension to Equations (17.1)–(17.3) where the adjustment term is a linear combination of the source consumption vector and that of the destination. Thus, the adjusted consumption vector at destination j due to a migration flow from source i is given by

$$
\mathbf{f}'_i = \left(1 - \frac{\bar{\mu}_{ij}}{\bar{\pi}_i} \right) \mathbf{f}_i
$$

$$
\mathbf{f}'_j = \mathbf{f}_j + \frac{\bar{\mu}_{ij}}{\bar{\pi}_j} \left[\alpha \mathbf{f}_j + (1 - \alpha) \mathbf{f}_i \right]
$$

(17.6)

where, as before, a bar indicates a value taken from data, $\mathbf{f}$ is a vector of per-sector final demand values from WIOD and α is the linear combination parameter.

Figure 17.1 shows the results of taking this parameter from 0, migrants consume exactly like the source country population, to 1, migrants consume exactly like the population of the destination country. The three countries chosen have markedly different experiences both in their sensitivity to α and in their GDP versus balance of trade changes. With $\alpha = 1$, migrants consume exactly as the destination population and Equations (17.1)–(17.3) reduces to Equations (17.1)–(17.3), our initial simple final demand scaling. Canada and Germany have very similar improvements to GDP, both around $US 5 billion, and the United States has a hugely greater improvement at around $US 35 billion. These figures are broadly in agreement with the demand channel discussion of Section 17.2.2.1, given in Table 17.2. The numbers do not match precisely since that analysis used migrants from every country in the data set, where this uses only migrants from the 40 countries of WIOD.

But when $\alpha = 0$ and migrants continue to consume in the exact pattern as they did in the source country, the picture looks extremely different. Here, Canada has a small advantage, increasing its GDP gains to just under $US 10 billion, and Germany stays broadly the same, falling to just under $US 5 billion. But the United States falls dramatically by more than $30 billion to just under $US 5 billion, almost identical to the gains of Germany. This huge difference suggests that a large proportion of the huge gains shown by the United States in the discussion of the familiarity effect of Section 17.2.2 is contingent on the migrants who arrive adapting their behaviour to match that of their new compatriots. Without this effect, the United States is in no better a position than Canada or Germany.

Between the extrema of α is what appears visually to be a straight line continuum of GDP benefits. In fact, this relationship is not quite linear. For the United States, the relationship is increasing in α, with $\frac{\partial^2 \text{GDP}}{\partial \alpha^2} = 0.6$, and for Canada it is decreasing, with $\frac{\partial^2 \text{GDP}}{\partial \alpha^2} = -0.6$. For Germany, the relationship is very close to linear.

The picture for balance of trade shows even more variation than that for GDP. At $\alpha = 1$, all three countries show a reduction in the balance of trade, which is a natural result of losing consumers in the foreign market and gaining domestic consumers. The US suffers by far the most with over $US 3 billion of losses, with Germany just over $US 1.5 billion and Canada at $US 1 billion. But at the other end of the scale, with $\alpha = 0$, the picture is completely reversed. Canada is now the worst off, with a small decrease in its situation to around $US 2 billion of losses. Germany improves slightly to around $US 1 billion. But for the United States, the difference

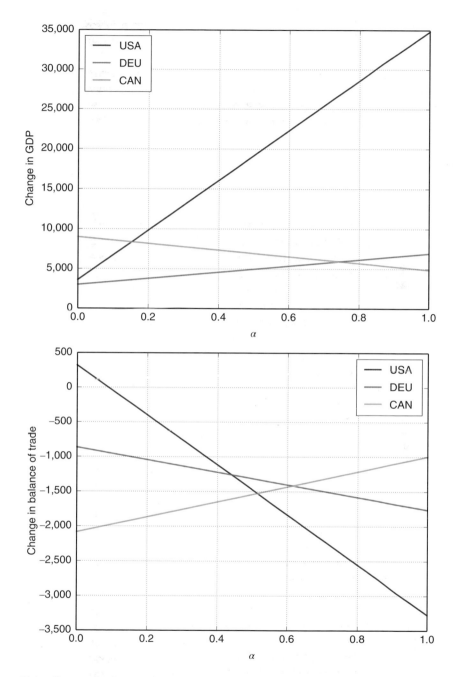

Figure 17.1 Changes in GDP and balance of trade for the three largest recipient of WIOD-country migrants in 2010 for various levels of the linear combination parameter α in Equations (17.1)–(17.3)

is again extremely marked, with the balance of trade now changing positively by around $US 300 million. At around $\alpha = 0.5$, the three countries are affected similarly in balance of trade terms.

This suggests that having migrants who behave like Americans is particularly bad for balance of trade in the United States, but having migrants who continue to behave like the population of their source countries may actually be beneficial to the balance of trade in the United States. Conversely, it is better for Canada's balance of trade if immigrants behave like Canadians. This is perhaps a reflection of the difference in immigration origins between Canada and the United States.

17.2.4   Conclusions

In this section, we have proposed two methods for the interaction of migration flow data with the global dynamics model. Firstly, we outlined the *familiarity effect*, which we split into a *demand channel* and a *trade channel*.

We showed how the demand channel has opposing effects on BoT and GDP and that countries with a greater reliance on imported goods will generally do worse in GDP terms from immigration. As might be expected, the biggest gains from this channel were in the United States. Mexico suffered the most from its emigration outflows.

The big winners from the trade channel in 2010 were China, Belgium and the Netherlands, although the effects were small compared to those countries which lost through this channel, notably the Czech Republic and Ireland both of which suffered losses of greater than 1% of GDP.

Combining these two effects showed that, globally, the United States, Canada and Australia are the biggest winners from the impact of migration on GDP, and the Czech Republic, Ireland and Mexico are the biggest losers.

Finally, an analysis of *consumption similarity* showed that the improvements in GDP due to net migration in both the United States and Germany are contingent on those new arrivals consuming like those people already in the country, and that the opposite is true for Canada: the gains Canada is modelled to get from its net migration are contingent on its arrivals continuing to consume in the patterns of their home country.

17.3   Adding Aid

17.3.1   Introduction

In this section, we estimate an exporter-specific measure which captures a country's overall ability or willingness to export, that is, all the unobserved factors which make a country a 'good' or a 'bad' exporter. These factors include, but are by no means limited to, export infrastructure, production capacity, trade tariffs, corruption, existing international relationships and currency strength.

By doing this, we will attempt to say something about the possible impact an 'Aid for Trade' policy might have not only on the recipient but also on the countries the recipient trades with and all the other countries in the trade network of the global dynamics model.

17.3.2   Estimating 'Exportness'

Rather than seeking explicitly to model the factors mentioned earlier which make a country a 'good' or a 'bad' exporter, we combine these effects into an arbitrary, unitless measure we might call 'exportness'. This exportness will be useful only as a comparison between countries (or between time periods).

We will use specification (5) from Chapter 4 of *Geo-Mathematical Modelling* with the point estimates of the country dummies acting as our measure of 'exportness'. Thus, the regression equation to estimate the flow of sector s from country i to country j is

$$\log\ y_{ijs} = \beta_0 + \sum_k (E_k \delta_k) + \beta_1 d_{ij} + \beta_2 \log\ x_j + \beta_3 \log\ f_{is} + \beta_4 \log\ v_{is} + \epsilon_{ijs} \qquad (17.7)$$

where $\delta_k = 1$ when $i = k$ and 0 otherwise, E_i is the exportness measure we are estimating, d_{ij} is the shortest distance between i and j (i.e. neighbouring countries have $d_{ij} = 0$), x_j is the total production of the importing country, f_{is} is the final demand for sector s in the exporting country and v_{is} is the per-unit value added.

Table 17.6 shows a summary of the unitless measure of exportness as estimated by Equation (17.3). The measure ranges from −0.6 to 6.5, with the world's least attractive trading partner being Bhutan and the most attractive being the United States. It is important to note that this is not merely a size measure: the inclusion of f_{is} absorbs the fact that the United States is the largest economy in the world.[3] China and Malaysia follow in the list of top exporters in Table 17.6a. This is perhaps evidence for what Stiglitz (2007) refers to as 'the Malaysian miracle' of successful poverty reduction, job creation and social cohesion. Germany is Europe's most attractive exporter, followed by France, the United Kingdom and Italy. Singapore, Japan and South Korea also score highly in exportness terms.

Table 17.6   A summary of the 'exportness' measure describing how attractive a country is an export partner beyond the effects included in the regression specified in Section 17.3.3. The measure is unitless and runs from 6.5 for the United States to −0.6 for Bhutan

(a) The 10 highest exportness countries		(b) The 10 lowest exportness countries	
Country	Exportness	Country	Exportness
United States	6.5	Bhutan	−0.6
China	6.2	Eritrea	−0.5
Malaysia	6.1	Iraq	−0.5
Germany	5.9	Cape Verde	−0.3
Singapore	5.6	Nepal	−0.3
France	5.6	Haiti	−0.2
United Kingdom	5.5	Angola	−0.1
Italy	5.4	Chad	−0.1
Japan	5.3	Maldives	−0.0
Korea	5.2	Mauritania	−0.0

---

[3] Note also that total production of the exporter cannot be included in the regression since it would be co-linear with the country indicator δ_k.

Bhutan is together with Nepal in being hugely dependent on a single market (India) for its exports. Other very low exportness countries are war-torn (Eritrea, Iraq, Angola), land-locked (Chad) or island nations (Cape Verde, Haiti, Maldives).

17.3.3   Modelling Approach

We will model aid for trade by assuming that aid has some kind of positive effect on the exportness measure estimated in Section 17.3.2. It is conceivable that we might estimate the effect of a dollar of aid for trade on exportness by using time series of aid for trade payments and an equivalent series of estimates for exportness, but the challenges of identifying aid for trade from total official development assistance (ODA) seem formidable and it might be hard to justify an assumption that *all* aid attempts to affect the ability of the recipient to export.

Instead, we will restrict ourselves to a comparison of the effect of an arbitrary, but fixed, increase in the exportness of each aid recipient country which might still allow us to say something about the countries which have the potential to most benefit from increased exportness (and, equivalently, which countries stand to lose the most from such increases elsewhere.)

Our analysis will focus on a subset of aid recipients. Namely, the recipients of 'ODA and official development aid' as a fraction of GDP greater than 2% in 2010, according to the World Bank's World Development Indicators, including only 'large countries', by which we mean those with a population of over 2 million[4]. This gives us 51 countries we will deem 'recipients' of which 31 are African[5]. We stress, though, that all the analysis of this section is done using all the countries in the global dynamics model, not just those deemed aid recipients.

For each of the 51 recipients, we will solve Equation (17.3) with the error term set to 0 to calculate estimated trade flows, $\hat{y}_{ijs}$, for each exporter, importer and sector. We will then increase the exportness by an arbitrary amount, ϵ, and solve the equation again. Then, using these $\hat{y}_{ijs}$, we will calculate import propensities using the same formula used to calculate them in Chapter 4 of Global Dynamics, which simply divides each flow by the total imports of a given importer. These import propensities are then inserted into the global dynamics model and the model is recalculated. This allows us to track the change in GDP in every country in the model for a change of ϵ to the exportness of each of our 51 recipient countries.

As outlined in the set-up of the global dynamics model, countries with no imports of a particular sector in the COMTRADE data upon which the model is based cannot calculate their import propensities in this way, and so it will be assumed that they import all their domestic requirements for that sector from the Rest of the World (RoW) entity.

17.3.4   Results

Throughout this section, we will consider the results of the experiment described in Section 17.3.2, using $\epsilon = 0.1$.

---

[4] Many of the largest recipients of ODA *per capita* are small Pacific islands and other countries with smaller populations.

[5] The non-African 'large' aid-recipient countries are Afghanistan, Albania, Armenia, Bolivia, Bosnia Herzegovina, Cambodia, Georgia, Haiti, Honduras, Jordan, Kyrgyzstan, Laos, Mongolia, Nepal, Nicaragua, Papua New Guinea, Moldova, Tajikistan, Vietnam and Yemen.

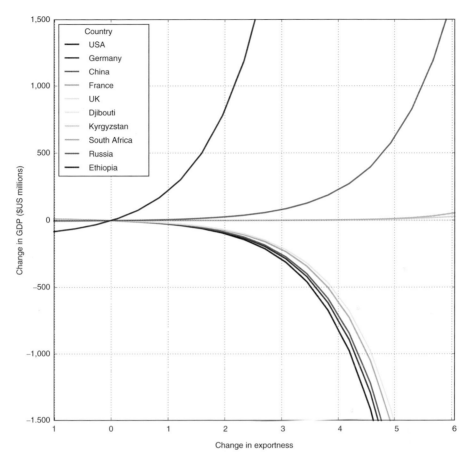

Figure 17.2 Changes in GDP for the five most positively affected and five most negatively affected countries due to a change in exportness of Ethiopia

17.3.4.1 Non-linearity in ϵ

As the choice of $\epsilon = 0.1$ was arbitrary, it is important to understand how the model response depends on this parameter choice. By picking a particular recipient and calculating the effects on the model at various values of ϵ, we can understand how the parameter choice affects the outcome.

In Figure 17.2, we pick Ethiopia (exportness 0.42) as our recipient country and vary ϵ such that Ethiopia's exportness varies across the whole scale of modelled countries from -0.6 of Bhutan to $+6.5$ of the United States. The changes to GDP of the five most increased and five most decreased countries are tracked as Ethiopia's exportness increases from that of Bhutan ($\epsilon = -1$) to that of the United States ($\epsilon = 6.1$). As expected, Ethiopia itself is the country most affected, followed by Russia, South Africa, Saudi Arabia and the United Arab Emirates. (These last two appear almost colinearly on the graph.) Negatively affected by Ethiopia's improved exportness are Germany, the Netherlands, the United Kingdom, France and Belgium. Clearly the effects on each of these countries are highly non-linear but, crucially for our purposes, appear to be monotonic with an increasing first derivative. This means that conclusions drawn about the relative effects of exportness at a particular level of ϵ will hold for any level of ϵ.

17.3.4.2 Own-Country Effects

We first consider the approach taken by the typical study of the effects of aid: the effect on the recipient country itself.

Table 17.7 shows the results of the experiment outlined in Section 17.3.3 on the country whose exportness was increased. Listed are the 10 most improved countries by percentage change to GDP. By far the most improved country is Vietnam, increasing its GDP by 5.6% in response to the increased exportness. The remainder of the 10 most improved countries are more closely spread, with just a single percentage point between Togo, at number 2, and Cambodia at number 10.

It is interesting to note that while two-fifths of the 51 countries in our recipient sample are African, only 3 of the top 10, Ivory Coast, Senegal and Togo, are in Africa. The others are Southeast Asian countries: Vietnam, Cambodia or Latin American: Honduras, Nicaragua, Bolivia.[6]

This raises the question of whether a recipient country must be geographically near to a regional power in order to benefit from an exportness increase. To test this, we regress the percentage change from Table 17.7 against the distance from either the United States, Germany, China or Australia. Figure 17.3 shows a scatter plot of the minimum distance, in thousands of kilometres, from one of these regional powers, against the own-country benefit from an exportness increase.

The relationship has the expected negative sign (-0.11) and is significant at the 5% level ($p = 0.037$). The relationship seems reasonably close, but there are some very noticeable outliers.

Table 17.7 The effect of increasing exportness by 0.1 on the affected country itself. Effect is measured in terms of change in GDP, both absolute and percentage. The top 10 by percentage are shown

Country	GDP ($ US millions)	Change	Change (%)
Vietnam	40,890	2,163	5.59
Togo	1,856	33	1.82
Honduras	6,209	92	1.50
Republic of Moldova	2,406	35	1.46
Papua New Guinea	2,668	37	1.40
Ivory Coast	10,652	123	1.17
Nicaragua	5,191	54	1.04
Senegal	7,871	73	0.93
Bolivia	7,183	65	0.91
Cambodia	2,754	23	0.86

---

[6] The top 10 also includes Moldova and Papua New Guinea.

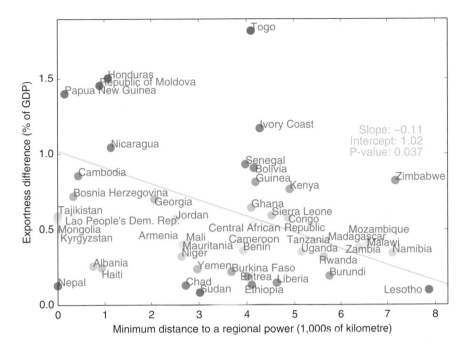

Figure 17.3   Scatter plot showing the relationship between the minimum distance of a recipient country to a regional power (the United States, China, Germany or Australia) and the results of the exportness experiment outlined in Section 17.3.3. Vietnam with an exportness difference of 5.6 and minimum distance of 0 is not shown on the graph, but does contribute to the regression line

Among those countries whose exportness results are less positive than we might expect given their proximity to a regional power, the largest outlier is Nepal. This country is land-locked between India and China and, as Jayaraman and Shrestha (1976) point out, Nepal's

> [...]excessive dependence on one market for its entire foreign trade places Nepal in a unique problematic situation. While the small size of the country impels Nepal to resort to international trade, its geographical position limits its option to trade with countries other than India.

Although improved since the 1970s, this situation is still severe, with India making up over 50% of Nepal's exports in 2012 (Simoes and Hidalgo, 2011). This reliance on a single export partner shows why Nepal is unable to benefit from increased exportness: no other market is really available, so being a more attractive export partner is of little use. This same effect may explain Haiti's presence as a negative outlier (85% of Haitian exports went to the United States in 2005), Chad (75% to the United States) and Sudan (83% to China and Japan combined).

On the positive side, the country to most benefit despite its comparative distance to a regional power is Togo. This country is such a stark outlier that it may deserve some closer inspection. Table 17.8 shows a summary of the results for Togo. It might be worth noting here that the numbers involved are extremely small. This is due to the small overall size of Togo's economy (just over $3 billion in 2010, around $500 per capita.) The analysis presented here is thus to

Table 17.8   Increases in trade flows from Togo, due to the exportness experiment, in which Togo's exportness was increased by 0.1

(a) Per importing country		(b) Per sector to the United States	
Importer	Trade ($US)	Sector	Trade ($US)
United States	1,280	Mining	202
France	716	Electricals	185
Germany	598	Vehicles	179
United Kingdom	447	Chemicals	153
Italy	445	Textiles	122

be taken as indicative of the kind of benefits a country with an economy and trading relationships such as those of Togo might experience due to changes in exportness, rather than as an empirical fact.

The biggest increase is to the United States with $1,280 additional trade. This trade with the United States is broken down into sector in Table 17.8b. The sector to benefit most is Mining, with $202 additional export (Togo exported $42,000 of phosphates to the United States in 2005), followed by Electricals (around $130,000 of integrated circuits was exported to the United States in 2005) and Vehicles ($1,500 of HS code 870590, 'special purpose motor vehicles', which seem primarily to be road manufacturing vehicles, such as bitumen spreaders.)

Togo is not a particularly large exporter to the United States, with only $11 million in 2005, making it the 35th largest of 53 African countries. Togo's large gains seem also not to be caused by inappropriate estimates for the value added per unit of Togo's sectors: the sector with the highest value added per unit is construction with 0.35 which has the same order of magnitude as Ethiopia's construction sector,[7] with 0.32, and that of the United States which has 0.48. But unlike the negative outliers, Togo's export profile was surprisingly diverse in 2005. The largest trade partner, Ghana, had only 15% of Togo's total exports.

To test the thesis that benefitting from exportness increases requires a broad spread of export partners, we calculate a Herfindahl index of export partner concentration based on the share, s_{ij}, which country j takes of country i's total exports. Throughout what follows, we will do all calculations based on 2005 data[8]. The concentration measure is then

$$H_{ij} = \sum_{j=1}^{C} s_{ij}^2 \tag{17.8}$$

where C is the number of countries in the data (in our case, there are 230 trade areas with an ISO3 country code, and we will use this as C throughout the following.) We can normalise H by a simple adjustment:

$$H_{ij}^* = \frac{H_{ij} - 1/C}{1 - 1/C} \tag{17.9}$$

---

[7] Ethiopia fares much worse in the exportness experiment than does Togo, which makes this a useful comparison.
[8] Note that here, uniquely in this work, a negative flow in the data source from Gabon to the United States has been corrected to a positive flow. Checks with online resources have been made to ensure the feasibility of this correction.

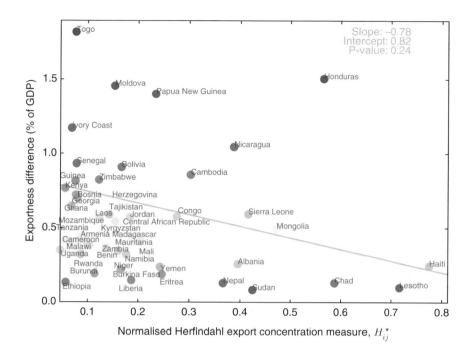

Figure 17.4 Scatter plot showing the relationship between the normalised Herfindahl index of export partner concentration and the results of the exportness experiment outlined in Section 17.3.3. Vietnam with an exportness difference of 5.6 and an H^* of 0.07 is not shown on the graph, but does contribute to the regression line

Large values of this measure indicate a highly concentrated export market, such as that of Haiti, Chad and Sudan, and smaller values indicate a more spread export market, such as that of Togo, as discussed earlier.

Figure 17.4 shows a scatter plot of the relationship between H^*_{ij} and the results of the exportness experiment. The relationship has the expected negative sign, indicating that countries with more concentrated export markets did worse on average, but the relationship is not significant. This may be because many countries have very similar values of H^*_{ij}, with the majority being in a cluster around 0.1 to 0.3. Such relationship as there is seems to be caused mainly by outliers such as Vietnam, Togo, Moldova, Haiti, Lesotho and Chad.

There is thus some evidence, albeit statistically weak, that Togo's position as a large bene-ficiary of exportness increases is related to its diverse export relationships.

17.3.4.3 Positive Other-Country Effects

Although the majority of effects were negative for countries other than those having their exportness increased (hereafter 'other' countries), a few other countries benefitted. Table 17.9 shows a summary of the positive results for other countries. The exportness of the country in the 'Recipient' column was increased as per the experiment of Section 17.3.2 and positive GDP effects on other countries were measured. The sum of these GDP increases is reported

Table 17.9   Other-country positive effects from the exportness experiment. The country in the 'Recipient' column had its exportness increased by 0.1. The 'Total effect' is the sum of all the positive changes in GDP from countries other than the recipient itself. The 10 largest by this measure are shown. The top three other-country effects are shown per recipient country

Recipient	Total +ve effect ($US millions)	Other country	Δ GDP ($US millions)
Vietnam	118.25	Korea	118.25
Honduras	18.92	Canada	13.52
		United States	4.98
		Israel	0.42
Kyrgyzstan	11.95	China	8.37
		India	1.75
		Turkey	0.34
Albania	9.65	Russia	9.62
		Kyrgyzstan	0.02
Kenya	9.01	Russia	7.95
		Kyrgyzstan	0.57
		South Africa	0.42
Papua New Guinea	8.97	Australia	7.92
		Indonesia	1.04
Georgia	8.64	Russia	8.54
		Azerbaijan	0.08
		Kyrgyzstan	0.01
Congo	8.31	Russia	3.60
		Israel	2.53
		South Africa	1.83
Jordan	6.67	Russia	5.70
		Kyrgyzstan	0.89
		Saudi Arabia	0.07
Bolivia	6.40	Chile	3.99
		Russia	2.41

in the 'Total effect' column, and the 10 largest by total effect are shown. The following two columns show the largest three other-country benefactors of the increased exportness in the recipient country.

The largest positive effect by a large margin is that associated with Vietnam, with a total effect of $118.25 million. The entire positive effect is due to an increase in GDP in Korea. This is an interesting result because Korea is only the fourth largest importer to Vietnam (11%), after China (18%), Singapore (13%) and Japan (12%) and because Vietnam makes up only 1.2% of Korea's total export market. Figure 17.5 shows a part of a network representation of all the changes resulting from an increase in Vietnam's exportness. The nodes of the network are either countries, labelled in light blue, or CSs, labelled in black as 'sector (country)'. All edges represent a change: those between CSs show changes in trade flows, and those between countries and CSs represent changes in total production of that sector in that country. Nodes are sized by total degree (in- plus out-degree). The width of the edges is proportional to the

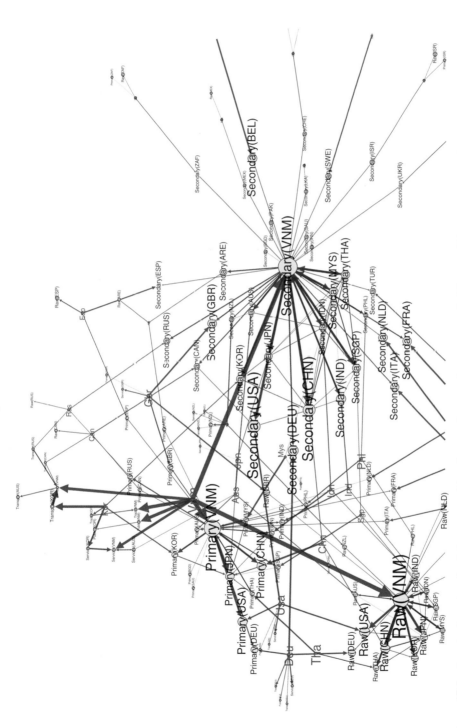

Figure 17.5 A part of a network of changes resulting from an increase in exportness in Vietnam. Blue arrows show increases and red arrows show decreases. Nodes are either countries (labelled in light blue) or country/sectors (CSs, labelled in black). Edges between CSs represent changes in trade flows. Those between countries and CSs represent changes in total production. All changes are logarithmic

logarithm of the magnitude of the change, with positive changes shown in blue and negative changes shown in red[9]. The sectors shown are the seven 'super-sectors' described in Chapter 5 of *Geo-Mathematical Modelling*.

We can see immediately that the change to total production in Vietnam (VNM, slightly left of centre of the image) is large and positive, as we would expect. There are large increases in the 'raw' and 'primary' sectors, which then trade with China, Japan and the United States. This leads to modest reductions in the production of these sectors in these countries. In Korea, in the top-left of the image, the 'primary' and 'secondary' sectors are reduced in output, but the 'trade', 'services' and 'transport' sectors all see comparatively large increases. These sectors all have increased trade with Vietnam. Russia and Canada also see increased output in the 'transport' sector due to increased trade with Korea. The 'raw' sectors in Thailand (THA, left-centre of the image) and Singapore (SGP, centre) also see increased production due to increased trade with Vietnam.

17.3.4.4 Negative Effects on the Entire Trade Network

As the discussion of Section 17.3.4.3 has indicated, the effect of an increase in exportness, beyond the effect on the recipient country itself, is largely (but not exclusively) negative. In this section, we will focus on the overall effect on global GDP, in contrast to the analysis of Section 17.3.4.3 where only those other-country effects which were positive were included. Additionally, we include the positive effect on the recipient country itself.

As with the positive effects, the largest effect by a large margin is that associated with an increase in exportness of Vietnam. Of the remaining nine countries, four are African: Ivory Coast, Kenya, Senegal and Ghana, and three are Latin American: Honduras, Bolivia and Nicaragua.

The largest other-country GDP reductions are suffered in every case by the United States, Germany and China, with the exception of Honduras, Bolivia and Bosnia Herzegovina, where the United Kingdom features among the three biggest reductions, and Jordan, where France does. This very brief piece of analysis gives us a glimpse into how the global dynamics model might be employed to assess the suitability of particular recipients from the perspective of a particular donor country.

17.3.5   Conclusions

In this section, we have outlined a simple method for using the global dynamics model to study questions relating to development aid, specifically 'aid for trade'. In order to do this, we assumed that aid was able to increase the export attractiveness, or 'exportness', of a recipient country. By running a simple multiplicative model, whose coefficients we estimated in Chapter 4 of *Geo-Mathematical Modelling*, we were able to estimate a set of flows resulting from an increased exportness. From these flows, we then calculate a new set of import propensities in the global dynamics model which allowed us to assess the impact of the change in exportness on the GDP of each country in the model (see Table 17.10).

---

[9] For visual clarity, edges with weight smaller than 3.25 have been filtered out. Additionally, only those nodes which are neighbours of the 'VNM' country node to a depth of 3 or less are shown.

Table 17.10   Total effects from the exportness experiment. The country in the 'Recipient' column had its exportness increased by 0.1. The 'Total effect' is the sum of all the changes in GDP from all countries in the model. The 10 largest by this measure are shown. The top three other-country effects are shown per recipient country

Recipient	Total effect ($US millions)	Other country	Δ GDP ($US millions)
Vietnam	−2,165	United States	−357.51
		Germany	−231.37
		China	−210.45
Ivory Coast	−123	China	−15.14
		Germany	−14.74
		United States	−11.87
Honduras	−94	China	−19.03
		Germany	−10.62
		United Kingdom	−8.53
Kenya	−93	United States	−12.26
		Germany	−10.93
		China	−10.37
Senegal	−74	China	−9.10
		Germany	−8.24
		United States	−6.30
Bolivia	−65	China	−9.34
		United States	8.30
		Germany	−6.27
Nicaragua	−59	China	−10.40
		Germany	−5.69
		United Kingdom	−4.62
Bosnia Herzegovina	−58	United States	−7.35
		China	−6.57
		United Kingdom	−5.33
Jordan	−54	United States	−7.57
		Germany	−6.10
		France	−5.00
Ghana	−52	China	−5.75
		Germany	−5.67
		United States	−5.20

We began by showing that the GDP responses were non-linear in the increase of exportness of a given country, but that their monotonicity allows us to choose an arbitrary increase to use as a comparison between recipient countries.

We noted that a disproportionate number of Southeast Asian countries were among the recipients to most benefit from an increase in exportness, showing that the benefits are correlated with closeness to a regional power (which we defined as the United States, Germany, China and Australia.) We also tested the thesis that a country must have a diverse set of export partners to benefit, but found that the export concentration measure we employed was too similar among too many countries for a statistically significant relationship to emerge.

We found that Vietnam was the recipient country most associated with positive GDP effects outside the recipient itself, with Korea being the biggest beneficiary from increased Vietnamese exportness, due to an increase in Korean services imports into Vietnam in response to the increased export demand in that country.

Finally, and unusually for an aid study, we looked at the negative effects on the rest of the trade network due to an increased export competitiveness in the recipient country. We found that the United States, Germany and China were most likely to lose out following an increase in an aid recipient's exportness.

17.4   Adding Security

17.4.1   Introduction

The domain of international security is different from the previous two domains, migration and development aid, because it contains no clearly defined concept which translates to bilateral relationships or flows. Since looking at the effects of our various domains in terms of pairs of countries has been crucial in the preceding sections, our approach here will be to use the dyadic concept of threat introduced in Chapter 11 of *Geo-Mathematical Modelling* and make an assumption about how a particular threat affects military expenditure in the threatened country, and hence the global economy.

17.4.2   Literature Review

Policies borne out of security considerations can have dramatic consequences on the economy. Funding of military activities, for instance, places a substantial burden on national budgets. In 2014, countries spent on average 2% of their GDP on their military, with this number rising up to 10% in some countries (SIPRI, 2015).

This sizeable contribution to the economy has not escaped the attention of scholars from the fields of economics and political science. A large literature has examined the determinants of military spending, often concluding that direct need for security (as evidenced by participation in ongoing international or civil wars), threats (both external and internal), neighbours' military expenditure and domestic characteristics (e.g. population, level of democracy, or GDP) are significant predictors globally (Collier and Hoeffler, 2002; Goldsmith, 2003, 2007; Nordhaus *et al.*, 2012; Böhmelt and Bove, 2014)

The impact of such expenditure has also been considered. Often using the state as a unit of analysis, a number of studies have considered whether large levels of military spending lead to economic growth or decline. A consensus, however, is yet to emerge (Aizenmann and Glick, 2006; Dunne *et al.*, 2005). Beyond considering spatial spillover effects (Shin and Ward, 1999; Yildirim and Öcal, 2014), the global economic impact of an increase in military expenditure in one specific country has rarely been considered. In other words, how does a change in military spending levels in one country affect other countries who trade with them?

In the remainder of this section, we consider how the global dynamics model might be used to address questions of international security. We show how the model offers benefits over approaches that only consider the economic impact of a state's military spending within that state. The global dynamics model enables us to examine the global impact of a domestic decision to increase military expenditure. In what follows, we outline an experiment to examine the effect an increase in threat between pairs of nations has on the economic performance of every country in the model. We explain how threat is operationalised, how it relates to military expenditure and how this in turn relates to the global dynamics model. Next, we model how trade flows vary with increasing demand for sectors of the national economy related to security. We then present an experiment with the global model before offering some tentative conclusions. The emphasis of this section is on a proof-of-concept study of how our the global model might be used to study security-related policy questions. Further research relating military expenditure to the sectors in the global model would lead to more concrete conclusions. These limitations are discussed in the conclusion to this section.

17.4.3   Measures of Threat and the Global Dynamics Model

Threats from the international security environment have been identified as significant predictors of military expenditure in recent global studies (Nordhaus *et al.*, 2012; Böhmelt and Bove, 2014). Attempts at measuring such threats can be traced back to the work of Lewis Fry Richardson, who investigated the influence of adversary spending abroad in increasing military spending at home using a simple mathematical framework (Richardson, 1919, 1960). In Richardson's model, the rate at which a nation increases their military expenditure is proportional to their adversary's military spending. In Chapter 11 of *Geo-Mathematical Modelling*, the Richardson model is extended by incorporating various factors that might alter this relationship. In particular, a measure of threat between any pair of nations, denoted by T_{ij}, is derived that captures the military resources of nation i that exist to target nation j. T_{ij} is defined as

$$T_{ij} = \frac{p_i p_j^\alpha e^{-\beta c_{ij}}}{\sum_k p_k^\alpha e^{-\beta c_{ik}}}, \tag{17.10}$$

where p_i is the level of military spending of nation i, c_{ij} is a measure of alliance between i and j, and α and β are fixed parameters. As explained in Chapter 11 of *Geo-Mathematical Modelling*, c_{ij} is constructed via a measure of alliance similarity (Signorino and Ritter, 1999) and geographic distance between countries i and j.

The model T_{ij} takes into account all other nations, their physical proximity to one another, the structure of the international alliance network, and some notion of collective security (in the sense that an ally will offset the threat posed by a mutual enemy).

Summing over all nations i obtains the total amount of threat faced by nation j. This measure is then incorporated in a global model of military expenditure and is shown to be a significant predictor of future expenditure. In particular, a 1% increase in the level of threat is found to increase military spending by 0.06%.

In this section, we assume that an increase in T_{ij} will lead to an increase in military expenditure in the global dynamics model. We systematically increase the threat of each pair of nations by 1%. We then assume that this corresponds to an increase in military spending of nation j of 0.06%.

One complication, however, is that the input–output models at the heart of the global model do not have a single sector for military expenditure. Such spending is likely to be spread over many different sectors in the model. As a simple proxy, we assume that an increase in military expenditure due to international threats will cause an increase in the demand for metals.

Furthermore, we suppose that the increase in demand for metals from an increase in threat occurs at the same rate as for military spending: an increase of 1% of threat corresponds to an increase of 0.06% of metals. In other words, we assume military expenditure consists entirely of expenditure on metals. This assumption can be relaxed in future work by using a typical 'basket' of sectors which make up the average unit of military expenditure.

The measures of threat and assumptions about how military expenditure relates to national expenditure enable us to relate dyadic measures of threat to changes in demand within a sector. This then enables us to explore the consequences of threat changes in the global model. In order to do this, a further set of assumptions are required regarding how this increase in demand is met. We next outline how we achieve this via an adjustment in import propensities.

17.4.4   Trade during Changing Security Conditions

In order to meet the increased demand of military expenditure that is due to increased threats, countries turn to imports. The way in which they obtain imports from existing trading partners is modelled by adjusting the modelled trade flow $y_{s,ij}$ as

$$\Delta y_{s,ij} = \Delta f_{s,i} \frac{s_{ij} y_{s,ij}}{\sum_k s_{ik} y_{s,ik}}, \tag{17.11}$$

where $f_{s,i}$ is the final demand of sector s in country i and s_{ij} is a measure of alliance similarity between countries i and j.

In this way, the change in levels of trade between two countries after a change of threat in the international system will depend on the level of alliance similarity between those two nations. The measure of alliance similarity s_{ij} is given by Signorino and Ritter's (1999) S-measure and is explained in more detail in Chapter 11 of *Geo-Mathematical Modelling*. It examines the alliance portfolio of the two countries with every other country and then measures their similarity. Two countries with a similar alliance portfolio will be considered allies and consequently suitable trading partners.

The adjustment to the trade flows defined in equation 17.11 is also dependent on the original trade flows, which ensures that trading partners are not created or destroyed. This model assumes that the suitability of the trading partner in the face of an uncertain security environment depends on both the level of current trade and their alliance structure.

We have included this feature to model the idea that increased spending on, for example, military hardware is more likely to come from allies than from potential threats. This makes particular sense in an environment where threat is increasing.

17.4.5   An Experiment of Increased Threat in the Global Dynamics Model

For the purposes of this experiment, we will restrict ourselves to studying only the largest portion of the dyadic threat measure. We arbitrarily restrict the threat relations to $T_{ij} \geq 1$ which restricts us to 399 dyadic relationships of a possible 16,770 in the data set.[10] Table 17.11 shows

---

[10] Dyads involving the neutral countries Costa Rica, San Marino, Switzerland and Turkmenistan are removed from the analysis. These nations were susceptible to generating low levels of alliance measures since they do not belong to many of the agreements that would otherwise be captured by this measure (e.g. Switzerland does not belong to the European Union or to NATO). They are also unlikely to be the source or targets of threat as a result of their neutrality.

Table 17.11   The 10 most commonly occurring sources and 10 largest targets of threat among the subset of $T_{ij} \geq 1$

(a) Threat sources		(b) Threat targets	
Country	Instances	Country	Instances
United States	127	United States	12
United Kingdom	61	Canada	10
France	55	Iraq	9
India	32	Libya	9
Saudi Arabia	32	Mauritania	9
Italy	27	Tunisia	8
Canada	21	Mali	8
Netherlands	11	Russia	8
Turkey	11	Algeria	8
Spain	9	Egypt	7

a summary of the 10 largest sources and targets among the 399 largest threats and the entire network of 399 relationships is shown in Figure 17.6. By far the most common source of threat comes from the United States which is the source of 127 of the 399 largest threats. The United Kingdom is a distant second with 61 followed by France (55), India and Saudi Arabia (32), and Italy (27). The United States is also the most common target for threat, although the targets are far more evenly distributed than the sources. Canada, Iraq, Libya, Mauritania, Tunisia, Mali, Russia, Algeria and Egypt are closely behind.

A simple measure of the effect of increased threat is which third-party country[11] benefits when threat is increased. This benefit will be brought about by an increased production (and export) of the metals sector due to increased demand from a threatened country which is either a large importer of metals or a close ally, according to Equation (17.4). Table 17.12 shows the 20 top countries by this measure, by total benefit summed across all 399 dyadic threat increases in Table 17.12a and by total benefit as a percentage of GDP in Table 17.12b. The two tables are markedly different. By total benefit, the top seven are among the world's largest economies: Canada, Italy, China, France, Russia, Japan and the United Kingdom. This reflects the fact that the large countries are more likely to be substantially existing exporters of metals and therefore have the most to gain from a third-party increase in demand. But this is not simply a measure of economy size: the United States, Brazil and Norway all come much further down the list. In the case of the United States, this is presumably because it is involved in so many of the 399 dyadic threat relationships and therefore cannot count as a beneficiary of any increase in demand as a result.

The list by percentage of GDP is, perhaps predictably, dominated by smaller countries, although several countries with a GDP of over $10 billion make it onto the list, including Canada, South Africa, Venezuela, Switzerland and Norway. Guinea taking first place in this table may be due to its large aluminium ore industry, which accounted for $600 million of exports in 2005, over a third of the global trade in this product. In the same year, Zambia

---

[11] Third party in the sense of being not the source or the target of the increased threat.

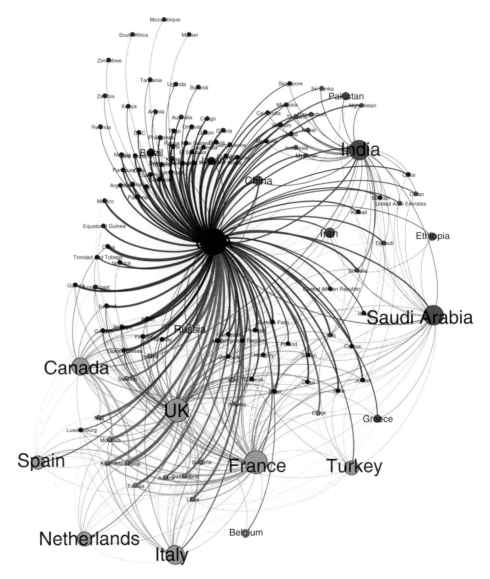

Figure 17.6 The 399 largest dyadic threat relationships (those with $T_{ij} \geq 1$). Node size is proportional to the weighted out-degree, meaning that larger nodes create more threat. The thickness of the edges is proportional to the logarithm of T_{ij}. Nodes are grouped into three clusters using a community detection algorithm

exported almost \$3.8 billion of copper, 72% of that country's total exports. Finally, Papua New Guinea's large exports of copper (3% of global exports) and gold (around 35% of its total exports) may account for its presence at the top of the table. Clearly gold is not a metal normally associated with military expenditure, but this is an inevitable drawback of the fairly high-level sector breakdown of the global dynamics model.

Table 17.12 The largest total beneficiaries of increased threat across the 399 dyadic threat relationships under experiment. Only third-party benefits are considered. Benefits are summed across all threat increases

(a) The top 20 by total benefit				(b) The top 20 by percentage of GDP			
Country	Benefit ($ M)	GDP ($ Bn)	Benefit (% GDP)	Country	Benefit ($ M)	GDP ($ Bn)	Benefit (% GDP)
Canada	22.45	914	0.0025	Guinea	0.22	3	0.0073
Italy	7.93	1,561	0.0005	Papua New Guinea	0.23	4	0.0054
China	7.43	1,511	0.0005	Zambia	0.31	6	0.0051
France	6.69	1,886	0.0004	Mozambique	0.24	6	0.0041
Japan	6.23	3,600	0.0002	Chile	2.36	95	0.0025
Russia	6.16	671	0.0009	Canada	22.45	914	0.0025
United Kingdom	5.91	2,162	0.0003	South Africa	3.13	186	0.0017
Mexico	4.73	642	0.0007	Venezuela	2.60	164	0.0016
Belgium	4.00	415	0.0010	Guyana	0.01	1	0.0014
Netherlands	3.86	676	0.0006	Brunei Darussalam	0.13	10	0.0013
Spain	3.79	877	0.0004	Malaysia	0.95	75	0.0013
Brazil	3.34	725	0.0005	Congo	0.11	8	0.0012
Switzerland	3.28	295	0.0011	Equatorial Guinea	0.12	11	0.0012
Korea	3.13	606	0.0005	Oman	0.31	27	0.0011
South Africa	3.13	186	0.0017	Switzerland	3.28	295	0.0011
Sweden	3.01	361	0.0008	Luxembourg	0.61	56	0.0011
Australia	2.94	563	0.0005	Yemen	0.14	13	0.0011
Venezuela	2.60	164	0.0016	Jamaica	0.10	9	0.0011
United States	2.48	11,230	0.0000	Norway	2.41	225	0.0011
Norway	2.41	225	0.0011	Libya	0.41	39	0.0011

Table 17.13 The threat dyads causing the biggest increase in the GDP of Guinea, following a fixed percentage increase in the threat level. The 10 largest effects are show

Threat from	Threat to	Δ GDP in Guinea
United States	Russia	0.1914
United States	France	0.0071
United States	United Kingdom	0.0057
China	United States	0.0029
India	United States	0.0015
United States	Korea	0.0014
United States	Spain	0.0011
United States	Italy	0.0008
United States	China	0.0007
United States	Ireland	0.0006

In order to examine in more detail how Guinea benefits from threat increases, we now look at which dyadic threats serve most to increase Guinea's GDP. The 10 most significant dyads are shown in Table 17.13. As with the overall dyadic threat dataset, threats from the United States dominate the list, with that between the United States and Russia coming at the top by a very wide margin. Almost 100% of Guinea's exports to Russia in 2005 were aluminium ore and although this flow accounted for only 0.2% of Guinea's total exports of this product, it represents 10% of all Russia's aluminium ore imports.

A more complete picture of the dyads which increase Guinea's GDP is shown in Figure 17.7 which is a network representation of the dyads, with edge weight being proportional to the increase in Guinea's GDP which a fixed-percentage increase in the dyadic threat measure induces. The 50 largest dyads are shown.

17.4.6   Conclusions

We have explored the impact of changing threat relations between pairs of countries on the global dynamics model. We have assumed that an increase in the threats faced by a nation leads to an increased final demand for metals and have explored the consequences of this. Via a simple experiment, Guinea was found to be the country most likely to benefit from increasing threat relations. It is conjectured that this is due to their substantial aluminium and copper industries, with copper being an essential component in the manufacturing of telecommunications and electronics.

This proof-of-concept study has demonstrated how the global dynamics model can be used to consider questions from different political arenas. There are, however, limitations with the present study which would benefit from further research. As emphasised in Chapter 11 of *Geo-mathematical Modelling*, there is scope for further development of the threat measure T_{ij}. There are, for instance, a number of alternative formulations of the alliance measure c_{ij} (Signorino and Ritter, 1999; D'Orazio, 2012) which may provide a better model of threats faced by different nations. In addition, further exploration of the parameter space may lead to more suitable values of the parameters α and β. In the context of changing threat flows, an increase in military expenditure is likely to influence many other sectors in the global model. Further research could investigate a typical profile of defence spending as it relates to the sectors within the global model. This would then enable a more thorough assessment of the impact of increasing military expenditure in different nations.

Although the experiment presented here represents a novel coupling of the global dynamics model with domestic security considerations, it should be borne in mind that the process by which they are coupled relies on a modelled quantity of threat. This is in contrast to the coupling of the global dynamics model with migration flows described in Section 17.2, where the influencing factor was migration flows obtained from data. Nevertheless, abstract quantities such as threat and its influence on military expenditure, if grounded in well-developed theory, can lead to models whose outputs can usefully contribute to policy discussions, whether it be in support of a policy decision or via the identification of aspects of the global system not previously considered but which may play an important role (such as Guinea's aluminium industry).

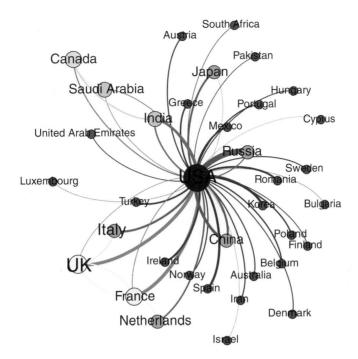

Figure 17.7 A network diagram of the 50 dyadic threat relationships to which the GDP of Guinea is most sensitive. Node size (and colour) is proportional to node degree. The weight of the edges is proportional to the increase in Guinea's GDP following a fixed-percentage increase in T_{ij}, the dyadic threat measure

17.5 Concluding Comments

In this chapter, we have pulled together threads from various parts of this book and its companion *Geo-Mathematical Modelling*. We have given examples of how the model of Chapter 4, extended to a larger set of countries with the techniques of Chapter 5 of *Geo-Mathematical Modelling*, might be used to study other global problem domains.

In the domain of migration, we used estimated migration figures from Chapter 7 to adjust final demand figures and trade relationships to estimate the impact a particular year's migration flows might have on the global economy.

We modelled the aid domain by focussing on the so-called 'aid for trade', assuming that aid was able to increase the export attractiveness of a recipient country. Running a simple trade model allowed us to assess the impacts, both positive and negative, of increasing one country's export attractiveness without increasing global demand. This allowed a new global insight into development aid spending and allowed us to make some tentative suggestions about how donors might select 'suitable' recipients.

We finally demonstrated, by building on the work of Chapter 12 of *Geo-Mathematical Modelling*, how the domain of international security might affect trade and the global economy. We assumed that increased threat led to increased military expenditure and assumed that this expenditure occurred entirely within the metals sector. By making a further assumption about how the additional demand for metals might be sourced according to military alliances, we were able to study the effects of an increase in threat between any pair of countries upon every country in the model. We then picked out and looked in more detail at those countries who stood to benefit from the largest number of threat pairs.

The assumptions and modelling approaches in this chapter are to be thought of as a guide to how the coefficients of the model introduced in Chapter 4 might be exploited to answer questions about a wide variety of global problem domains. The reader is encouraged to see these as starting points for approaches to problems in their own global problem domain and perhaps as inspiration for further research.

References

Aizenmann, J. and Glick, R. (2006) Military expenditure, threats, and growth. *The Journal of International Trade and Development: An International and Comparative Review*, **15** (2), 129–155.

Askegaard, S., Arnould, E.J., and Kjeldgaard, D. (2005) Postassimilationist ethnic consumer research: qualifications and extensions. *Journal of Consumer Research*, **32** (1), 160–170, doi: 10.1086/426625.

Beine, M., Docquier, F., and Özden, C. (2011) Diasporas. *Journal of Development Economics*, **95** (1), 30–41, doi: 10.1016/j.jdeveco.2009.11.004.

Bertoli, S. and Fernández-Huertas Moraga, J. (2013) Multilateral resistance to migration. *Journal of Development Economics*, **102**, 79–100, doi: 10.1016/j.jdeveco.2012.12.001.

Böhmelt, T. and Bove, V. (2014) Forecasting military expenditure. *Research and Politics*, **1**, 1–8.

Clark,X., Hatton, T.J., and Williamson, J.G. (2007) Explaining U.S. Immigration, 1971–1998. *Review of Economics and Statistics*, **89** (2), 359–373, doi: 10.1162/rest.89.2.359.

Collier, P. and Hoeffler, A. (2002) Military expenditure: threats, aid and arms races. World Bank Policy Research Working Paper 2927.

D'Orazio, V. (2012) Advancing Measurement of Foreign Policy Similarity, APSA 2012 Annual Meeting Paper, Available at SSRN: http://ssrn.com/abstract=2105547 (accessed 31 December 2015).

Dunne, J., Smith, R., and Willenbockel, D. (2005) Models of military expenditure and growth: a critical review. *Defence and Peace Economics*, **16** (6), 449–461.

Felbermayr, G.J. and Jung, B. (2009) The pro-trade effect of the brain drain: sorting out confounding factors. *Economics Letters*, **104** (2), 72–75, doi: 10.1016/j.econlet.2009.04.017.

Felbermayr, G.J. and Toubal, F. (2012) Revisiting the trade-migration nexus: evidence from new OECD data. *World Development*, **40** (5), 928–937, doi: 10.1016/j.worlddev.2011.11.016.

Genc, M., Gheasi, M., Nijkamp, P., and Poot, J. (2012) The impact of immigration on international trade: a meta-analysis, in *Migration Impact Assessment: New Horizons* (eds P. Nijkamp, J. Poot, and M. Sahin), Edward Elgar, Cheltenham, pp. 301.

Girma, S. and Yu, Z. (2002) The link between immigration and trade: evidence from the United Kingdom. *Weltwirtschaftliches Archiv*, **138** (1), 115–130, doi: 10.1007/BF02707326.

Goldsmith, B. (2003) Bearing the defense burden, 1886–1989: why spend more? *Journal of Conflict Resolution*, **47**, 551–573.

Goldsmith, B. (2007) Arms racing in 'space': spatial modelling of military spending around the world. *Australian Journal of Political Science*, **42** (3), 419–440.

Grogger, J. and Hanson, G.H. (2011) Income maximization and the selection and sorting of international migrants. *Journal of Development Economics*, **95** (1), 42–57, doi: 10.1016/j.jdeveco.2010.06.003.

Jayaraman, T.K. and Shrestha, O.L. (1976) Some trade problems of landlocked Nepal. *Asian Survey*, **16** (12), 1113–1123, doi: 10.2307/2643448.

Karamba, W.R., Qui nones, E.J., and Winters, P. (2011) Migration and food consumption patterns in Ghana. *Food Policy*, **36** (1), 41–53, doi: 10.1016/j.foodpol.2010.11.003.

Mayda, A.M. (2009) International migration: a panel data analysis of the determinants of bilateral flows. *Journal of Population Economics*, **23** (4), 1249–1274, doi: 10.1007/s00148-009-0251-x.

Nordhaus, W., Oneal, J., and Russett, B. (2012) The effects of the international security environment on national military expenditures: a multicountry study. *International Organization*, **66** (3), 491–513.

Ortega, F. and Peri, G. (2013) The effect of income and immigration policies on international migration. *Migration Studies*, **1** (1), 47–74, doi: 10.1093/migration/mns004.

Pedersen, P.J., Pytlikova, M., and Smith, N. (2008) Selection and network effects—Migration flows into OECD countries 1990–2000. *European Economic Review*, **52** (7), 1160–1186, doi: 10.1016/j.euroecorev.2007.12.002.

Richardson, L. (1919) Mathematical psychology of war, in *The Collected Papers of Lewis Fry Richardson, Quantitative Psychology and Studies of Conflict*, vol. **2** (eds I. Sutherland, O. Ashford, H. Charnock, P. Drazin, J. Hunt, and P. Smoker), Cambridge University Press, Cambridge, pp. 61–100.

Richardson, L. (1960) *Arms and Insecurity*, The Boxwood Press, Pittsburgh, PA.

Shin, M. and Ward, M. (1999) Lost in space: political geography and the defense-growth trade-off. *Journal of Conflict Resolution*, **43**, 793–817.

Signorino, C. and Ritter, J. (1999) Tau-b or not Tau-b: measuring the similarity of foreign policy positions. *International Studies Quarterly*, **43** (1), 115–144.

Simoes, A.J.G. and Hidalgo, C.A. (2011) The economic complexity observatory: an analytical tool for understanding the dynamics of economic development. Scalable Integration of Analytics and Visualization.

SIPRI (2015) Sipri Military Expenditure Database 2015, http://milexdata.sipri.org (accessed 7 September 2015).

Stiglitz, J. (2007) The Malaysian Miracle.

Wallendorf, M. and Reilly, M.D. (1983) Ethnic migration, assimilation, and consumption. *Journal of Consumer Research*, **10** (3), 292–302.

Yildirim, J. and Öcal, N. (2014) Military expenditures, economic growth and spatial spillovers. *Defence and Peace Economics*, doi: 10.1080/10242694.2014.960246.

Index
